COMPUTER APPLICATIONS IN TECHNOLOGY

KENNETH C. MANSFIELD, JR.

JAMES L. ANTONAKOS

BROOME COMMUNITY COLLEGE

Pearson
Custom
Publishing

Cover Art: "Dawn of Computer Age," by Guy Crittenden/Stock Illustration
Source, Inc.

Excerpts taken from:

Introduction to Microsoft Windows for Engineering & Technology,
by James L. Antonakos and Kenneth C. Mansfield, Jr.
Copyright © 2000 by Prentice-Hall, Inc.
A Pearson Education Company
Upper Saddle River, New Jersey 07458

Microsoft Windows is a registered trademark ® of the Microsoft Corporation.

This special edition published in cooperation with Pearson Custom Publishing

Printed in the United States of America

10 9 8 7 6 5 4 3

Please visit our web site at www.pearsoncustom.com

ISBN 0–536–60908–X

BA 992076

PEARSON CUSTOM PUBLISHING
75 Arlington Street, Boston, MA 02116
A Pearson Education Company

Preface

Students graduating from school today must be equipped with increasingly sophisticated skills. In addition to the fundamental core of knowledge, students should know how to use the proper tools to enhance their job performance. Within the last fifteen years, the ability to use a personal computer has become one of the most important additions to a graduate's portfolio.

It is our belief as educators that being comfortable with the personal computer, its peripherals, the various Windows operating systems, and application software is required for students graduating from two- and four-year science, technology, and engineering programs. Students who have never used a computer before, and students who have a working knowledge of the personal computer but want more details, will find this textbook a valuable resource. One student may ask, "What is HTML?" whereas another might ask, "How do I defragment my hard disk?" In this text there are exercises designed to answer both of these questions and more.

After using this text, the students will be able to explain the internal operation of the personal computer, including its microprocessor, memory, and input/output systems. The Windows operating systems, including Windows 95/98 and NT, are covered in detail. Students will be able to perform various tasks with each operating system, such as word processing, preparing new disks, spreadsheets, printing, installing and upgrading software, and troubleshooting, to name a few. There are numerous exercises designed to provide hands-on experience in these areas.

The social aspects of using computers and their technical challenges are illustrated through the exploits of Joe Tekk, a fictitious computer specialist working at a fictitious software company called RWA Software. Joe encounters the successes and failures commonly associated with computers and their operation, and also interacts with many different individuals regarding their computer experiences. Students are encouraged to consider the social implications of computers whenever possible.

OUTLINE OF COVERAGE

The textbook is divided into three major parts.

UNIT I: INTRODUCTION TO COMPUTER TECHNOLOGY

The Windows 95/98, and Windows NT operating systems are all presented, with numerous screen shots provided to illustrate exactly what students should see as they try new commands and procedures. They are shown how to manage the graphical user interface provided by Windows, how to install new software and upgrade existing software, how to use the Control Panel to customize Windows, and much more. Important topics, such as networking, computer security, and the Internet are also presented.

UNIT II: PRODUCTIVITY SOFTWARE

This unit covers a number of popular software packages, showing the student how to use word processors, spreadsheets, databases, presentation software, and electronic mail.

UNIT III: TECHNICAL APPLICATIONS

This unit describes the use of two technical applications: Electronics Workbench and Actrix Technical. Electronics Workbench allows you to design and simulate electronic circuits. Actrix Technical is a CAD-based drawing application.

The exercises in each part all have the same format.

Each exercise begins with a short example of why the exercise is applicable to microcomputer use. A fictitious employee is typically used to convey the problem or situation. Here is the first example:

Joe Tekk has just been hired by RWA Software as its new computer specialist. His manager was impressed during Joe's job interview by the fact that Joe had read RWA's company literature before the interview. When asked why, Joe said, "I like to be prepared, to know in advance what to expect of a situation."

Every exercise starts with **performance objectives.** The performance objectives indicate what new skills and knowledge will be learned in the course of completing the exercise.

The instructor will usually administer the requirements of the performance objectives.

The **background information** section presents all the information needed in order to perform the exercise and pass the review quiz. The background information section, which is usually the longest section in the exercise, contains important and detailed information.

The background information section also contains a **troubleshooting** area. Tips, techniques, and actual problems and their solutions are presented.

The next section in each exercise is the **self-test.** The self-test is there to help verify understanding of the material covered in the background information section.

The self-test is divided into several types of test questions. These are True/False, Multiple Choice, Completion, and Open-Ended. This is done to make the test more interesting and more reflective of what needs to be reviewed, and to help present the different types of questions asked during job interviews. Answers to odd-numbered self-test questions are given at the end of the book.

The next section in each exercise is the **familiarization activity.** This is the "hands-on" application of what was just learned.

The familiarization activity is usually performed in the lab. However, there may be some exercises that the instructor will assign as outside work. This is usually the case for exercises on software. In these exercises, the familiarization activity usually consists of a series of software interactions with a computer. These interactions can be performed as a homework assignment or done in some other place such as a computer room that provides access to computers for all students.

The next section found in all the exercises is the **questions/activities** section. In this section there are questions about the familiarization activity just completed. These questions are designed to help reinforce important concepts picked up during the familiarization activity. There may be times when other activities are suggested, and the instructor may or may not assign them. These other activities usually include outside assignments and are selected to give the opportunity to broaden understanding of the subject of the exercise.

The last section, the **review quiz,** restates the performance objectives. Note that some exercises may be skipped, read at home, or even covered out of sequence, depending on the educational situation.

ANCILLARIES

These ancillaries are available:

- **Instructor's Manual:** This manual includes solutions to selected text problems, teaching suggestions, and sample syllabi.
- **PowerPoint® Slides:** Figures from the text are designed to help instructors with classroom/lecture presentations. The slides are contained on a CD packaged with the Instructor's Manual.

SPECIAL NOTE

Exercises 20 and 21, which are specific to electronic and CAD applications, contain references to Thought Projects contained in the textbook *Strategies for Problem Solving Workbook*. You are encouraged to refer to the indicated Thought Project as you encounter new material.

ACKNOWLEDGMENTS

We would like to thank our editors, Frank Burrows and Charles Stewart, for their encouragement and assistance during the development of this project. Thanks also go to our copyeditor, Sandra Hutchinson, and our production supervisor at Pearson Custom Publishing, Kristen Colman.

We also thank everyone at ITT for their advice and assistance, especially Tom Bledsaw, Bill Perkins, Wen Liu, and Sunand Bhattacharya.

These companies and the following individuals were especially helpful to us regarding permission requests, and we deeply thank them:

- Scott Duncan of Interactive Image Technologies, Ltd. for providing Electronics Workbench (multiSIM). Contact Interactive at www.interactiv.com.
- Elizabeth Compton of America Online, Inc. for allowing screen shots of Netscape Communicator and the Netscape browser window, © 1999 Netscape Communications Corporation. Used with permission.
- Microsoft® Corporation for allowing screen shots of their operating systems and software applications, particularly Microsoft Office. Screen shots reprinted by permission of Microsoft Corporation.
- Network Associates, Inc. and McAfee.com Corp., for allowing screen shots of VShield and VirusScan.
- Farnoush Deylamian of Autodesk, Inc. for allowing screenshots of Actrix™ Technical.
- Prentice-Hall/Pearson Education for allowing screen shots of MathPro.
- Power Quest Corp., for allowing screen shots of PartitionMagic.

Kenneth C. Mansfield Jr.
mansfield_k@sunybroome.edu
http://www.sunybroome.edu/~mansfield_k

James L. Antonakos
antonakos_j@sunybroome.edu
http://www.sunybroome.edu/~antonakos_j

To Norma and Vince.

James L. Antonakos

To Betty and Bob.

Kenneth C. Mansfield Jr.

Brief Contents

Contents

UNIT ■ Introduction to Computer Technology

1 PC Technology Revolution

Don, Joe Tekk's manager at RWA Software, asked Joe if he would come with him to pick up some new computer equipment. They drove to a local computer store and loaded the company van with several large boxes.

When they returned, Don asked Joe to unpack everything and set up the new computer. Thirty minutes later, Joe turned power on and the new computer booted up. He showed it to Don with admiration. "It's a real nice system, Don. It has everything: 450 MHz Pentium III, 64 megabytes of RAM, 3-D hardware acceleration, a 12-gigabyte hard drive, and lots of other goodies."

Don smiled at Joe. "I'm glad you like it. It's your new computer."

PERFORMANCE OBJECTIVES

Upon completion of this exercise, you will be able to

1. Discuss the history of the personal computer.
2. Identify the major parts of a microcomputer system.
3. Explain the purpose of each of these major parts.
4. Browse the World Wide Web.

BACKGROUND INFORMATION

The personal computer has come a long way since its introduction in 1981. Back then, the 5 MHz 16-bit Intel 8086 microprocessor seemed fast, and 640 KB was plenty of memory. The first PC did not even have a hard drive. At power-on it loaded BASIC (a simple programming language) from on-board ROM, or booted DOS from a 5¼ inch floppy drive.

Today, Pentium III microprocessor clock speeds have hit 1000 MHz, making them two-hundred times faster than the 8086. In addition, the Pentium III is a 64-bit microprocessor, with internal cache (high speed memory), floating-point unit (high speed math), and many other special hardware features designed to enhance its performance.

Hard drives for the PC have also evolved, from the initial 20 MB drives to the 10 (or more) GB drives in use today. The price of one bit of hard drive storage has gone from 4.29 cents per bit to less than 0.000001 cents per bit over the past 20 years. System memory has increased from 640 KB to 32 MB or 64 MB (minimum RAM for reasonable performance).

Along with the hardware evolution of the PC there was a change in the software used to control the PC (the operating system). The first operating system, DOS, had a text-based, command-line oriented interface that required the user to remember and enter commands such as

```
DIR *.TXT /S /P
```

and

```
EDIT A:NAMES.LST
```

When Intel came out with more advanced microprocessors (the 80386, 80486, and finally the Pentium), DOS was unable to take advantage of the new features built-in to each processor (through a special mode of operation called *protected mode*). This paved the way for a new operating system called Windows. The Windows operating system, unlike DOS, runs in protected mode and utilizes a graphical user interface (along with a mouse) to provide an easy-to-use environment.

Let us take a look at a typical PC.

A TYPICAL MICROCOMPUTER SYSTEM

Figure 1.1 illustrates the major parts of a microcomputer system. Table 1.1 lists the major parts shown in Figure 1.1, as well as the purpose of each part.

Each of these major parts is called a **peripheral device** because the device (such as the printer) is separate from the microcomputer.

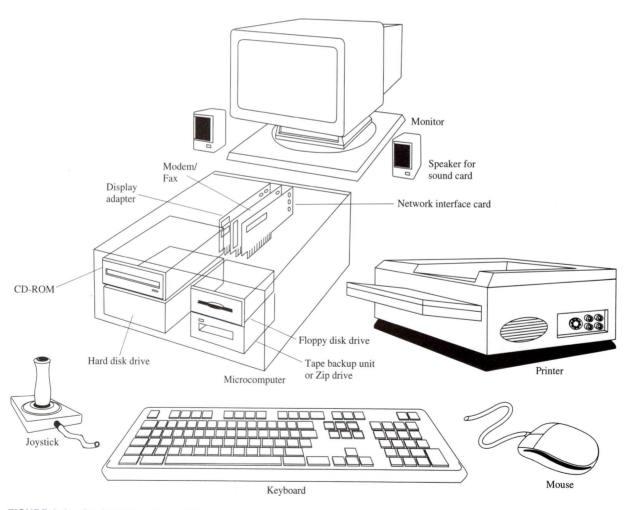

FIGURE 1.1   A microcomputer system

4

LAN (Local area network)

Figure 1.2 shows the relationship of each peripheral device to the microcomputer. As shown in Figure 1.2, some of the peripheral devices serve only as input devices. An **input device** is one that can only input information to the microcomputer.

Other devices serve as output devices. An **output device** is one that can only get information from the microcomputer.

The third type of peripheral device is the kind that serves as both an input and an output device. These devices are capable of putting information into the microcomputer as well as getting information from the microcomputer. Peripheral devices that are capable of both inputting information and getting information from the microcomputer are called **I/O (input/output) devices.**

TABLE 1.1   Major parts of a microcomputer system

Part	Purpose
Microcomputer	Central component of the system. Performs all the calculations and logic functions. Also called the CPU (central processing unit).
Keyboard	Consists of miniature switches with alphanumeric and other labels. Allows the program user to enter information directly into the computer.
Monitor	Contains a viewing screen. Gives the program user temporary information useful in the operation of the microcomputer. Requires a display adapter card, such as a graphics accelerator.
Hard disk drive	Serves as a storage place for information. Consists of one or more rigid magnetic disks used to store programs and other items useful to the user. These disks cannot be changed by the user.
Tape backup unit or Zip drive	Used to back up files to/from the hard drive. Tapes and cartridges are removable.
Floppy disk drive	Will copy information from or place information on small disks consisting of magnetic material. These disks are an easy and quick way of getting information into the microcomputer and can be changed by the user.
CD-ROM	Provides very large storage capability. Reads compact disks provided by the user. Rewritable CD-ROMs are also available.
Printer	Consists of a printing head and paper mechanism for the purpose of making permanent copies of useful information contained in the microcomputer.
Mouse	A small device moved by hand across a smooth surface. Used with information on the screen to control the microcomputer quickly and easily.
Joystick	Used for quick interaction with the monitor. Usually used for interacting with computer games.
Speakers	Provide left and right audio output from a soundboard.
Telephone modem	A device for transferring information between computers by use of telephone lines.
Network interface card	Used to connect the computer to a network.

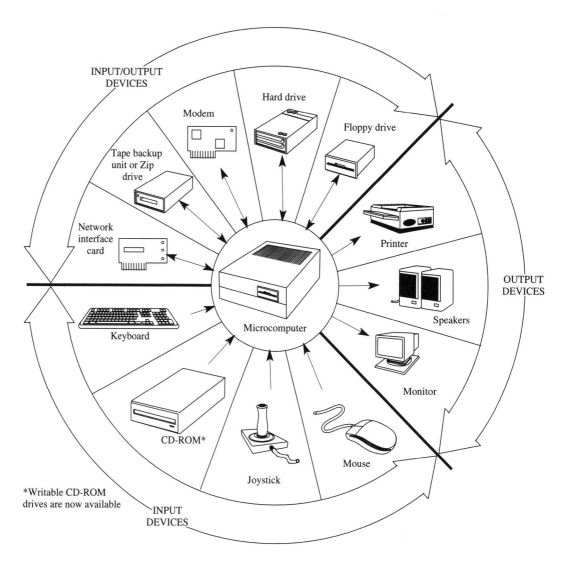

INPUT/OUTPUT
DEVICES

Hard drive

Modem

Floppy drive

Tape backup
unit or Zip
drive

Network
interface
card

Printer

OUTPUT
DEVICES

Keyboard

Microcomputer

Speakers

Monitor

CD-ROM*

Joystick

Mouse

*Writable CD-ROM
drives are now available

INPUT
DEVICES

FIGURE 1.2   Relationships of peripheral devices to the microcomputer

THE WINDOWS DESKTOP

As mentioned earlier, the operating system for the PC has evolved into a more powerful, graphical, point-and-click operating system which is much easier-to-use than DOS.

Figure 1.3 shows a screen shot of the Windows desktop. This is what you see after the machine boots up. Though the desktop is the subject of Exercise 4, a few points deserve mention now. Note that the time is displayed in the lower right corner. This is convenient, and eliminates having to constantly enter the TIME command in DOS.

The Start button at the lower left is a very important starting point for many Windows operations. Left-clicking the Start button once will bring up its menu of items.

Along the left side of the screen are the 18 icons representing applications or folders (directories in DOS terminology) that can be started or opened using a left double-click on the mouse when the mouse pointer is over the icon. For example, to start the MathPro application, place the mouse pointer (typically an arrow) over the icon that reads "mathpro" and left double-click.

If an application does not have an icon on the desktop you can still access it by using the Start button and selecting Programs from the menu. A list of all installed programs will appear, allowing you to select the MathPro application. This is typical of Windows, where one task can usually be performed several different ways.

FIGURE 1.3   Windows desktop with open application

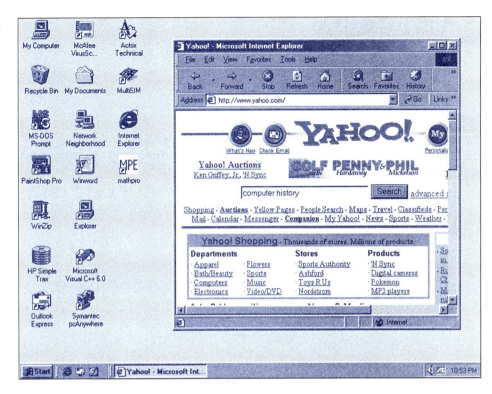

Next, note the large square window covering most of the desktop. This is the Internet Explorer World Wide Web browser window. The web site for the Yahoo search engine is currently being displayed. This was accomplished by entering 'www.yahoo.com' in the Address field near the top of the browser window. The information contained in the Yahoo web site contains images, text with different fonts and sizes, and interactive fields and buttons in a WYSIWYG format. WYSIWYG stands for what-you-see-is-what-you-get. This is a big improvement over the plain text interface used by DOS.

Last, notice that the user has entered the words 'computer history' in the Search field of the Yahoo page. When the Search button is left clicked, Yahoo will return a page of links to other web pages containing information about computer history. More information about the World Wide Web and the Internet can be found in Exercise 13.

The graphical interface of Windows, the existence of the World Wide Web, and the impressive power of today's computers have fundamentally changed the way we use our computers. Truly, we are in the middle of a PC technology revolution.

TROUBLESHOOTING TECHNIQUES

A brand-new computer fresh out of the box should assemble and boot up with a minimum of fuss. Many of the external connectors (keyboard, printer, mouse) only allow one way to plug in the appropriate cable, so there is no need to worry about accidentally plugging the printer cable into the mouse port.

If all external connections are correct and the machine will not boot, it could be the result of vibration damage that could have occurred during shipment. For example, an already-loose peripheral card may have simply popped out of its slot during transit. It is good to take a look inside the chassis (even before powering on for the first time, if necessary). A good visual inspection might turn up the problem.

This self-test is designed to help you check your understanding of the background information presented in this exercise.

True/False

Answer *true* or *false*.

1. A complete microcomputer system consists of more than just the microcomputer itself. *False*
2. All the calculations and logic functions are performed by the microcomputer. *True*
3. The keyboard can be viewed as an example of an output device. *False*
4. The monitor screen can be thought of as an input device. *False*
5. The hard disk drive is a storage place for information. *True*

Multiple Choice

Select the best answer.

6. An output device
   a. Copies information from the microcomputer.
   b. Puts information into the microcomputer.
   c. Can copy information from or put information into the microcomputer.
   d. Is the keyboard.
7. An input device
   a. Copies information from the microcomputer.
   b. Puts information into the microcomputer.
   c. Can copy information from or put information into the microcomputer.
   d. Is the monitor.
8. An I/O device
   a. Copies information from the microcomputer.
   b. Puts information into the microcomputer.
   c. Can copy information from or put information into the microcomputer.
   d. Is the printer.
9. A joystick is classified as an
   a. Output device.
   b. Input device.
   c. I/O device.
   d. None of the above.
10. The monitor is used
    a. For storing permanent information.
    b. When the printer fails.
    c. For putting information into the computer.
    d. As a temporary storage for immediately useful information.

Matching

Match each phrase on the left with the correct peripheral device or devices on the right.

11. Copies information from or places information into the microcomputer	a. Monitor
12. Used with the monitor for quick control of the microcomputer	b. Keyboard
13. Can transfer information between computer systems	c. Printer
14. Makes a permanent copy of information in the microcomputer	d. Mouse
15. Makes a temporary copy of information for immediate use	e. Joystick
	f. Modem
	g. Microcomputer
	h. Hard drive
	i. Floppy drive
	j. CD-ROM

12. D
13. F
14. H
15. A

8

Completion

Fill in the blanks with the best answers.

16. Any device that copies information into the microcomputer is called a(n) _input_ device.
17. Any device that copies information from the microcomputer is called a(n) _Output_ device.
18. I/O devices are capable of copying information _From_ the microcomputer as well as putting information into it.
19. A(n) _mouse_ is a small device moved by the hand across a smooth surface.
20. The disk drive that can have its disks changed by the user is called the _____ disk drive.

FAMILIARIZATION ACTIVITY

Your instructor may give a laboratory demonstration showing a complete microcomputer system, or your lab station may be equipped with a complete microcomputer system. In either case, you should know how to identify the following components of your system:

1. Microcomputer
2. Floppy, hard, and CD-ROM drives
3. Keyboard
4. Monitor
5. Printer
6. Modem
7. Network interface card
8. Mouse
9. Joystick
10. Tape backup unit or Zip drive

As an aid in familiarizing yourself with each of these peripheral devices, answer the following questions as they apply to the system used in the demonstration or at your lab station:

1. Who is the manufacturer of the microcomputer?
2. How many disk drives does the system contain?
3. Describe the monitor used in this system.
4. Who is the manufacturer of the printer?
5. How many keys are contained on the keyboard?
6. What is the capacity of the tape backup?

QUESTIONS/ACTIVITIES

Using a current computer magazine, list two manufacturers of each of the following:

1. Microcomputer
2. Modem
3. Printer
4. Monitor
5. Mouse/joystick
6. Disk drive
7. CD-ROM
8. Tape backup unit

Using a current computer catalog, list the prices of each of the following:

1. Telephone modem
2. Printer
3. Monitor
4. Disk drive
5. Mouse/joystick
6. CD-ROM
7. Network interface card

Under the supervision of your instructor, Start Up Internet Explorer, go to Yahoo, and search for "computer history." Look at 5 of the sites that show up in the search results.

Under the supervision of your instructor,

1. Discuss the history of the personal computer.
2. Identify the major parts of a microcomputer system.
3. Explain the purpose of each of these major parts.
4. Browse the World Wide Web.

2 An Introduction to Windows

INTRODUCTION

Joe Tekk was just about to leave for his lunch break when he was stopped by Don, his manager. "Hey, Joe, what happened with the laser printer?"

Joe replied that he had no luck trying to install the old drivers on the printer installation disks, so "I downloaded the new driver from the Web and installed it. Now the printer works fine."

"You did what?" Don asked, seeking more of an explanation.

"I brought up Netscape and used Yahoo to search for the printer manufacturer. Yahoo gave me a link to their home page. I went to it and found new drivers on their support page. I just had to click on the right one to download a copy to my hard drive."

Don was impressed. "Joe, you fixed a problem the customer has had for months with that laser printer. Good job!"

After Joe left, Don smiled to himself. "I would have searched with Alta Vista."

PERFORMANCE OBJECTIVES

Upon completion of this exercise, you will be able to

1. Explain the various items contained on the Windows 95 desktop.
2. Discuss the differences between Windows 3.x and Windows 95.
3. Briefly list the new features of Windows 95.
4. Identify the differences between Windows 95 and Windows 98.

BACKGROUND INFORMATION

The Windows operating system has gone through many changes since it first appeared in the mid-1980s. It has evolved from a simple add-on to DOS to a multitasking, network-ready, object-oriented, user-friendly operating system. As the power of the underlying CPU running Windows has grown (from the initial 8086 and 8088 microprocessors through the Pentium III, as well as other microprocessors), so too have the features of the Windows operating system. For users familiar with Windows 3.x, the good news is that many of the operating system features are still there. For example, a left double-click is still used to launch an application. The purpose of this exercise is to familiarize you with many of the features that are new in Windows 95. Where possible, comparisons will be made to Windows 3.x to help you gain an appreciation for how things have changed. In addition, many of the new features of Windows 98 will be introduced.

FIGURE 2.1 Windows 95 desktop

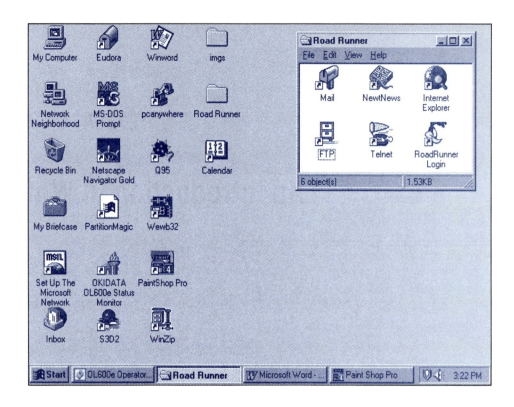

THE DESKTOP

Once Windows 95 has completely booted up, you may see a display screen similar to that shown in Figure 2.1. This graphical display is called the *desktop,* because it resembles the desktop in an office environment. The desktop may contain various *folders* and *icons,* a *taskbar,* the *current time,* and possibly *open folders* containing other folders and icons. A folder is more than just a subdirectory. A folder can be shared across a network, and *cut* and *pasted* just like any other object. You can even e-mail a folder if you want.

Typically, the bottom of the display will contain the *taskbar,* which contains the Start button, icons for all applications currently running or suspended, open desktop folders, and the current time. You can hide, resize, or move the taskbar to adjust the display area for applications. Simply left-clicking on an application's icon in the taskbar makes it the current application.

A new application may be launched by left double-clicking its desktop icon. The desktop may contain a picture, centered or tiled, called the *background image.* The desktop itself is an object that has its own set of properties. For example, you can control how many colors are available to display the desktop. Everything is controlled through the use of easily navigated pop-up menus. The desktop is the subject of Exercise 4.

LONG FILE NAMES

File names in DOS were limited to eight characters with a three-letter extension (commonly called 8.3 notation). Because Windows 3.x ran on top of DOS, it, too, was limited to file names of the 8.3 variety, even though Program Manager allowed longer descriptive names on the program icons.

Windows 95 eliminates the short file name limitation by allowing up to 255 characters for a file name. As shown in Figure 2.2, a *long file name* has two representations. One is compatible with older DOS applications (the old 8.3 notation using all uppercase characters). The other, longer representation is stored exactly as it was entered, with uppercase and lowercase letters preserved. To be compatible with older DOS applications, Windows 95 uses

FIGURE 2.2 Two examples of long file names

```
Volume in drive D is FIREBALLXL5
Volume Serial Number is 245F-15E6
Directory of D:\repair3e\e15

.                <DIR>        12-31-97 12:58a .
..               <DIR>        12-31-97 12:58a ..
LIST     TXT             0    01-06-98  2:01p list.txt
E15      DOC        31,232    01-06-98  2:01p e15.doc
WOWTHI~1 DOC        20,480    01-06-98  1:38p Wow This is a LONG filename.doc
WOWTHI~2 DOC        20,992    01-06-98  2:01p Wow This is LONG too.doc
         4 file(s)        72,704 bytes
         2 dir(s)     23,674,880 bytes free
```

the first six characters of a long file name, followed in most cases by ~1. When two or more long file names appear in the same directory, Windows 95 will enumerate them (~1, ~2, etc.), as you can see in the directory listing of Figure 2.2. To specify a long file name in a DOS command, use the abbreviated 8.3 notation, or enter the entire long file name surrounded by double quotes. For example, both of these DOS commands are identical in operation:

```
TYPE    WOWTHI~1.DOC
TYPE    "Wow This is a LONG filename.doc"
```

The great advantage of long file names is their ability to describe the contents of a file, without having to resort to cryptic abbreviations.

CONTEXT-SENSITIVE MENUS

In many instances, right-clicking on an object (a program icon, a random location on the desktop, the taskbar) will produce a *context-sensitive menu* for the item. For example, right-clicking on a blank portion of the desktop produces the menu shown in Figure 2.3. Right-clicking on the time in the lower corner of the desktop generates a different menu, as indicated by Figure 2.4. Note that the two example menus are different. This is what the "context-sensitive" term is all about. Windows 95 provides a menu tailored to the object you right-click on. This is a great improvement over Windows 3.x, which rarely did anything after right-clicking.

FIGURE 2.3 Context-sensitive desktop menu

FIGURE 2.4 Context-sensitive time/date menu

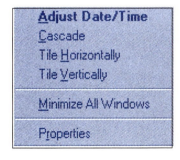

FIGURE 2.5   Help menu

IMPROVED HELP FACILITY

The built-in help available with Windows 95 is significantly different from that provided by Windows 3.x. To get help, go to the Start menu. Figure 2.5 shows the Help Topics window that is displayed when Help is selected.

Three tabs appear at the top of the display. The Index tab allows the user to enter keywords that might be found by looking in an index. As each letter is entered, the display is updated to show all matching items. Help for the highlighted selection (left-click to choose a different help topic) is then displayed by left-clicking the Display button or by double-clicking on the help topic.

The two other tabs, Contents and Find, provide additional support in the form of a guided tour of Windows 95, troubleshooting methods, tips, and alternate ways of finding specific help topics.

WINDOWS EXPLORER

Windows 3.x provided two main applications that made life bearable: Program Manager and File Manager. In Windows 95, the services provided by these two applications, as well as many new features, are found in the new Windows Explorer program. Figure 2.6 shows a typical Explorer window. Although we will cover Windows Explorer in detail in Exercise 6, it is worth taking a quick look at now. The small box in the upper left corner containing W95 (C:) indicates the current folder selected. Clicking the down arrow produces a list of folders to choose from. The two larger windows display, respectively, a directory tree of drive C: (folders only) and the contents of the currently selected folder (which also happens to be drive C:). Note the different icons associated with the files shown. Windows Explorer allows you to change the icon, or associate it with a different file. In general, as with Windows 3.x, double-clicking on a file or its icon opens the application associated with it.

Windows Explorer also lets you map network drives, search for a file or folder, and create new folders, among other things. It is truly one of the more important features of Windows 95.

THE REGISTRY

The Registry is the Windows 95 replacement for the SYSTEM.INI and WIN.INI configuration files used by Windows 3.x. The Registry is an internal operating system file maintained

FIGURE 2.6   Windows Explorer

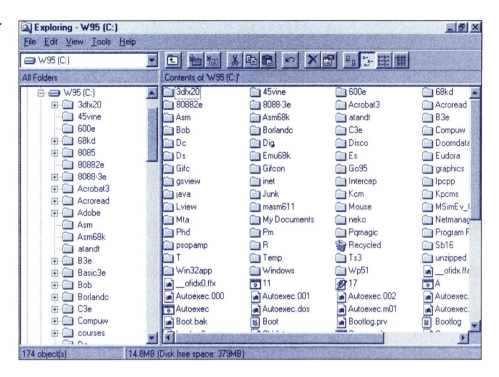

FIGURE 2.7   REGEDIT window

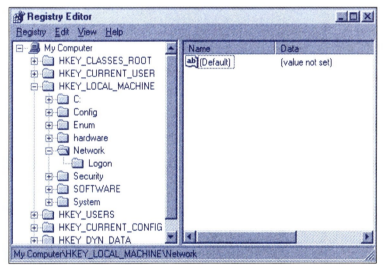

by Windows 95. As each application is installed, the installation program makes "calls" to the Registry to add configuration information, storing similar information to what was previously stored in the .INI files. In this way, the Registry is protected and therefore is harder to corrupt. The Registry is accessed using the REGEDIT program as illustrated in Figure 2.7. The Registry is nothing to fool around with simply because you feel like experimenting. A corrupt Registry can prevent Windows 95 from booting and could possibly require a complete reinstallation of Windows 95.

The Registry contains all the information Windows 95 knows about both the hardware and software installed on the computer.

NETWORKING

Windows 95 offers a major improvement in networking capabilities. Windows 3.1 had no built-in networking support and required network software to be loaded and maintained by DOS. This situation was improved slightly with the release of Windows for Workgroups 3.11, which provided limited networking via a network protocol called **NetBEUI** (NetBIOS Extended User Interface), which allows e-mail and file sharing in small peer-to-peer networks.

15

NetBEUI is still available in Windows 95 (providing the Network Neighborhood feature, network drives, shared printers), along with other additions. Two important protocols have been added, PPP (point-to-point protocol) and TCP/IP (transmission control protocol/Internet protocol). PPP is used with a serial connection (such as a modem) and is the basis for the dial-up networking provided by Windows 95. TCP/IP is the protocol used by the Internet. You can use TCP/IP applications such as Netscape or Internet Explorer to browse the World Wide Web, connect to a remote computer and share files, send and receive e-mail, and much more. Typically, TCP/IP is used in conjunction with a network interface card for fast data transmission, although it can also be used over a modem connection (by encapsulating it inside the PPP).

DOS

Yes, DOS is still a part of Windows. However, the role of DOS in Windows 95 is different in many ways from what was required under Windows 3.x. For example, Windows 3.x relied on the file system set up and maintained by DOS. This is commonly referred to as running "on-top" of DOS. Windows 95 does things differently, providing its own improved file system (long file names) and a *window* to run your DOS application in. You can open several DOS windows at the same time if necessary, and run different applications in each. Figure 2.8 shows a typical DOS window.

Furthermore, in Windows 3.x, it is possible for a wayward application running in a DOS shell to completely hang the system. Applications are not allowed that much control in Windows 95. If a DOS application (or any other, for that matter) hangs up, all you need to do is press Ctrl-Alt-Del (oddly enough) to bring up the Close window. Figure 2.9 illustrates the Close window.

FIGURE 2.8   DOS window

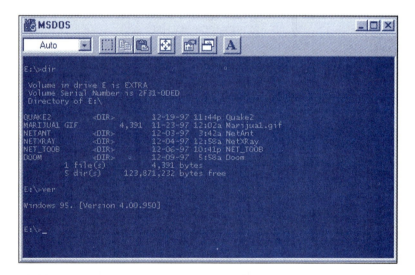

FIGURE 2.9   Close window

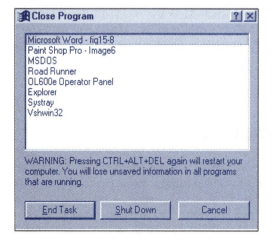

It is important to note that Windows 95 is always in control somewhere in the background, monitoring everything. A DOS application cannot hang the system like its Windows 3.x counterpart.

PREEMPTIVE MULTITASKING

A significant change in the way multiple programs are executed was made when Windows 95 was developed. Windows 3.x used a technique called *cooperative multitasking* to run more than one 16-bit application at a time. Each application would get a slice of the processor's computation time, with periodic switches between applications. The problem with this form of multitasking is that one task can take over the system (by requesting all available memory or other resource) and prevent the other tasks from running.

Windows 95 uses a method called *preemptive multitasking* to run multiple 32-bit applications while still running the older 16-bit Windows 3.x applications in cooperative mode (in a single shared block of memory). Preemptive multitasking means that the current application can be interrupted and another application started or switched to. This provides a degree of *fairness* to the set of applications competing for processor time. For example, referring back to Figure 2.1, it is possible to quickly double-click on several icons (Netscape, Eudora, Winword) before the first application has completely come up. All applications clicked will start up, with the first one clicked getting the initial shot at the desktop.

In Windows 3.x, when you double-clicked an application to start it, you had to wait until the hourglass went away before you could do anything else. In Windows 95, the hourglass still indicates that the operating system is busy with a chore, but it may be preempted to begin a new task at any time. Once again, we see that both the user and the operating system have more control under Windows 95 than was possible with Windows 3.x.

RECYCLE BIN

The Recycle Bin shown in Figure 2.10(a) is a holding place for anything that is deleted in Windows 95. The nice thing about the Recycle Bin is that *you can get your files back* if you want to, using the Undelete option. Double-clicking on the Recycle Bin icon brings up the window shown in Figure 2.10(b). Any or all of the files shown in the window may be recovered. *Warning:* If you delete files while in DOS, they are not deposited in the Recycle Bin, and may be impossible to recover at a later time (with the old UNDELETE command).

FIGURE 2.10   Recycle Bin

Recycle Bin

(a) Program icon

Name	Original Location	Date Deleted	Type	Size
~$fig	D:\repair3e\e15	1/9/98 3:44 PM	Microsoft Word Doc...	1KB
~$list	D:\repair3e\e15	1/9/98 3:44 PM	Microsoft Word Doc...	1KB
ENGLISH.RES	C:\GRAVIS\GRAVU...	1/4/98 12:06 AM	Intermediate File	102KB
GRAVUTIL	C:\GRAVIS\GRAVU...	1/4/98 12:06 AM	Application	365KB
GRAVUTIL	C:\GRAVIS\GRAVU...	1/4/98 12:06 AM	Help File	9KB
GRAVUTIL	C:\GRAVIS\GRAVU...	1/4/98 12:06 AM	Shortcut to MS-DOS...	1KB
OKIDATA OL600e...	C:\WINDOWS\Des...	1/7/98 9:24 PM	Shortcut	1KB

Recycle Bin — File Edit View Help

7 object(s)   475KB

(b) Window

OTHER NEW FEATURES

A summary of several other new features is included here to help you get a good idea of how much of an improvement Windows 95 is over Windows 3.x.

- *Remote Procedure Calls*   RPCs allow computers in a network to share their processing capabilities. For example, a 486 machine could issue an RPC to a Pentium-based machine to execute code *on that machine* and send the results back to the 486.
- *Support for OLE 2*   Object Linking and Embedding 2 (OLE 2) provides powerful features supporting a dynamic application environment. Cutting and pasting text, images, and other types of objects are just the beginning of OLE 2. Windows 95 provides a large set of enhanced OLE 2 features, such as drag-and-drop, nested objects, and optimized object storage.
- *Accessibility Options*   Many new features have been added to allow differently abled users to customize their computers to meet individual needs. The keyboard, display, sounds, and mouse can be set to a variety of combinations to suit most needs.
- *Microsoft Exchange*   E-mail is now a standard feature on the Windows 95 operating system.

WINDOWS 95 VERSION B

A number of bugs encountered in version A (the first release) of Windows 95 can be fixed by downloading Service Pack 1 from Microsoft's Web site (http://www.microsoft.com). The service pack updates Windows 95 by modifying various portions of the operating system (such as the kernel), and provides additional drivers and other improvements.

Version B of Windows 95 (called OEM Service Release 2) offers additional improvements, such as FAT32 (a better way of organizing files on your hard drive to reduce lost storage space), many new drivers, and other enhancements. Unfortunately, you cannot upgrade Windows 95 A to Windows 95 B.

THE WINDOWS 98 DESKTOP

Many changes in Windows 98 are visible right on the desktop. A *channel bar* provides one-click access to your favorite or often-used Internet links. A channel is a connection to a Web site that allows you to schedule automatic updates of the information you need. The desktop can be organized and operated as a Web page if you desire, even displaying a specific Web page as its background.

Figure 2.11 shows a sample Windows 98 desktop. The channel bar is to the right, and contains several preassigned Internet links. The taskbar has a new area that holds often-used shortcuts. This is a nice feature because desktop shortcuts are hidden from view in Windows 95 whenever a window uses the full screen.

WHAT ELSE IS NEW IN WINDOWS 98?

The Welcome to Windows tour (Programs, Accessories, System Tools) contains details for users of Windows 3.x, Windows 95, and new users. All the new features of Windows 98 are highlighted and presented in an easy-to-view multimedia format. Figure 2.12 shows the main feature screen. Clicking on a feature opens up a short slide show with narration and a quick run-through of the menus required to use the feature. Let us take a brief look at each feature group.

WINDOWS 98 IS EASIER TO USE

In addition to the Web-style features we will cover shortly, many other improvements are found in Windows 98. Two or more monitors (with associated display adapter cards) can

FIGURE 2.11   Windows 98 desktop

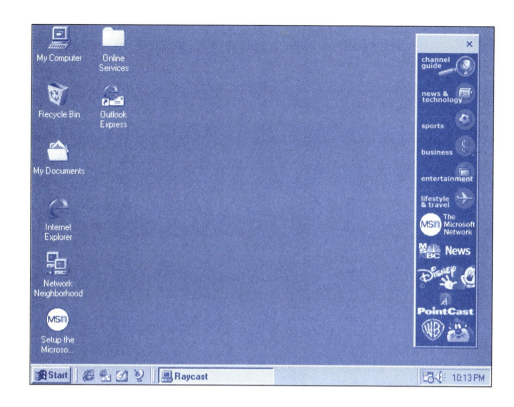

Figure 2.12   New features in Windows 98

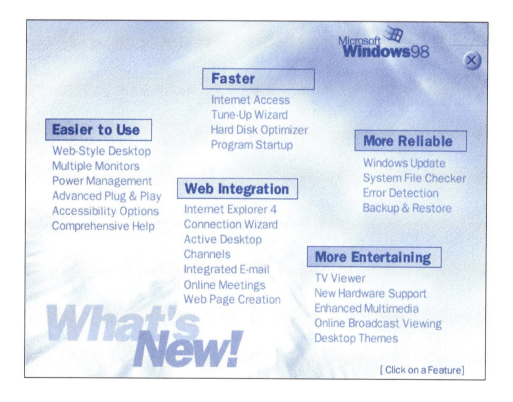

be used at the same time. Special OnNow hardware can be managed to reduce power use. Advanced Plug & Play includes support for the Universal Serial Bus. A new utility called Microsoft Magnifier uses a portion of the desktop as a magnifying glass. The portion of the desktop closest to the mouse is displayed with an adjustable level of magnification. In addition, a substantial amount of online help is provided.

FIGURE 2.13   Web-style desktop

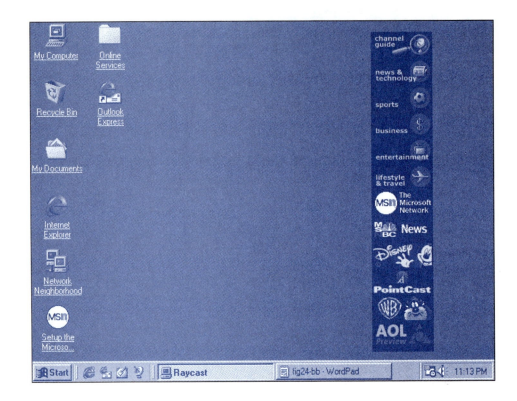

WINDOWS 98 IS ENHANCED FOR THE WEB

Windows 98 comes with Internet Explorer 4.0 included, a Connection Wizard to help make your connection to the Internet a smooth process, desktop Web page capabilities, Channels (favorite Internet links), Outlook Express, an all-purpose e-mail application, NetMeeting (use a digital whiteboard to sketch proposals, use Chat to confer with group members), and even FrontPage Express for creating your own Web pages. Figure 2.13 shows the Windows 98 desktop setup for Web-style appearance. If a Web page is used as the background, you can elect to hide the desktop icons to get a true browser appearance.

WINDOWS 98 IS FASTER AND MORE RELIABLE

Microsoft has added a number of features to Windows 98 to enhance its operating speed and increase reliability. The Hard Disk Optimizer will convert FAT16 file systems into the newer FAT32 file system, which allows for larger drives and more efficient use of drive space. In addition, applications that are used frequently can be relocated to a special portion of the disk to allow quicker launching.

Windows 98 can be updated automatically over the Web. For reliability, a System File Checker application checks the integrity of the files required by the operating system. At boot time, ScanDisk runs automatically if the previous shutdown was not completed correctly.

WINDOWS 98 IS MORE ENTERTAINING

A number of multimedia improvements are found in Windows 98. With special hardware, Windows 98 allows you to view broadcast television from the airwaves or gather and present information from the Internet (using NetShow). The Universal Serial Bus is also supported; this allows devices such as DVD (digital video disk) players and *force-feedback* joysticks to be connected.

DirectX support is built-in, enabling real-time high-quality 3-D graphics for supported graphics adapters. There are also many new screen savers (which take advantage of DirectX for stunning visual effects).

Overall, Windows 98 is another big step forward, as Windows 95 was compared to Windows 3.x.

TROUBLESHOOTING TECHNIQUES

There are so many new features in Windows 95/98 it is easy to lose track of them and not use one or more of them to make your life easier. For instance, rather than getting out a large, heavy Windows reference book, simply search the topics in Help for the answers you need.

Also, remember that it is usually OK to right-click on *anything,* any object (file, program icon, taskbar, display), and a context-sensitive menu will typically pop up. Sometimes you might get a What's This? button, which provides a short description of whatever you right-clicked on.

SELF-TEST

This self-test is designed to help you check your understanding of the background information presented in this exercise.

True/False

Answer *true* or *false.*

1. Long file names are limited to 500 characters. *False*
2. NetBEUI is a networking protocol. *True*
3. The Registry is just a database of .INI files. *False*
4. You can open only one DOS window at a time. *True*
5. Once a file enters the Recycle Bin it is gone forever. *False*
6. FAT32 is available on all Windows 95 computers. *False*
7. Windows 98 contains a channel bar used to watch television. *False*
8. The Windows 98 desktop can be organized and operated as a Web page. *True*
9. The Windows 98 operating system requires an 80486 processor or better. *False*

Multiple Choice

Select the best answer.

10. The desktop contains
    a. The taskbar, application icons, and folders.
    b. The taskbar, running applications, and a communications console.
    c. A set of folders for each hard drive.
    d. All of the above.
11. A context-sensitive menu is displayed when
    a. Double-clicking on a program icon.
    b. Pressing both mouse buttons at the same time.
    c. Right-clicking almost anything.
    d. None of the above.
12. Windows 95 supports
    a. 16-bit applications.
    b. 32-bit applications.
    c. Both a and b.
    d. DOS applications only.
13. Windows Explorer
    a. Replaces Program Manager and File Manager.
    b. Explores the Internet.
    c. Searches for viruses on the hard drive.
    d. None of the above.

14. Windows 98 includes Outlook Express, which provides:
    a. A digital whiteboard and a chat feature.
    b. An HTML editor to create custom Web pages.
    c. An all-purpose e-mail application.
    d. Reduced system power consumption during peak system activity.
15. The new Windows 98 utility program Microsoft Magnifier is used to:
    a. Show all channels on the desktop.
    b. Show a portion of the display as a magnifying glass.
    c. Send e-mail messages on the Internet.
    d. Access the devices on a Universal Serial Bus.
16. FAT32 is better than FAT16 because:
    a. Hard drive space is used more efficiently.
    b. Memory resources are reduced and the system runs faster.
    c. It cannot be used on small disk drives.
    d. 32 is higher than 16 and it is a multiple of 2.

Completion

Fill in the blank or blanks with the best answers.

17. Two new Windows 95 protocols are PPP and _HT ML_.
18. Windows 95 runs 32-bit applications using _several_ multitasking.
19. The _Desktop_ contains folders, icons, and the taskbar.
20. Windows Explorer can be used to map a(n) _specific_ drive.
21. A centralized location storing all configuration information about a Windows 95 system is called the _Bus_.
22. Windows 98 can use multiple _Tasking_ at the same time.
23. ScanDisk runs automatically on Windows 98 if the previous _scan_ was not successful.

1. Boot up Windows 95/98.
2. On the taskbar, left-click the Start button.
3. Move the mouse pointer up to the Help icon and left-click it.
4. In the Help Topics window, left-click the Index tab.
5. Enter the word "network" in the text box (left-click inside the box if the cursor is not visible).
6. From the list of network topics, chose Dial-Up Networking by left-clicking on it.
7. Left-click the Display button.
8. Left-click the Cancel button.
9. Double left-click the Dial-Up Network topic.
10. Left-click the Cancel button.
11. Left-click the Cancel button to return to the desktop.
12. Right-click on an empty portion of the desktop.
13. Move the mouse pointer to the New menu selection.
14. Move the mouse pointer to the Shortcut selection on the submenu.
15. Left-click on the Shortcut item.
16. Click the Browse button.
17. Locate the Windows folder icon (using the scroll bar).
18. Double left-click on the Windows folder icon.
19. Locate the Calendar icon (using the scroll bar).
20. Left-click on the Calendar icon.
21. Left-click the Open button.
22. Left-click on the Next button.
23. Click on the Finish button.
24. Double left-click on the Calendar icon.
25. Close the Calendar application.

QUESTIONS/ACTIVITIES

1. Does Windows 95 ever lose control of the system?
2. How are long file names backward compatible with the older DOS 8.3 notation?
3. What types of networking does Windows 95 provide?
4. Name five features found in Windows 98 that are not found in Windows 95.

REVIEW QUIZ

Under the supervision of your instructor,

1. Explain the various items contained on the Windows 95 desktop.
2. Discuss the differences between Windows 3.x and Windows 95.
3. Briefly list the new features of Windows 95.
4. Identify the differences between Windows 95 and Windows 98.

3 Windows NT

Joe Tekk was examining the computer book section of a local bookstore. Shelf after shelf contained books about the Windows operating system. There were books about Windows 95, Windows 98, Windows NT, and Windows 2000.

Joe thought about where the future of the Windows operating system is headed. Just skimming through the new features of Windows 98 showed him that a major push toward integrating the Web into the operating system was undertaken.

Next, Joe turned his attention to the Windows NT operating system. With support for multiple processors, improved security and system management tools, and full 32-bit code utilization, Windows NT has significant differences from Windows 95/98, and is well suited for network server operation.

A voice from behind broke into Joe's thoughts. "I remember when the entire operating system for the PC fit into 32KB of RAM."

Joe turned to see who had spoken to him. It was an old man, well into his 80s, hunched over and standing with the help of a cane. Joe noticed that the cane had a microchip encased in clear plastic on the handle.

"Now you need 16MB just to run the install program," Joe replied. He spent the next two hours talking to the old man about operating systems, and learned a great deal from him.

PERFORMANCE OBJECTIVES

Upon completion of this exercise, you will be able to

1. Identify the differences between Windows 95/98 and Windows NT.
2. Identify the key features of Windows NT.
3. Identify any common features of Windows 95/98 and Windows NT.
4. Describe the characteristics of Windows NT domains.

BACKGROUND INFORMATION

Windows NT is another operating system developed by Microsoft. It was developed to create a large, distributed, and secure network of computers for deployment in a large organization, company, or enterprise. Windows NT actually consists of two products: Windows NT Server and Windows NT Workstation. The server product is used as the server in the client-server environment. Usually a server will contain more hardware than the regular desktop-type computer, such as extra disks and memory. The workstation product is designed to run on a regular desktop computer (consisting of an 80486 processor or better).

25

Windows NT provides users a more stable and secure environment, offering many features not available in Windows 95/98 such as NTFS, a more advanced file system than FAT 16/32. The newest version of Windows NT is called Windows 2000.

We will use the Windows NT Server product to illustrate the user interface into the Windows NT environment. Let us begin by looking at the Windows NT login process.

WINDOWS NT OPERATING SYSTEM LOGON

One of the first things a new user will notice about the Windows NT environment is the method used to log in. The only way to initiate a logon is to press the Ctrl-Alt-Delete keys simultaneously as shown in Figure 3.1. This, of course, is the method used to reboot a computer running DOS or Windows 95/98. Using Windows NT, the Ctrl-Alt-Delete keys will no longer cause the computer to reboot, although it will get Windows NT's attention.

If the computer is not logged in, Windows NT displays the logon screen, requesting a user name and password. During the Windows NT installation process, the Administrator account is created. If the computer (Windows NT Server, Windows NT Workstation, Windows 95/98, or Windows for Workgroups 3.11) is configured to run on a network, the logon screen also requests the domain information. After a valid user name, such as Administrator, and the correct password is entered, the Windows NT desktop is displayed, as shown in Figure 3.2.

WINDOWS NT SECURITY MENU

After a Windows NT Server or Workstation is logged in, pressing the Ctrl-Alt-Delete keys simultaneously results in the Windows NT Security menu being displayed as illustrated in

FIGURE 3.1   Windows NT Begin Logon window

FIGURE 3.2   Windows NT desktop

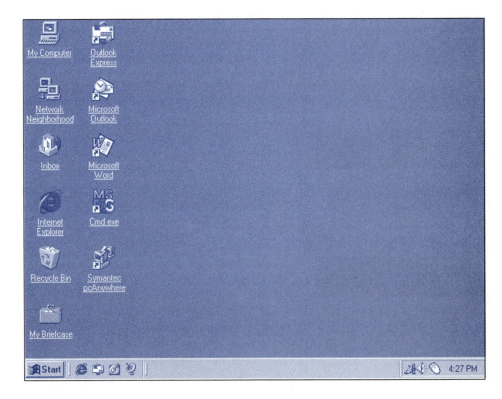

Figure 3.3. From the Windows NT Security menu it is possible for the operator to select from several different options, including Cancel to return to Windows NT.

The Lock Workstation option is used to put the Windows NT Server or Workstation computer in a *locked* state. The locked state is usually used when the computer is left unattended, such as during lunch, dinner, nights, and weekends. When a computer is locked, the desktop is hidden and all applications continue to run. The display either enters the screen saver mode or displays a window requesting the password used to unlock the computer. The password is the same one used to log in.

The Logoff option is used to log off from the Windows NT computer. The logoff procedure can also be accessed from the shutdown menu by selecting the appropriate setting. The logoff procedure terminates all tasks associated with the user but continues running all system tasks. The system returns to the logon screen shown in Figure 3.1. The Shutdown option must be selected before system power can be turned off.

The Change Password option is used to change the password of the currently logged-in user.

The Task Manager option causes the Windows NT Task Manager window to be displayed. The Task Manager is responsible for running all the system applications, as indicated by Figure 3.4. Notice that individual applications may be created, selected, and ended or switched by using the appropriate buttons. It is sometimes necessary to end tasks that are not functioning properly for some reason or another. In these cases, the status of the application is usually "not responding."

FIGURE 3.3   Windows NT Security menu

FIGURE 3.4   Task Manager applications

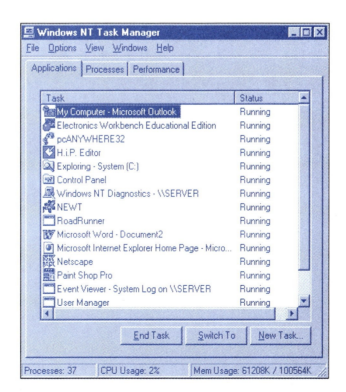

FIGURE 3.5   Windows NT processes

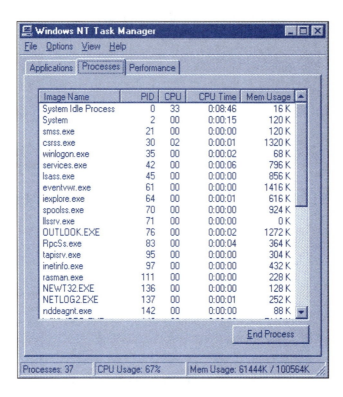

FIGURE 3.6   Task Manager Performance display

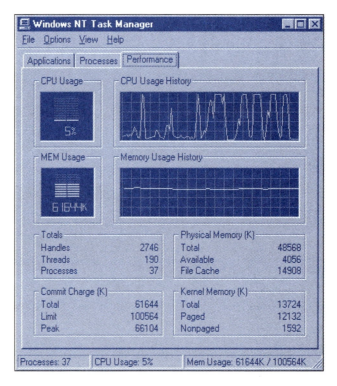

Each application controls processes that actually perform the required tasks. Figure 3.5 shows a number of processes being executed by the Task Manager. Applications may create as many processes as necessary. Extreme caution must be exercised when ending a process shown on the Processes display. The processes used to control Windows NT can also be ended, causing the computer to be left in an unknown state. If processes must be terminated, it is best to use the Applications tab.

The Task Manager can also display the system performance. Figure 3.6 shows a graphical display of current CPU and memory utilization. It also shows a numeric display of other critical information.

FIGURE 3.7 Windows NT Start menu

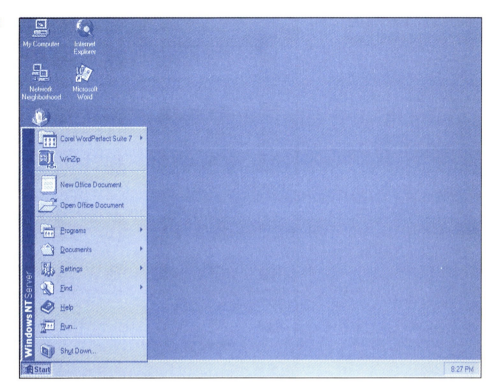

WINDOWS NT DESKTOP, TASKBAR, AND START MENU

The look and the feel of the Windows 95/98 desktop have been incorporated into the Windows NT desktop. At first glance, it might be hard to tell the difference between the two operating systems. Figure 3.7 shows the Start menu of a typical Windows NT desktop. Aside from the Windows NT text display along the left margin of the Start menu, it contains some of the same categories found on Windows 95/98 computers. As you will see, there are many similarities to Windows 95/98 (such as the desktop, Start menu, and the taskbar), and many differences.

WINDOWS NT CONTROL PANEL

Let us continue our investigation of Windows NT by looking in the Control Panel as illustrated in Figure 3.8. Within the Control Panel, there are several changes. For example, the System icon in Windows NT brings up a completely different display than Windows 95/98, as shown in Figure 3.9. This is no longer the place to go when examining system hardware components. The Devices icon provides that function in Windows NT. This is illustrated in Figure 3.10. Each individual device may be configured from this window.

The Services icon is used to control the software configuration on Windows NT. Usually, a service is started when the computer is booted, but it can also be started or stopped when a user logs in or out of Windows NT. Note that the Administrator account must be used when making changes to the Windows NT service configuration. A typical Services window is shown in Figure 3.11. You are encouraged to explore the different icons in the Windows NT Control Panel.

WINDOWS NT DOMAINS

Windows NT computers usually belong to a computer network called a *domain*. The domain will collectively contain most of the resources available to members of the domain. Computers running Windows NT Server software offer their resources to the network clients (Windows NT workstations, Windows 95/98, and Windows for Workgroups 3.11).

29

FIGURE 3.8 Items in the
Windows NT Control Panel

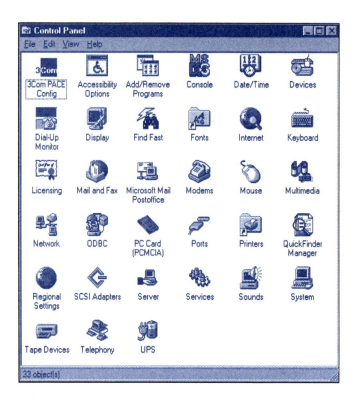

FIGURE 3.9 Startup and
shutdown system properties

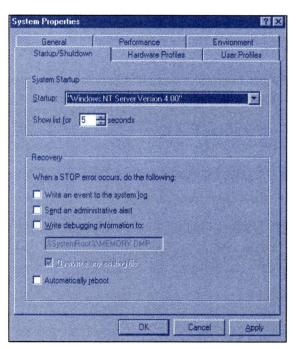

For example, during the logon process, a Windows NT Server responsible for controlling a domain will verify the user information (a user name and password) before access to the computer is allowed. The Network Neighborhood allows access to the resources available on other computers in the domain. Figure 3.12 shows the Network menu, where network components are configured.

The *network administrator* determines how the network is set up and how each of the components are configured. It is always a good idea to know whom to contact when information about a network is required. If the setting is not correct, unpredictable events may occur on the network, creating the potential for problems.

Any group of personal computers can be joined together to form either a workgroup or a domain. In a workgroup, each computer is managed independently, but may share some of

FIGURE 3.10   A list of devices

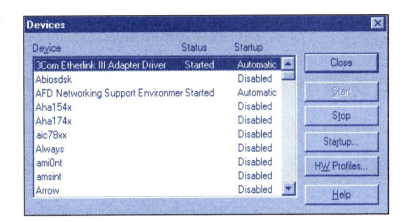

FIGURE 3.11   A list of services

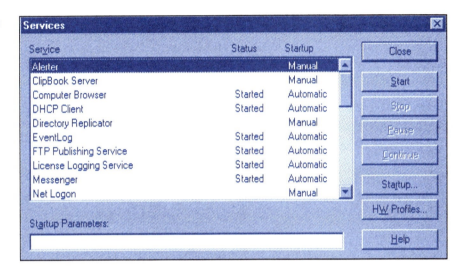

FIGURE 3.12   Currently installed Network Protocols

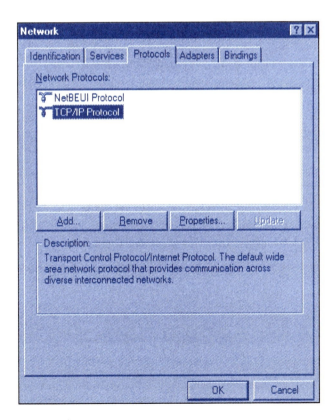

its resources with the other members of the network, such as printers, disks, or a scanner. Unfortunately, as the number of computers in the workgroup grows, it becomes more and more difficult to manage the network. This is exactly the situation where a Windows NT domain can be used. A domain offers a centralized mechanism to relieve much of the administrative burden commonly experienced in a workgroup. A domain requires at least one computer running the Windows NT Server operating system. Table 3.1 illustrates the characteristics of a workgroup and a domain.

DOMAINS

Each Windows NT domain can be configured independently or as a group where all computers are members of the same domain. Figure 3.13 shows two independent domains. Each domain consists of at least one Windows NT primary domain controller (PDC) and any number of backup domain controllers (BDC). One shared directory database is used to store user account information and security settings for the entire domain.

A BDC can be promoted to a PDC in the event the current PDC on the network becomes unavailable for any reason. A promotion can be initiated manually, causing the current PDC to be demoted to a backup. Figure 3.14 shows a domain containing two Windows NT server computers. One computer is the PDC, and the other computer is the BDC.

Windows NT can administer the following types of domains:

- Windows NT Server domains
- Windows NT Server and LAN Manager 2.x domains
- LAN Manager 2.x domains

A LAN Manager 2.x domain is a previous version of Microsoft networking software used by older MS-DOS and Windows computers.

The different types of activities that can be performed on a domain include the following:

- Create a new domain
- Modify an existing domain
- Join a domain
- Add a computer to a domain
- Remove a computer from a domain
- Synchronize files in a domain

TABLE 3.1 Comparing a workgroup and a domain

Workgroup	Domain
Small networks	Large networks
Peer-to-peer	Client–server
No central server	Central server
Low cost	Higher cost
Decentralized	Centralized

FIGURE 3.13 Independent Windows NT domains

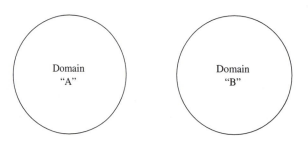

Domain "A" Domain "B"

- Promote a BDC to a PDC
- Establish trust relationships

When a system is set up as a PDC, the new domain name is required in order to proceed through the Windows NT installation process. This domain name is required by all other computer users who want to join the domain. Note that each domain can contain only *one* primary domain controller. All other Windows NT Server computers can be designated as backups or ones that do not participate in the domain control process at all.

A computer can be configured to join a domain during the Windows NT installation process, using the Network icon in the system Control Panel, or by using the Server Manager tool. A computer can be removed using the Network icon in the system Control Panel or the Server Manager tool.

Synchronizing a domain involves exchanging information between a primary domain controller and any secondary or backup domain controllers as previously shown in Figure 3.14. The synchronization interval for a Windows NT computer is five minutes. This means account information entered on the primary domain controller takes only five minutes to be exchanged with all secondary computers. This synchronization is performed automatically by Windows NT.

Domains can also be set up to offer *trust relationships*. A trust relationship involves either providing or receiving services from an external domain, as shown in Figure 3.15. A trust relationship can permit users in one domain to use the resources of another domain. A trust relationship can be a one-way trust or a two-way trust, offering the ability to handle many types of requirements.

A one-way trust relationship as shown in Figure 3.15(a) identifies domain "B" as a trusted source for domain "A." A two-way trust, shown in Figure 3.15(b), involves two separate domains sharing their resources with each other. Each domain considers the other to be a trusted source. Extreme caution must be exercised when setting up trust relationships. If the trusted domain is really untrustworthy, valuable information can be lost using the "trusted" accounts.

FIGURE 3.14   Domain "A" configuration

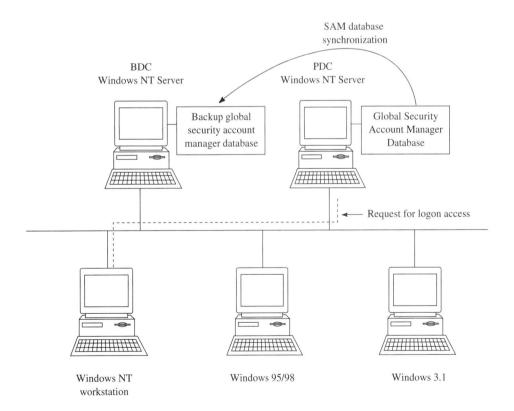

FIGURE 3.15   Domain trust
relationships

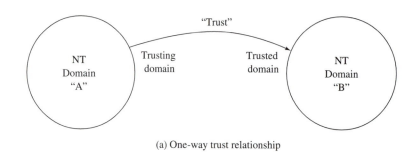

(a) One-way trust relationship

(b) Two-way trust relationship

DOMAIN CLIENTS

A Windows NT domain can support many different types of clients, such as:

- Windows NT servers
- Windows NT workstations
- Windows 95/98 clients
- Windows 3.11 clients
- Windows 3.1 clients
- MS-DOS clients
- OS/2 workstations

LOGGING ONTO A NETWORK

When a computer is configured to run in a network, each user must be authorized before access to the computer can be granted. Figure 3.16 shows a typical Windows 95/98 logon screen. Each user must supply a valid user name and a valid password in order to gain access to the computer and any network resources. In a *workgroup* setting, all password information is stored locally on each computer in PWL files. The PWL files are named using the following format: the first eight letters of the user name entered in the logon screen followed by the .PWL file extension. The PWL files contain account and password information stored in encrypted form. These files are typically stored in the Windows directory. Figure 3.17 shows the concept of a workgroup where each computer is administered independently.

FIGURE 3.16   Windows 95/98
logon screen

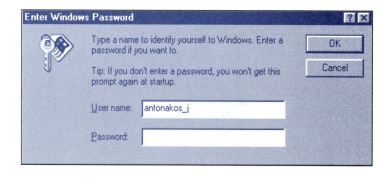

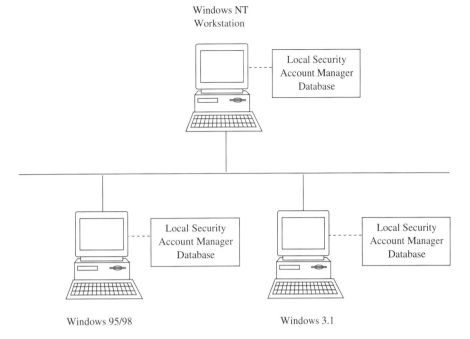

FIGURE 3.17   Workgroup concept where each computer is administered independently

Windows NT Workstation

Local Security Account Manager Database

Local Security Account Manager Database

Local Security Account Manager Database

Windows 95/98

Windows 3.1

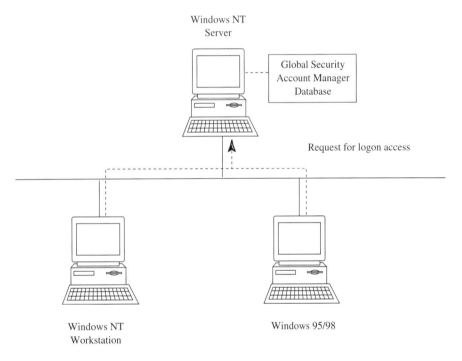

FIGURE 3.18   Domain concept

Windows NT Server

Global Security Account Manager Database

Request for logon access

Windows NT Workstation

Windows 95/98

In a *domain* setting, a centralized computer running Windows NT is contacted to verify the user name and password. If the information provided to the server is valid, access is granted to the local machine. If either the user name or password is invalid, access to the local computer is denied. As you might think, this method offers tremendously more flexibility as far as the administration is concerned. This concept is illustrated in Figure 3.18.

RUNNING A NETWORK SERVER

Running a network server involves installing the Windows NT operating system and then configuring it to run as a primary or secondary domain controller during the installation process. After a PDC is created during the installation, the domain exists on the network.

Windows NT computers can then join the domain by changing the Member of domain as shown in Figure 3.19. Windows computers join the domain by changing individual settings on each computer. Figure 3.20 shows the Primary Network Logon selecting the Client for Microsoft Networks option. Then, by selecting the properties for Client for Microsoft Networks, the specific domain can be identified as illustrated in Figure 3.21. After making these changes, a system reset is necessary to make the changes active.

Network server computers are also assigned the task of running more applications to manage both the server and network. For example, a Windows NT Server may be used to add fault tolerance to disks using a Redundant Array of Inexpensive Disks (RAID) technology. A server may also run the WWW server application, Windows Internet Naming System (WINS), Dynamic Host Configuration Protocol (DHCP), and Remote Access Server (RAS). These services are usually required 24 hours a day, seven days per week.

Windows NT Server computers are designed to handle the computing workload for entire organizations, corporations, or any other type of enterprise. In these cases, many

FIGURE 3.19   Configuring a Windows NT Server

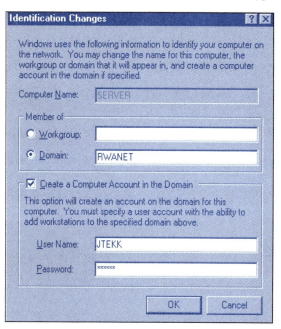

FIGURE 3.20   Windows 95/98 Network settings

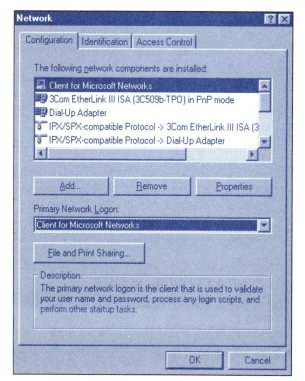

servers (including a PDC and several BDCs) are made available to guarantee the availability of any required services.

USER PROFILES

In a domain, the primary domain controller maintains all user profiles. This allows for centralized control of the Security Accounts Manager (SAM) database. Two programs are provided to update the SAM database. One of the programs is used in a stand-alone (no domain) environment and the other is for use where a domain is specified. Otherwise the programs operate in the same way. Let us examine what is involved when setting up a user account as illustrated in Figure 3.22. Information must be specified about each user account including user name, full name, a description of the account, and the password setting. The check boxes are used to further modify the account, such as requiring a password change during the first logon, restricting changing the password, extending the life of a password, and, lastly, disabling the account.

The three buttons at the bottom of the New User window (Figure 3.22) allow for each new account to be added to different *groups* as shown in Figure 3.23. It is a good idea to

FIGURE 3.21 Configuring Windows 95/98 to log on to a domain

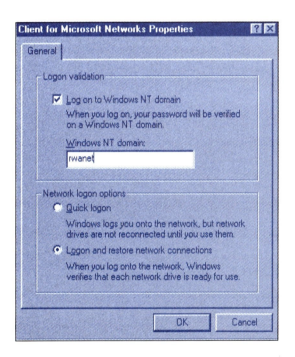

FIGURE 3.22 New User dialog box

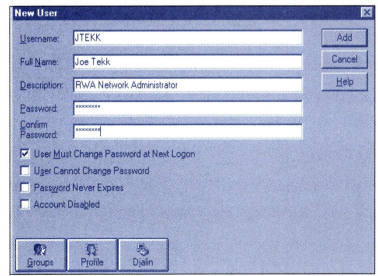

grant access to groups on an individual basis as certain privileges are granted by simply belonging to the group, such as administrator.

The User Environment Profile screen specifies the path to an individual profile and any required logon script name. Additionally, the home directory may be specified as shown in Figure 3.24.

Lastly, the Dialin Information window determines if a Windows NT account has access to Dial-Up Networking. The Call Back option may also be configured to require the computer to call the user back. This is an additional security feature that may be implemented if necessary. Figure 3.25 shows these settings.

SECURITY

Windows NT is a C2 compliant operating system, when it is configured properly as defined by the National Computer Security Center (NCSC). C2 compliance involves properly con-

FIGURE 3.23 Group Memberships selection screen

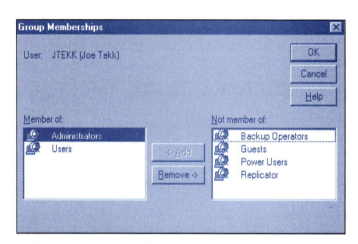

FIGURE 3.24 User Environment configuration screen

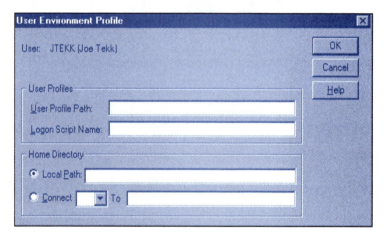

FIGURE 3.25 Dialin Information settings

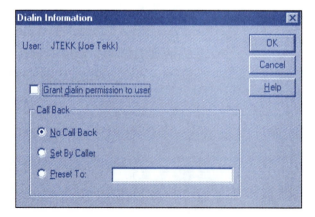

FIGURE 3.26 System events display

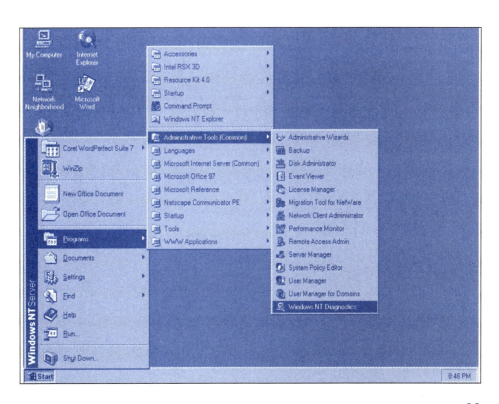

Event Viewer - System Log on \\SERVER

Log View Options Help

Date	Time	Source	Category	Event	User	Computer
5/5/98	11:26:08 PM	Service Control Mar	None	7000	N/A	SERVER
5/5/98	11:24:50 PM	BROWSER	None	8033	N/A	SERVER
5/5/98	10:55:30 PM	Srv	None	2013	N/A	SERVER
5/5/98	10:55:30 PM	Srv	None	2013	N/A	SERVER
5/5/98	10:55:30 PM	Srv	None	2013	N/A	SERVER
5/5/98	10:50:29 PM	Service Control Mar	None	7001	N/A	SERVER
5/5/98	10:50:29 PM	Service Control Mar	None	7001	N/A	SERVER
5/5/98	10:50:27 PM	Service Control Mar	None	7000	N/A	SERVER
5/5/98	10:50:27 PM	Service Control Mar	None	7000	N/A	SERVER
5/5/98	10:50:14 PM	EventLog	None	6005	N/A	SERVER
5/5/98	10:50:27 PM	Service Control Mar	None	7000	N/A	SERVER
5/5/98	3:12:37 PM	BROWSER	None	8033	N/A	SERVER
5/5/98	3:10:18 PM	Srv	None	2013	N/A	SERVER
5/5/98	3:10:18 PM	Srv	None	2013	N/A	SERVER
5/5/98	3:10:18 PM	Srv	None	2013	N/A	SERVER
5/5/98	3:05:07 PM	Service Control Mar	None	7001	N/A	SERVER
5/5/98	3:05:07 PM	Service Control Mar	None	7001	N/A	SERVER
5/5/98	3:05:05 PM	Service Control Mar	None	7000	N/A	SERVER
5/5/98	3:05:05 PM	Service Control Mar	None	7000	N/A	SERVER
5/5/98	3:04:51 PM	EventLog	None	6005	N/A	SERVER
5/5/98	3:05:05 PM	Service Control Mar	None	7000	N/A	SERVER
5/5/98	3:03:46 PM	BROWSER	None	8033	N/A	SERVER
5/5/98	3:01:55 PM	Service Control Mar	None	7001	N/A	SERVER

figuring Windows NT to use the built-in safeguards. An application tool supplied with the operating system (C2CONFIG.EXE) examines the operating system setting against a recommended setting. Any exceptions are noted.

Windows NT provides security-logging features designed to track all types of system activities, such as logon attempts, file transfers, Telnet sessions, and many more. Typically the System Administrator will determine which types of events are logged by the system. Figure 3.26 shows the system log. The icons along the left margin are color coded to draw attention to more serious events. Event logs should be reviewed daily.

TROUBLESHOOTING TECHNIQUES

Windows NT provides many different troubleshooting aids to tackle a wide variety of common problems. The Administrative Tools menu shown in Figure 3.27 shows a list of common applications designed to properly configure a Windows NT computer. One of the

FIGURE 3.27 Administrative Tools menu for Windows NT

most common administrative tools is the Windows NT Diagnostics application shown in Figure 3.28. The Diagnostics menu contains several different tabs, each showing a specific area of the system. When experiencing problems, it is always a good idea to examine all the information shown on the diagnostics window.

A networked computer environment (especially when using Windows NT) can become somewhat complex, requiring the system or network administrator to have many technical skills. Fortunately, Windows NT also provides many resources designed to tackle most networking tasks. For example, the Administrative Tools menu contains the Administrative Wizards option shown in Figure 3.29. Most of these wizards perform the activities that are necessary to get a network up and running.

FIGURE 3.28 Windows NT Diagnostics system information

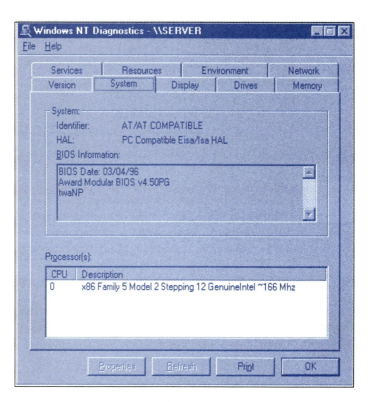

FIGURE 3.29 Administrative Wizards display

FIGURE 3.30   Windows NT Help display

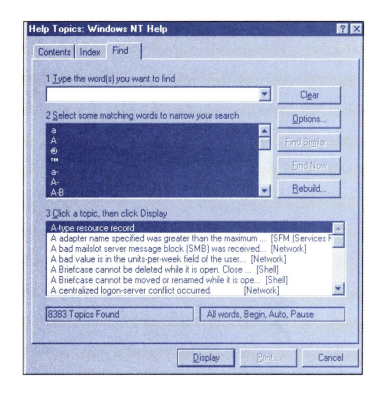

It is also a good idea to examine the online help system to get additional information, which may simplify any task. Figure 3.30 shows a Help screen that contains a total of 8383 topics, many of which contain information about networking.

SELF-TEST

This self-test is designed to help you check your understanding of the background information presented in this exercise.

True/False

Answer *true* or *false*.

1. Logoff can be selected from the Windows NT Security menu. *False True*
2. A task that is not responding must be stopped manually. *True False*   *False*
3. The Windows NT operating system has the look and feel of a Windows 98 computer. *True*
4. The Windows NT operating system requires an 80486 processor or better. *False*
5. The Devices icon in the Windows NT Control Panel replaces the Add New Hardware icon in Windows 98. *True False*
6. Each Windows NT domain can be configured independently. *True*
7. A primary domain controller can be demoted to a backup domain controller. *True*
8. A backup domain controller is updated every 10 minutes. *False 5min*
9. Windows 95 computers function only marginally in a Windows NT domain. *False*

Multiple Choice

Select the best answer.

10. Windows NT consists of two products:
    a. Client and server.
    b. Client and workstation.
    c. Server and workstation.
    d. Server and client workstation.
11. Windows NT uses the
    a. FAT32 file system.
    b. NTFS file system.

41

    c. RBFS file system.

    d. All of the above.

12. The Task Manager is used to

    a. Put the computer in a locked state.

    b. Log off from Windows NT.

    c. Create new system applications.

    d. None of the above.

13. When a computer is in a locked state,

    a. The desktop is hidden and applications are paused.

    b. The desktop is hidden and applications continue to run.

    c. The desktop is not hidden and applications are paused.

    d. The desktop is not hidden and applications continue to run.

14. The Windows NT Security menu is displayed when the

    a. Computer is booted the first time.

    b. Computer is accidentally turned off and then turned back on.

    c. User presses the Ctrl-Alt-Delete keys simultaneously.

    d. User presses Ctrl-C or Ctrl-Break repeatedly.

15. Windows computers are added to a Windows NT domain by

    a. Double-clicking on the Windows NT computer in the Network Neighborhood.

    b. Modifying the properties of the TCP/IP network settings.

    c. Modifying the properties of the Client for Microsoft Network settings.

16. A Windows NT Server can be a

    a. Parent domain controller and a child domain controller.

    b. Secondary domain controller and a backup domain controller.

    c. Primary domain controller and a secondary domain controller.

17. A trusted domain

    a. Contains only one primary domain controller and no secondary controller.

    b. Is granted special access to all Windows computers in the trusted domain.

    c. Permits users in one domain to use the resources of another domain.

18. Running a network server involves

    a. Installing and configuring a Windows NT Workstation computer.

    b. Installing and configuring a Windows NT Server computer.

    c. Connecting Windows 95 computers to Windows NT workstation computers.

Completion

Fill in the blank or blanks with the best answers.

19. The _shutdown_ option will cause a Windows NT computer to perform a system reset operation and restart the operating system.

20. The Windows NT task can be stopped when a process is _____ _____.

21. If a Windows NT task must be stopped, it is best to use the _aplication_ tab.

22. Disk _____ is an example of an administrative tool.

23. When experiencing problems, it is a good idea to examine the _Diagnostic_ window.

24. A backup domain controller is _____ to a primary domain controller.

25. A large number of computers cannot be managed effectively in a(n) _Workgroup_ setting.

26. Computers administered centrally are part of a(n) _____.

27. Each domain must contain _one_ primary domain controller.

Using a Windows NT computer, perform the following activities:

1. Open the Control Panel.

2. Run each of the Control Panel applications to determine what type of settings may be controlled.

3. Close the Control Panel.

4. Examine the administrative tools.

5. Review online help information.

6. Configure a Windows NT computer to function as a primary domain controller.

7. Configure a Windows NT computer to function as a backup domain controller.
8. Change the name of the domain.
9. Add some client computers to the Windows NT domain.
10. Examine networking topics using the online help system.
11. Run the Administrative Wizard applications to become familiar with account and network information.

QUESTIONS/ACTIVITIES

1. Create a new Windows NT user using the User Manager administrative tool.
2. Look at the Windows NT Event Viewer application.
3. Under what circumstances can a Windows NT workstation computer become a primary domain controller? A Windows NT Server computer?
4. When should a Windows NT domain be used instead of a workgroup?
5. What is necessary for an operating system to become a network client?
6. What is C2 security?
7. Where can additional C2 information be found?

REVIEW QUIZ

Under the supervision of your instructor,

1. Identify the differences between Windows 95/98 and Windows NT.
2. Identify the key features of Windows NT.
3. Identify any common features of Windows 95/98 and Windows NT.
4. Describe the characteristics of a Windows NT domain.

4 The Desktop

Jeff Page, who was in charge of all Web development at RWA Software, ran into Joe Tekk's office. He was covered with sweat. "Joe, you've got to come and help me. My machine is trashed."

Joe followed as Jeff ran back to his office. "I've rebooted my machine five times and get the same result each time. My desktop is completely blank. There are no icons for anything."

Joe looked at the display. The taskbar was visible at the bottom of the display screen, but the rest of the screen was a large empty patch of green background. Joe tried to reassure Jeff. "Believe it or not, Jeff, I've seen something like this before. I think I might know what is wrong."

Joe sat down and grabbed the mouse. Before long, he had clicked his way into the Recycle Bin, which contained, among other things, all of the missing desktop icons. Within moments the icons were restored.

Jeff wondered how Joe had identified the problem so quickly. Joe explained, "Last week one of the network guys brought his daughter with him while he did some work. She ran over to an open laptop and deleted every icon on the desktop. I think she figured out what happens when you right-click, and learned how to delete things to make them invisible. Apparently she knows just enough about Windows to cause trouble with it."

"Do you think she's still here?" Jeff asked, looking worried. "I have to leave for a basketball game in 20 minutes and I need to leave my machine on."

Joe laughed and shook his head. "Password protect your screen saver. That will keep her out."

PERFORMANCE OBJECTIVES

Upon completion of this exercise, you will be able to

1. Demonstrate the different features of the Start menu.
2. Identify items on the taskbar.
3. Change the appearance and properties of the desktop.
4. Create and use shortcuts.
5. Explain the function of the standard desktop icons.

Since the desktop is common to all three advanced Windows operating systems (95, 98, NT), we will refer to it simply as the Windows desktop. Desktop features unique to a single operating system will be discussed as necessary. The desktop is the centerpiece of the Windows operating system. The desktop typically contains a number of standard icons provided by the operating system as well as user-defined icons. As with the Windows 3.x operating system, an application is launched by double-clicking the program icon, although single-clicking may also be used in Web-style desktop settings.

The desktop also contains the taskbar, a system tray, any open folders and applications, and an optional background image called a wallpaper. Figure 4.1 shows the contents of a typical desktop. All the icons down the left-hand side of the desktop are automatically created and placed on the desktop when Windows is installed. These icons are entry points to the many built-in features of Windows. All the other icons are *shortcuts,* user-defined links to programs or other files that are used frequently.

In the upper right corner of the desktop shown in Figure 4.1 is an open *folder* called "Road Runner," which contains icons for six objects. Recall that Windows is an object-oriented operating system. Double-clicking on any of the icons in the Road Runner folder will start their associated applications. Note that the taskbar at the bottom of the desktop contains an entry for the Road Runner folder.

The lower right-hand corner of the desktop contains the *system settings area,* where several background tasks (virus protection, speaker volume, and the time-of-day clock) are represented by small icons. The icons may be hidden if necessary, to make more room on the taskbar.

Finally, the background of the desktop itself is an object that has properties that we can alter, such as what type of image to display, the overall display resolution, and the number of colors available. In the following sections we will discuss each of the many desktop components in greater detail.

THE START BUTTON

Microsoft made it easy to begin using Windows by placing the Start button at the lower left-hand corner. Everything it is possible to do in Windows is accessible through the Start button. Left-clicking on the Start button generates the menu shown in Figure 4.2. The Shut Down, Run, and Help items are selected by left-clicking on them. The other four items

FIGURE 4.1 Sample Windows desktop

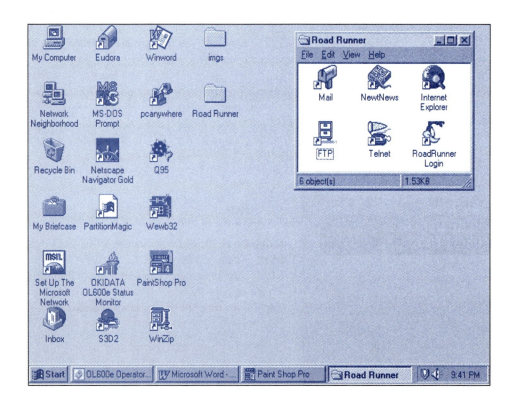

FIGURE 4.2   Start menu

FIGURE 4.3   Windows 95 Shut Down menu

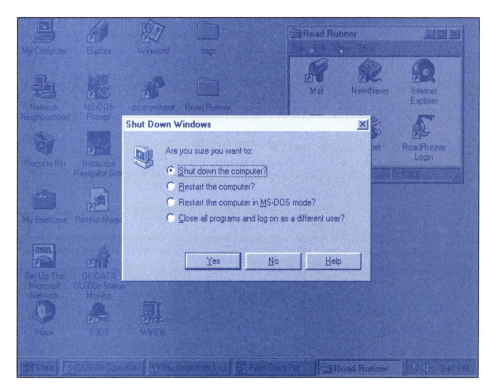

produce submenus when selected by the mouse (which only needs to pass over the menu item to create the associated submenu). Let us look at each menu item and its operation.

The Shut Down item, when left-clicked, produces the display shown in Figure 4.3. The display is darkened, except for a bright Shut Down menu. You may back out of Shut Down by left-clicking the No button (or the Close box). You may also choose to restart the computer in the MS–DOS mode of operation.

The Run menu item allows a program to be executed by typing the file name in the text box or by selecting it from a graphical file menu accessed through the Browse button. Figure 4.4 shows a Run dialog box with the file name "scandisk" entered in the text box. The Run menu gives you one method to access programs that are not contained on the desktop.

The next Start menu item, Help, provides access to the help facility, which is vastly improved over that of Windows 3.x. The Help menu is shown in Figure 4.5. Do not let the short list of topics mislead you. There is so much help available that you could spend days reading the various topics provided. There is help for Dial-Up Networking, managing printers, customizing your desktop and Windows environment, and much more. You are encouraged to spend some time looking through the Help information; you may pick up some valuable tips on the way.

FIGURE 4.4   Run dialog box

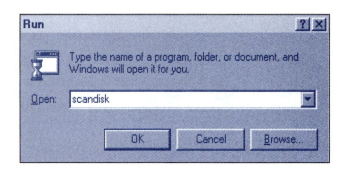

FIGURE 4.5   Windows 95 Help menu

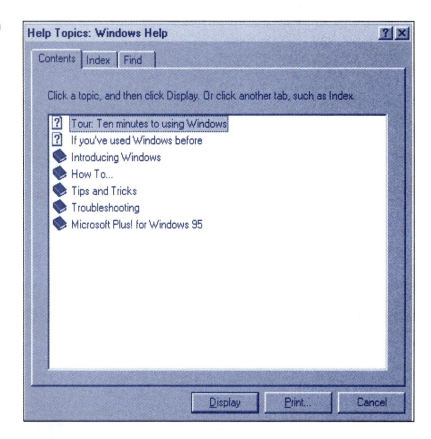

FIGURE 4.6   Windows 95 Settings submenu

The Settings menu item opens up a submenu to three other applications, as illustrated in Figure 4.6. The Control Panel is a powerful set of utilities that are used to control how the hardware and software work under Windows. The Control Panel is the subject of Exercise 5. The Printers item brings up the Printers folder, which we will examine in Exercise 7. The third item, Taskbar, opens up the Taskbar Properties window, which we will look at in the next section. The contents of the Start menu are controlled using the Taskbar option from the Settings submenu.

The Documents menu item produces a submenu containing links to the last 15 documents opened. As shown in Figure 4.7, files such as Microsoft Word documents appear on the Documents submenu, placed there automatically when opened by the user.

FIGURE 4.7 Documents
submenu

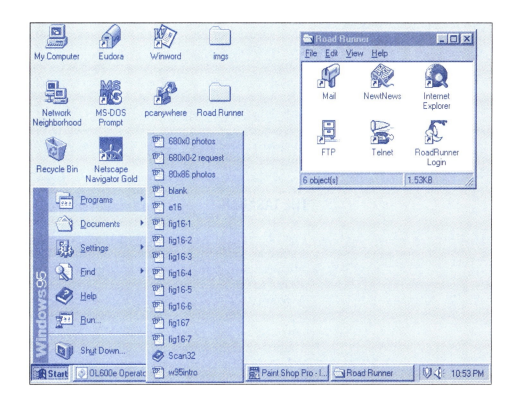

FIGURE 4.8 Programs
submenu

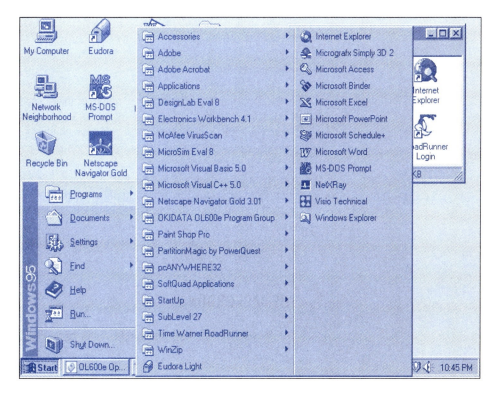

The last item on the Start menu is the Programs menu item, the starting point when starting a new application. Placing the mouse over the Programs menu item generates a submenu similar to that illustrated in Figure 4.8. The Programs submenu consists of folders and program icons. Each folder icon and associated folder name shown in the list of applications indicates that a submenu exists for that particular item. Also note the small black triangles that point to the right on many of the menu items. These indicate that submenus exist for the menu items. As the mouse is dragged over this type of menu item, a new submenu will appear.

When the mouse is dragged over a program icon (NetXRay, for example), the icon becomes highlighted, and its associated application can be launched by simply left-clicking on it.

As new applications are installed, they are automatically added to the list of applications on the Programs submenu.

If you change your mind and do not want to run an application from the Programs submenu, simply left-click anywhere on the desktop outside the submenu area or choose a different menu item from the Start menu.

THE TASKBAR

The taskbar is used to switch between tasks running on the desktop. If no tasks are running, the taskbar contains only the Start button and the System Settings area. When other applications are running, or folders are open on the desktop, their icons will be displayed on the taskbar. In Figure 4.1 the taskbar contains four entries. The first three are for applications, the fourth is for the open Road Runner folder. Program and folder entries in the taskbar appear from left to right in the order they were started or opened. The taskbar can be resized, moved, and set up to automatically hide itself when not in use. Figure 4.9(a) shows a taskbar with so many applications displayed on it, it is impossible to read the names of any of them. In Figure 4.9(b) the same taskbar has been resized to allow more space for each application to be identified. To resize the taskbar, move the mouse near the top of the taskbar. It will change into an up/down arrow. Left-click and hold to drag the taskbar to a larger or smaller size, and then release.

The taskbar can also be moved to any of the four sides of the display. Figure 4.10 shows the taskbar moved to the right-hand side of the display. To move the taskbar, left-click and hold on any blank portion of the taskbar, then drag the mouse to the desired side of the display and release.

The taskbar has its own set of properties (such as its Auto hide feature) that can be adjusted. Recall that the Settings menu item of the Start menu displays a submenu containing a taskbar item. This is the entry point to the Taskbar Properties window. You can also right-click on a blank area of the taskbar and get a context-sensitive menu containing a Properties item you can click. Figure 4.11 shows the contents of the Taskbar Properties window.

You should experiment with each of the settings to see what they do and to settle on a format for the taskbar that you find pleasing. For example, if the Always on top box is unchecked, it is possible for an application to completely cover up the taskbar when it is opened up to full screen size. The only way to get the taskbar back is to minimize all applications or move applications around on the desktop to uncover the taskbar. Checking the Always on top box allows Windows to do the housecleaning for you, and keep the taskbar visible no matter how many applications are open (unless the Auto hide feature is enabled).

THE BACKGROUND

The look of the desktop background can be adjusted for a pleasing appearance. Right-clicking anywhere on an empty portion of the desktop brings up the Display Properties

FIGURE 4.9 Two views of the taskbar

(a) The taskbar with many applications displayed on one row

(b) The same taskbar resized to two rows

FIGURE 4.10   A different location for the taskbar

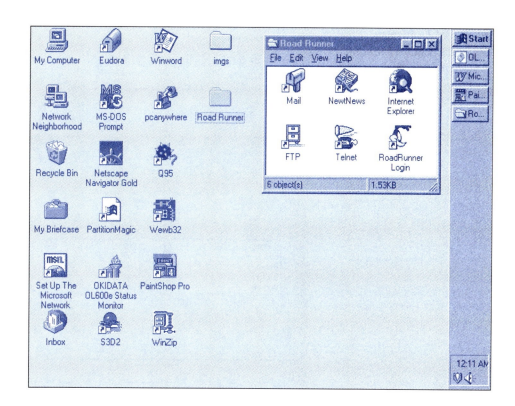

FIGURE 4.11   Taskbar Properties window

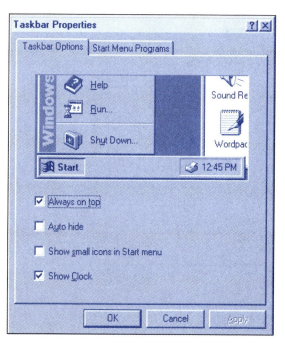

window shown in Figure 4.12. The Background controls allow you to choose a pattern or wallpaper to paint over the background. Both may be selected by double-clicking on their entry. Custom wallpapers can be selected using the Browse button, which allows selection of a *bitmap* file (.BMP extension), a standard Windows file format used for graphic images. The bitmap files displayed in the list are located in the directory where the Windows operating system is installed.

When a pattern or wallpaper type is selected with a single click, the display icon shows an example of what you will see. This is illustrated in Figure 4.13. Note that the pattern chosen will completely cover the background of the desktop. If a wallpaper is

FIGURE 4.12 Display Properties window showing Background controls

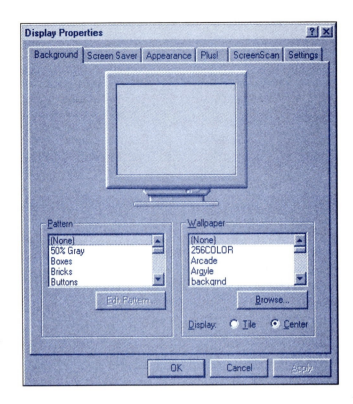

FIGURE 4.13 Choosing the Boxes pattern

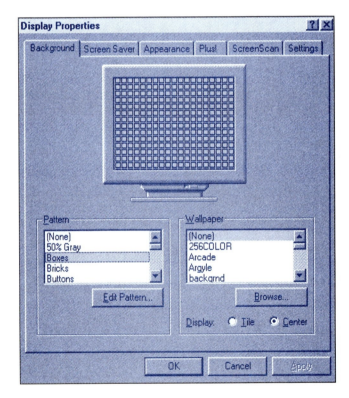

selected instead, you may choose to tile the wallpaper (repeat the wallpaper pattern over and over to cover the entire desktop) or simply center the wallpaper pattern, leaving the rest of the desktop filled with a blank background or a patterned background. All of these options are shown in Figure 4.14.

In Figure 4.14(a), notice that icons are drawn on top of the background pattern in a special way so that you can still read their names. In Figure 4.14(b), icons and folders are drawn on top of the background wallpaper, which is centered. The desktop area not

FIGURE 4.14 (a) Boxes pattern, no wallpaper, (b) Forest wallpaper (centered), no pattern

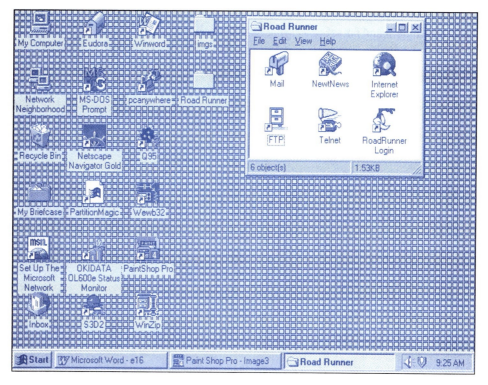

(a)

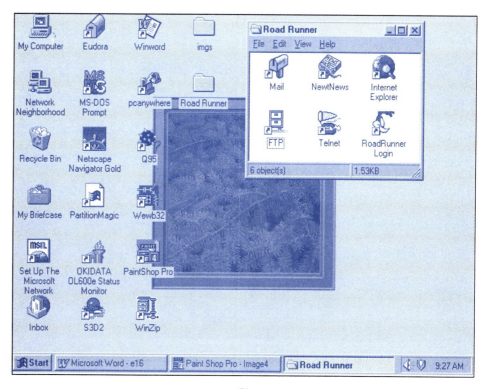

(b)

covered by a small wallpaper image can be patterned, as shown in Figure 4.14(c). If the wallpaper is tiled, the entire surface of the desktop is covered by repeating the wallpaper image horizontally and vertically as many times as necessary. This last option is shown in Figure 4.14(d).

53

FIGURE 4.14 *(continued)*
(c) Bricks pattern and Forest wallpaper (centered), and (d) Forest wallpaper (tiled)

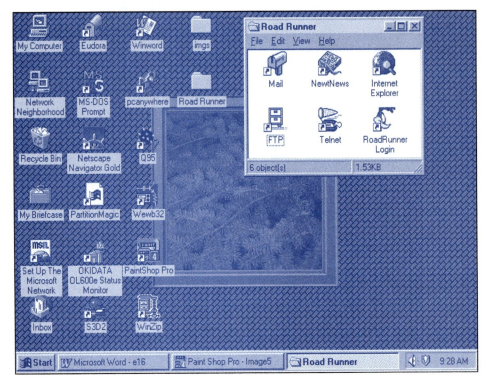

(c)

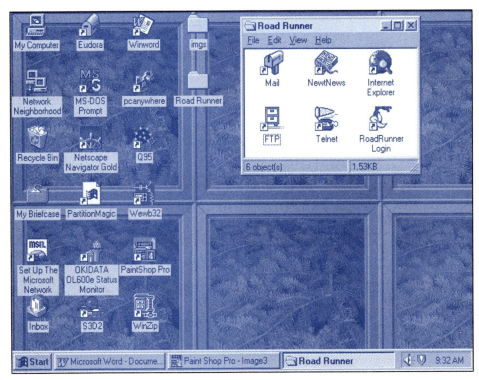

(d)

THE SYSTEM SETTINGS AREA

The system settings area of the taskbar typically contains a speaker icon (for controlling the speaker volume) and a 24-hour clock. Just placing the mouse near the time display causes a small pop-up window to appear with the full date displayed. This is illustrated in Figure 4.15.

FIGURE 4.15 Pop-up date
window

FIGURE 4.16 Date/Time
Properties window

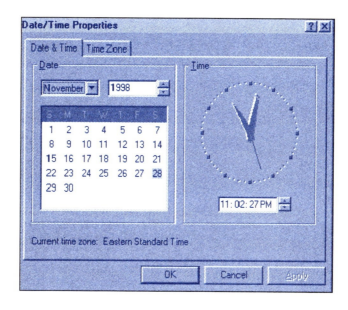

Double-clicking on the time display brings up the Date/Time Properties window illustrated in Figure 4.16. It is easy to change any part of the time or date with a few left-clicks or keystrokes. Windows will even adjust its clock automatically twice a year to properly "spring ahead" or "fall back" as required.

Some applications place their icons in the system area displayed to the left of the time. By placing the mouse over these icons, their current status is displayed. The applications are accessed by single-clicking or double-clicking them. We will examine these items in detail when we discuss accessories in Exercise 8.

CREATING AND USING SHORTCUTS

Some applications add their shortcut icons to the desktop automatically during installation. A shortcut to any application installed on a computer can be added to the desktop by simply right-clicking on an empty portion of the desktop and selecting Shortcut from the New submenu. The Create Shortcut dialog window is displayed as shown in Figure 4.17.

Simply type the name of the file to be added, or click the Browse box to locate the desired file. When the Browse box is selected, the Browse window is displayed. Figure 4.18 shows a sample Browse window for the C: drive. From this dialog box, any program located on the local hard drive or network hard drive can be selected. For example, suppose we want to add a shortcut to the Calculator program. Search for the application using the Browse window. After the Calculator program is located and selected, simply click the Open button to return to the Create Shortcut dialog box. Notice that the correct path and program name is displayed in the Command line text box shown in Figure 4.19.

By selecting the Next button, Windows provides an opportunity to enter a name for the icon (which is displayed under the icon on the desktop). The name of the file is used by default but can be changed by simply typing a new name in the text box, as illustrated in Figure 4.20.

When the Finish button is selected, Windows creates the shortcut and adds the program icon to the desktop. Figure 4.21 shows the desktop with the new Calculator icon.

FIGURE 4.17   Create Shortcut dialog window

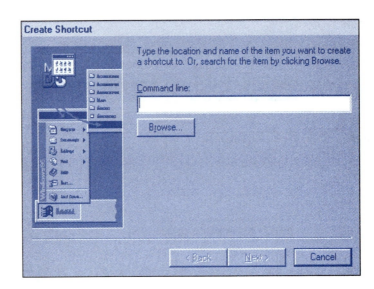

FIGURE 4.18   Browse window

FIGURE 4.19   Create Shortcut window with file selected

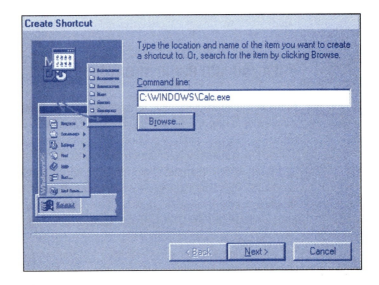

CHANGING OTHER DESKTOP PROPERTIES

Several other desktop properties you may want to experiment with are the Screen Saver, Appearance, and Settings items. Screen savers are programs that run in the background, only performing their screen saving when there has been no mouse or keyboard activity for a predetermined time period. A typical screen saver draws an interesting shape on the

FIGURE 4.20 Naming the shortcut

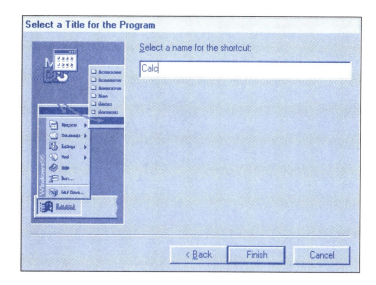

FIGURE 4.21 The desktop with the new shortcut

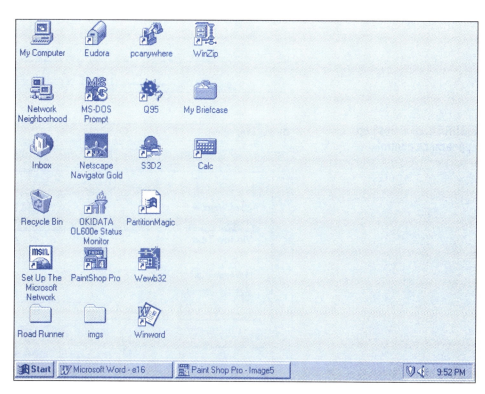

screen or performs some other graphical trick, keeping most of the screen blank. This is done to prevent an image from being burned into the phosphor coating on the display tube. When a key is pressed, or the mouse is moved, the screen saver restores the original display screen and goes back into the background.

Screen savers can also be password protected, to prevent unauthorized access to your desktop if you are away from your computer. Figure 4.22 shows the Screen Saver control window, with a small example of the selected screen saver being displayed.

The appearance of the desktop, from the color used in the title bar of a window, to the font of the desktop text, can be adjusted using the Appearance controls in the Display Properties window. Practically everything you see in the Appearance window illustrated in Figure 4.23 is clickable, including the scroll bar shown in the active window, which allows its width to be set. If the defaults for each item are not acceptable, spend some time tailoring them to suit your needs.

The Settings tab on the Display Properties window provides access to another important set of controls. As shown in Figure 4.24, the Settings controls allow you to change the size

FIGURE 4.22   Selecting a screen saver

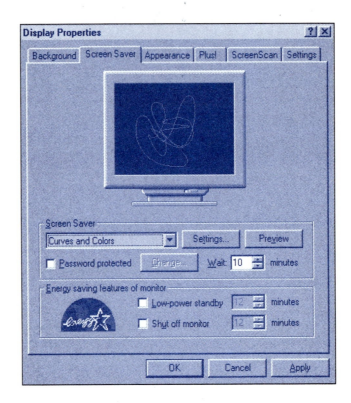

FIGURE 4.23   Desktop appearance controls

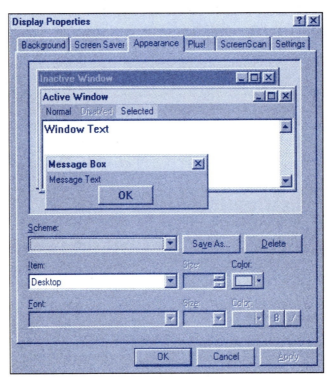

(also called *resolution*) of the desktop, the number of colors available, and the type of display connected to the computer. A small sample of the desired desktop is displayed to assist you when adjusting the settings.

MANAGING THE DESKTOP CONTENTS

The degree of control you have over the appearance of the desktop can be adjusted in many ways. For example, if you like to place new desktop icons (shortcuts or newly installed

FIGURE 4.24 Display
Properties window

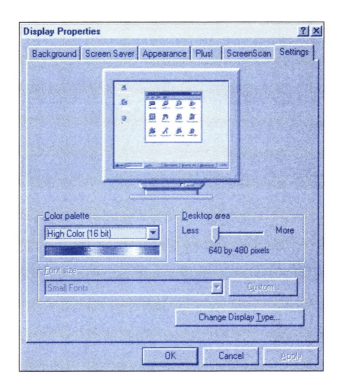

FIGURE 4.25 Arrange Icons
submenu

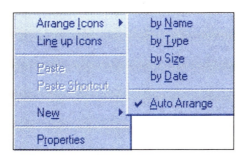

application folders) in specific locations, you may do so by left-clicking on them and holding the mouse button down as you drag the icons to their new positions on the desktop (drop them by releasing the mouse button). If you want Windows to manage your icons for you, or if you need to prevent inexperienced users from messing up the appearance of the desktop, select the Auto Arrange option from the Arrange Icons submenu shown in Figure 4.25. Right-clicking on a blank portion of the desktop brings up the context-sensitive menu.

When Auto Arrange is enabled, icons dragged from their positions will snap back in place when released. The desktop icons will remain in an orderly arrangement.

Many applications install folders on the desktop (such as the Road Runner folder shown in many of the figures in this exercise). The desktop itself is actually a special folder, and you can view its contents using Windows Explorer. This is demonstrated in Figure 4.26. In Exercise 6 you will see exactly how to do this, and many other useful operations, such as creating folders and moving folders to different locations.

MY COMPUTER

Double-clicking the My Computer icon opens up the folder shown in Figure 4.27. The contents will vary from machine to machine, depending on the actual hardware installed on each one. Double-clicking on any of the drive icons will open a directory window for the selected drive. The properties of a disk can be displayed by right-clicking on any of the drive icons and selecting Properties from the pop-up menu. The Disk Properties window shown in Figure 4.28 shows the disk label and current usage. The used space, free space, and capacity are displayed in bytes.

FIGURE 4.26   Viewing the desktop with Windows Explorer

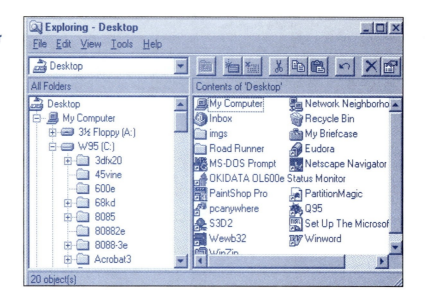

FIGURE 4.27   Contents of My Computer

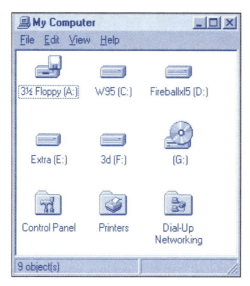

FIGURE 4.28   Disk Properties window

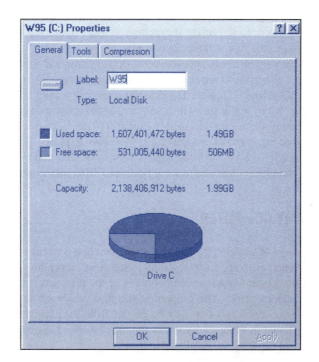

60

FIGURE 4.29   Disk tools

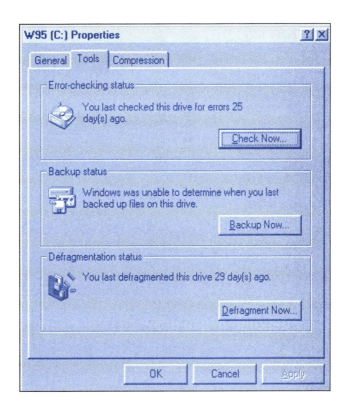

The Tools tab illustrated in Figure 4.29 shows the current status of each of the built-in disk tools provided by Windows. Each tool displays how long it has been since the utility has been run.

The Compression tab controls disk compression options. These utility programs will be examined in Exercise 8. The other system features are accessed and controlled through the Control Panel, Printers, and Dial-Up Networking folders, all of which we will cover in detail in Exercises 5, 7, and 11, respectively.

THE NETWORK NEIGHBORHOOD

If your computer is connected to a network, via a network interface card or a serial PPP (point-to-point protocol) connection, the Network Neighborhood icon, when clicked, allows you to examine other machines connected to the same network. Note that the other machines must be running a protocol called NetBEUI for them to be included in the network neighborhood. NetBEUI is part of Windows for Workgroups 3.11 and one of many protocols used with Windows.

As Figure 4.30 shows, the Network Neighborhood looks similar to a directory tree. In fact, files and printers may be shared among computers participating in the Network Neighborhood. Thus, one laser printer can serve the needs of a small laboratory or office.

THE RECYCLE BIN

Whenever anything is deleted in Windows, from executable programs to text files, images, and desktop icons, it is not yet gone for good. The exception to this rule are files deleted while running inside a DOS window. They are simply gone, unless the old UNDELETE command is still active. When you delete an item, Windows prompts you with a question, allowing you to change your mind, if necessary (as indicated in Figure 4.31).

The first destination of a deleted item is the Recycle Bin. Double-clicking the Recycle Bin icon on the desktop brings up the Recycle Bin window, which lists all files, if any, that have been deleted. Figure 4.32 shows the Recycle Bin window containing several deleted

FIGURE 4.30   Network Neighborhood

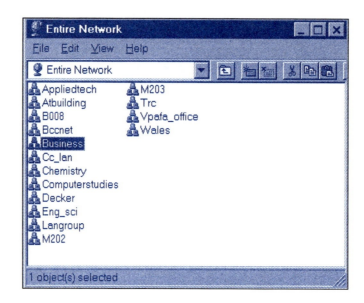

FIGURE 4.31   One of the Windows deletion safeguards

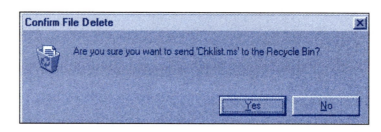

FIGURE 4.32   Contents of the Recycle Bin window

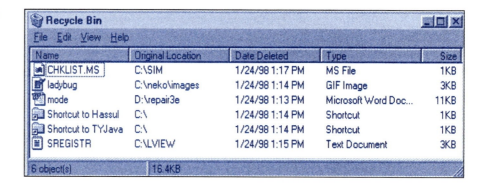

files. Under the File menu, the Empty Recycle Bin option is used to finally delete the files from the hard drive. Windows provides one final confirmation window, to make sure you really want to delete the items in the Recycle Bin. To get a file back, simply select it and choose the Undo Delete option from the Edit menu (or Restore from the File menu).

One thing to keep in mind is that the amount of free disk space does not change until the Recycle Bin is emptied. This is because a copy of the deleted file is stored in the Recycle Bin, which stores its contents on the hard drive.

MY BRIEFCASE

When you need to work on one or more files on two different computers (home and office machines), the My Briefcase utility can be used to organize and properly update the files. Files are copied into the Briefcase by dragging them onto the My Briefcase icon while in Windows Explorer. The entire Briefcase is copied onto a floppy disk by dragging its icon onto the icon for the floppy drive.

After the Briefcase files have been modified on a different computer, it is necessary to update the original files. Dragging the Briefcase icon from the floppy drive back to the desktop will cause Windows to evaluate the contents of the Briefcase and determine which files need updating.

THE INBOX

The Inbox is the central location for all e-mail activity. The first time the Inbox is selected, the Inbox Setup Wizard will begin the configuration process. The Microsoft Network, Microsoft Mail, and Microsoft Fax selections are available. Figure 4.33 shows the Inbox Setup Wizard window.

When the Inbox is selected after the services have been configured, the Microsoft Exchange application will start and begin to perform all e-mail and fax services. Figure 4.34 shows a Microsoft Exchange mail window with several e-mail messages.

E-mail is more than just letters and numbers. You can send and receive all types of files, from graphic images to executable programs. Typically, these types of files are called *attachments*. The paper clip on the fourth e-mail message in Figure 4.34 indicates an attachment exists for that message. In general, double-clicking the attachment brings up the application associated with the attachment's file type. Or, if the attachment is an executable program, it is executed when double-clicked.

All in all, e-mail provides services that are essential to modern-day computing, business, and education. Spend some time learning to use this feature of the Windows operating system.

FIGURE 4.33   Inbox Setup Wizard window

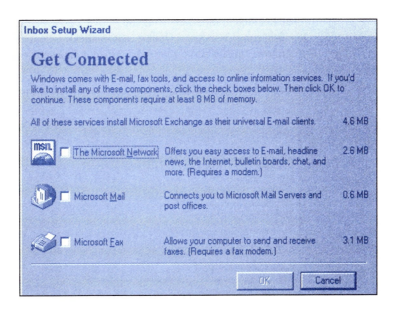

FIGURE 4.34   Microsoft Exchange mail window

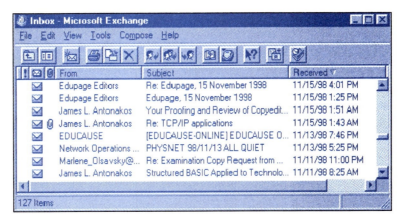

GETTING BACK TO THE DESKTOP FROM DOS

If you are running in a full-screen MS-DOS window, you can exit back to the desktop by entering EXIT at the DOS prompt. If you want to keep the DOS session active, but also to be able to access the desktop, pressing ALT-ENTER will automatically resize your DOS environment to a window and bring back the desktop.

WINDOWS 98 HELP

One of the first differences between Windows 95 Help and Windows 98 Help is the appearance of the Help window. As shown in Figure 4.35, the Windows 98 Help window displays the Windows 98 logo alongside the Contents window. When a Help item is clicked in the Contents window, the associated Help information replaces the Windows 98 logo. This is a nice change from Windows 95 Help, which opened a new Help-specific window, making it awkward to return to the Contents window.

WINDOWS 98 SETTINGS MENU

Figure 4.36 shows the Settings menu found in the Windows 98 Start menu. Three new selections are available: Folder Options, Active Desktop, and Windows Update. The Folder Options selection allows you to choose how folders appear and function (single-click to open versus double-click), and handles file associations. Additional desktop properties can be examined or modified using the Active Desktop selection.

Selecting Windows Update automatically connects your computer to Microsoft's Windows update site on the Web. Once connected, you can pick and choose the updates

FIGURE 4.35 Windows 98 Help window

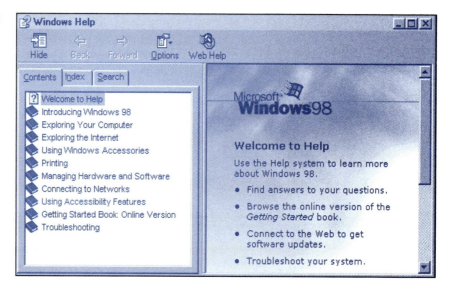

FIGURE 4.36 Windows 98 Settings menu

your Windows system requires. Software patches are downloaded and executed to make the necessary changes.

WINDOWS 98 EXPLORER

Under Windows 98, the Windows Explorer has been updated to work with the Web in addition to the desktop. Figure 4.37 illustrates how Explorer integrates the Web as just another item on the desktop. In addition to displaying lists of files and folders, Explorer is also designed to display Web content. Look along the left-hand side of the Explorer screen to see that the Voltage Regulator HTML document displayed is found in the Internet Explorer folder. The address of the HTML document is shown in the Address field.

WINDOWS 98 WEB-STYLE DESKTOP

To further integrate the Web into the Windows 98 computing environment, you have the option of configuring your desktop to act like a Web page. Folders can be opened or applications launched with a single left-click, instead of the usual double-click. When applications are launched in this way, the application is selected by simply moving the mouse over the desktop icon. As Figure 4.38 shows, the names under each icon have the familiar underline associated with clickable links on a Web page. The Web-style desktop can also be configured so that the underline is displayed when the icon is selected. With all the available choices, no two desktops look the same.

WINDOWS NT DESKTOP EVOLUTION

The Windows NT desktop has evolved along the same path as the Windows desktop. The first version, Windows NT 3.5, took on the appearance of the Windows 3.1 desktop. This is because the first version of Windows NT was an industrialized version of Windows 3.1.

The desktop of the second version of Windows NT, version 4.0, looks like the Windows 95 desktop. The next release of Windows NT, Windows 2000, is expected to look a lot like Windows 98, with additional improvements.

FIGURE 4.37 Windows 98 Explorer window displaying a Web page

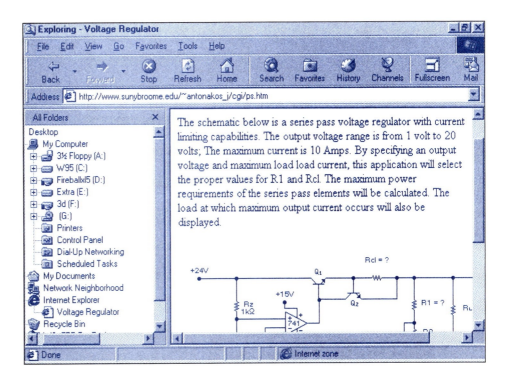

**FIGURE 4.38 Windows 98
Web-style desktop**

The release of Windows Explorer Version 4.0 offered the Windows 98 look and feel to the Windows 95 and Windows NT desktops. To determine the current operating system, look at the Start menu. Along the left-hand side is a graphic indicating the operating system version.

**TROUBLESHOOTING
TECHNIQUES**

It is important to be patient with Windows. For example, suppose you have a number of applications open, and have just clicked the Close button on one of them, expecting to see the application close instantly; instead, you see nothing happen for a long time.

An impatient user may decide that Windows has died and simply turn the computer off. This is not the recommended approach (there may be important information that needs to be backed up to the hard drive). In this case, where Windows has "gone away" for a long time, attempt to switch to another application on the taskbar (or press Ctrl-Alt-Del to bring up the Close Program window). Just seeing the mouse pointer move when you use the mouse is a good sign that the operating system is still listening.

You may find that, eventually, Windows finishes closing the "dead" application, or reports an error message like the one shown in Figure 4.39. There is not really much that can be done for the application if an illegal operation is detected, but at least Windows catches these problems and does not crash because of them.

**FIGURE 4.39 Illegal operation
detected**

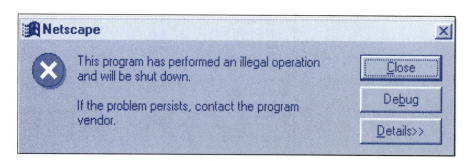

This self-test is designed to help you check your understanding of the background information presented in this exercise.

True/False

Answer *true* or *false*.

1. The taskbar contains only the Start button, running applications, and the system settings area. *False*
2. A shortcut for any application can be added to the desktop. *True*
3. Context-sensitive help is displayed whenever the mouse is simply moved on top of any desktop icon. *True*
4. The Documents menu item on the Start menu contains important help information for Windows. *False*
5. When looking for help, enter HELP on the keyboard. *False*
6. Ctrl-Alt-Del instantly reboots the computer. *True*
7. The background may contain a pattern and a wallpaper. *True*
8. The size of the desktop is fixed. *False*
9. Right-clicking a desktop icon will cause its application to start up. *True*
10. The name and position of each desktop icon may be changed. *True*

Multiple Choice

Select the best answer.

11. The Windows desktop is
    a. Part of the installation process.
    b. An essential component of the operating system.
    c. An optional application program.
12. When files are deleted in a DOS window, they are
    a. Gone for good.
    b. Sent to a temporary directory.
    c. Sent to the Recycle Bin.
13. When files are deleted in Windows, they are
    a. Gone for good.
    b. Sent to a temporary directory.
    c. Sent to the Recycle Bin.
14. The My Computer icon displays information about
    a. Each disk drive.
    b. Folders for the printer and control panel.
    c. Both a and b.
15. Wallpaper files are image files called
    a. Pixmaps.
    b. Bitmaps.
    c. Pic files.
16. To quickly bring up a context-sensitive display menu
    a. Left-click anywhere on the desktop.
    b. Right-click anywhere on the desktop.
    c. Press M on the keyboard.
17. Left double-clicking a disk icon in My Computer
    a. Formats the disk.
    b. Displays the disk directory.
    c. Displays the disk properties.
18. If Windows goes away for a long time
    a. Just turn the computer off.
    b. Press every key on the keyboard and move the mouse.
    c. Press Ctrl-Alt-Del and see what happens.

19. When Auto Arrange is enabled, the desktop icons
    a. Periodically move themselves to new locations.
    b. Snap back into place if moved.
    c. Converge at the center of the screen.
20. Windows help is
    a. Minimal due to lack of disk space.
    b. Only available on CD-ROM.
    c. Online and significantly better than Windows 3.x.

Completion

Fill in the blank or blanks with the best answers.

21. The Start button is located on the _____.
22. The desktop contains the _____ Neighborhood icon.
23. Wallpaper may be centered or _____.
24. The Programs submenu is located on the _____ menu.
25. The _____ is used to maintain a set of files that are used on more than one computer.
26. When a file is deleted, it goes to the _____ _____.
27. A(n) _____ takes over the display screen when there is no user activity for a length of time.
28. Electronic mail and fax services are provided by selecting the _____ icon.
29. The size of the display is set in the Display Properties _____ menu.
30. To exit from a DOS shell but keep the session running, press the _____ and _____ keys.

FAMILIARIZATION ACTIVITY

1. Create a shortcut on the desktop to the Calendar application in the Windows directory and demonstrate its use.
2. Experiment with the desktop appearance by choosing five different patterns and five different wallpapers.
3. Compare the centered and tiled options for three different wallpapers.
4. Move the taskbar to all four corners of the display and vary the size. Explain your preference for the normal position of your taskbar.
5. Create several small text files with a text editor or word processor. Put a copy of each file into the Briefcase. Drag the My Briefcase icon onto the icon for your floppy drive. Take your floppy to a different computer and drag the Briefcase from the floppy onto the desktop. Edit the files in the Briefcase. Drag the Briefcase back to the floppy. Put the floppy in the original computer and drag the Briefcase back onto the desktop. Double-click the My Briefcase icon and examine its contents. Update the files in any manner you choose.
6. Click the time display on the taskbar. Use the up and down arrows to determine the first and last year recognized by Windows. Can you grab the minute hand and move it around the clock?
7. Enable the Auto hide feature of the taskbar and experiment with the mouse to determine how and when the taskbar will reappear. Resize the taskbar. How large and small can it be?
8. Create a folder on the desktop and name it Temp. Drag some of the other desktop icons into it. Open the Temp folder and drag the icons back to the desktop one at a time. Do the icons snap into place or can you put them anywhere you choose?
9. Open the Recycle Bin. If there are already files in it, determine when they were put there and whether it is safe to delete them. Open a DOS window. Do a directory listing of the current drive and make note of the free space on the drive. Now empty the Recycle Bin. Has the free space changed?
10. Slowly move your mouse over the items in the Programs submenu as if choosing an application to start up. Do you encounter any difficulty trying to select any of the applications?

1. If you were a Windows 3.x user, how would you explain the similarities between Program Manager and the Windows desktop?
2. Why is it good to be patient with Windows?
3. If the operating system looks dead, what might you do to check for signs of life?
4. Does the taskbar have to be at the bottom of the screen? Does it have to be on the screen at all?
5. What is the Network Neighborhood?
6. Explain how to produce the Display Properties menu.

Under the supervision of your instructor,

1. Demonstrate the different features of the Start menu.
2. Identify items on the taskbar.
3. Change the appearance and properties of the desktop.
4. Create and use shortcuts.
5. Explain the function of the standard desktop icons.

5 The Control Panel

INTRODUCTION

Joe Tekk was sitting at his desk, eating lunch in front of his computer. His Windows 98 desktop showed the open Control Panel folder. Joe selected an icon from the bottom row, ODBC, and examined all of its various submenus and options, never actually selecting any of them.

Don, Joe's manager, saw Joe's screen and was concerned. "What are you doing in Control Panel? Is there something wrong with your system?"

"No, Don," Joe answered. "I'm just checking out some of the operations I never use. Every now and then I discover one that has some new feature I am looking for."

PERFORMANCE OBJECTIVES

Upon completion of this exercise, you will be able to

1. Identify several of the many useful icons in the Control Panel and explain their functions.
2. Use the Control Panel to modify system properties such as time/date, sounds, the display format, and keyboard/mouse functionality.
3. Show how printers and modems are configured.

BACKGROUND INFORMATION

The Control Panel is one of the most important tools included in Windows. Just like the Control Panel included with Windows 3.x, the Windows Control Panel allows you to fine-tune and customize the hardware and software of your system. The degree of control and the number of features available is much improved over Windows 3.x. Figure 5.1 shows three views of the Control Panel. Note the similarities in all three Control Panels. Windows 95 and Windows 98 are the most similar. The Windows NT 4.0 Control Panel shares many of the same icons, but has several icons that are unique to the Windows NT 4.0 environment, such as Devices, Licensing, and Services. As indicated in Figure 5.1, nearly every aspect of the operating system can be controlled, examined, or configured through one of the many Control Panel icons. Let us examine the operation of several important icons.

FIGURE 5.1 (a) Windows 95 Control Panel, (b) Windows 98 Control Panel, and (c) Windows NT 4.0 Control Panel (Web-style format)

(a)

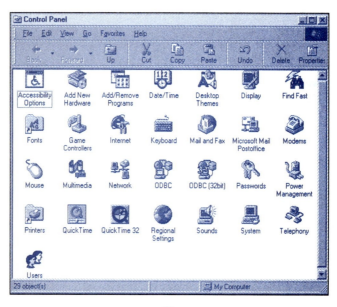

(b)

(c)

**FIGURE 5.2 Accessibility
Properties menu**

ACCESSIBILITY

Figure 5.2 shows the Accessibility Properties menu. For users who want to customize how the keyboard, sound, display, and mouse operate, this menu provides the way. As indicated in Figure 5.2, a lot of attention has been paid to how each key on the keyboard behaves. Other accessibility options involve generation of short tones when Windows performs an operation, helpful pop-up messages, using the arrow keys to control the mouse, and changing the display colors to high-contrast mode for easier viewing.

ADD NEW HARDWARE

The process used to add new hardware to Windows 95 or 98 usually involves two steps. First, the new hardware device must be physically added to the computer. This requires removing the cover from the computer, identifying the proper location to install the new hardware, and then performing the actual installation procedure. The second step involves configuring the software to properly communicate with the new device. This is accomplished using the Add New Hardware Wizard shown in Figure 5.3. Adding new hardware in Windows NT is different.

**FIGURE 5.3 Add New
Hardware Wizard window**

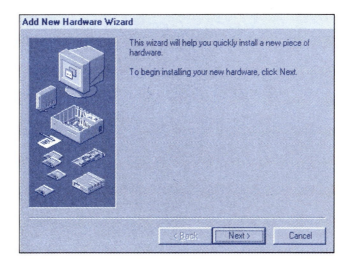

FIGURE 5.4 Add/Remove
Programs Properties menu

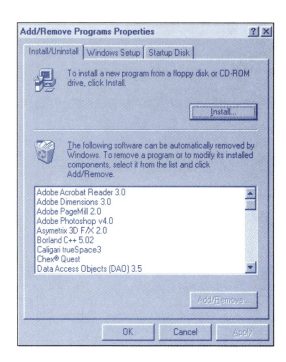

ADD/REMOVE PROGRAMS

When installing software applications in Windows, many software vendors require the installation to be performed from the Add/Remove Programs Properties menu illustrated in Figure 5.4. When the software is installed, it is registered with Windows. If at some point it is necessary to remove an application from the system, the user can simply return to the menu, select the application, and press the Add/Remove button. In addition to user application programs, the Windows operating system can also be modified by adding or removing components. This topic is discussed in Exercise 9.

DATE/TIME

This menu can be opened from inside Control Panel or by double-clicking the time display in the taskbar. The Date/Time Properties menu is shown in Figure 5.5. The user can easily change the time or date with a few left clicks. The Time Zone submenu allows you to select the time zone your computer is located in. Windows will then automatically adjust the clock for daylight saving time when required.

FIGURE 5.5 Date/Time
Properties menu

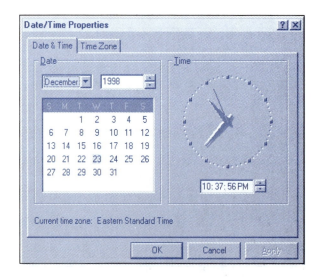

FIGURE 5.6 Display Properties menu

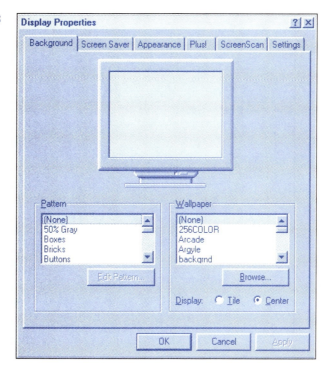

DISPLAY

The Display Properties menu is shown in Figure 5.6. Recall from Exercise 4 all of the various ways we can adjust the display. The menu in Figure 5.6 is also reachable by right-clicking on a blank area of the desktop and selecting Properties.

KEYBOARD

The Keyboard Properties menu is displayed in Figure 5.7. The Speed tab is used to set the repeat delay and repeat rate when a character is pressed and held down on the keyboard and

FIGURE 5.7 Keyboard Properties menu

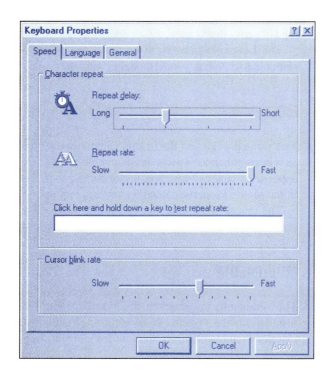

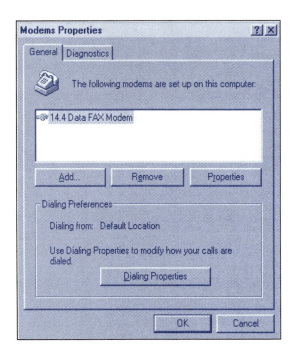

FIGURE 5.8 Modems Properties menu

the rate at which the cursor blinks in a data entry field. Many people need to change these settings because of individual keyboarding styles. We will investigate the consequences of changing these settings during the activities at the end of this exercise.

The Language tab is used to indicate the keyboard language, and the General tab is used to identify the specific type of keyboard used, such as the Microsoft Natural keyboard.

MODEMS

Modems are very easy devices to work with in Windows. The Modems Properties menu (Figure 5.8) identifies the specific type of modem installed in your computer.

Windows can be set up to dial from many different dialing locations. For example, many offices may require a 9 to be pressed to access an outside line; other dialing locations may not require a 9. All the places frequently called can be given a name and can be selected very easily. Even specific communication details, such as the number of data bits and the type of parity that is used, can be adjusted.

The Diagnostics submenu is useful for interrogating the modem and examining its response to commonly used modem commands.

MOUSE

This menu provides all the functional control over the mouse. The mouse can be set up for left- or right-handed operation, its double-click speed adjusted, mouse trails enabled, and appearance changed by choosing one of several scenes (3-D pointer, for example). The initial Mouse Properties menu is shown in Figure 5.9.

The mouse is a significant component of the Windows operating system. The mouse properties you choose for yourself can make your system easier to use.

MULTIMEDIA

The Multimedia Properties menu is used to control the multimedia hardware and software installed on the system. Windows comes equipped with the ability to display MPEG (Motion Pictures Experts Group) video in real time, and provides virtual device drivers for

FIGURE 5.9 Mouse Properties menu

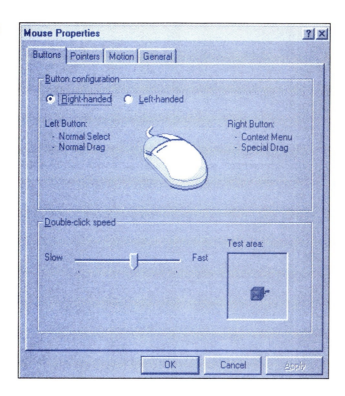

FIGURE 5.10 Multimedia Properties menu

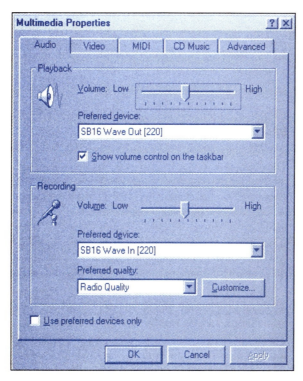

many popular sound cards and graphics accelerators. Windows can also take advantage of the new MMX technology available in the Pentium family of microprocessors. Figure 5.10 shows the initial Multimedia Properties menu.

NETWORK

The Network menu allows you to add, modify, or remove various networking components, such as protocols (NetBEUI, TCP/IP), drivers for network interface cards, and Dial-Up

FIGURE 5.11   Network menu

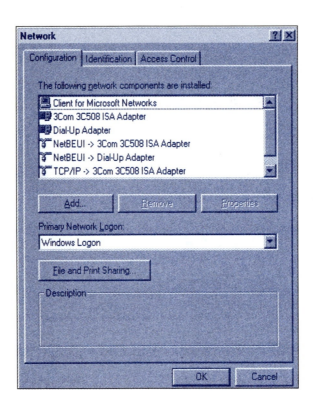

Networking utilities. You can also specify the way your machine is identified on the network, as well as various options involving file and printer sharing, and protection. Figure 5.11 shows a sample Network menu. Selecting any of the network components allows its properties to be examined.

Networking will be covered in detail in Exercise 11.

PASSWORDS

The Passwords Properties menu found in Windows 95 and Windows 98 is used to maintain information about passwords and security when a computer is shared among different people. Passwords are maintained in files with a .PWL extension if the computer does not participate in a Windows NT domain. The file name portion consists of the first eight letters of the user name entered at the Windows logon screen. If you are required to supply passwords for services used on the network, the passwords may be saved in a .PWL file. Unfortunately, you are allowed to delete any .PWL file you want, effectively removing all password protection for the affected user. This may be a factor in a security-conscious network.

Windows 95, Windows 98, and Windows NT can be configured to allow each computer user to maintain individual desktop settings plus other related preferences all protected by an initial password. Figure 5.12 illustrates the Passwords Properties menu. Windows NT Server can be configured as a domain controller responsible for maintaining passwords for all computers in the network.

PRINTERS

The Printers window shown in Figure 5.13 shows all the printers currently installed on a particular computer. The Add Printer icon is used to add a brand-new local or network printer to Windows. Although printers are covered in detail in Exercise 7, it is important to note the starting point for printer operations. The system tray portion of the taskbar will indicate when a printer is in use and the status of the current print job. Right-clicking on an installed printer will allow you to bring up the Properties window and change printer parameters as necessary.

FIGURE 5.12   Passwords Properties menu

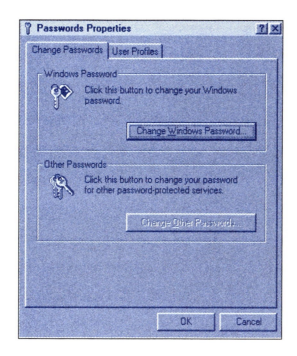

FIGURE 5.13   Printers window

REGIONAL SETTINGS

The Regional Settings menu is used to configure Windows to conform to the many different international standards. For example, the display format for numbers, currency, date, and time can all be modified. Figure 5.14 illustrates the initial Regional Settings menu. To make the job simple, simply select the specific region, and all the individual items are automatically configured. If you travel internationally, your system can automatically adjust to the new region with a few clicks of the mouse.

SOUNDS

Windows uses sounds to accompany many typical operations, such as closing an application or shutting down the computer. The sound, if any, that is played for an event is specified using the Sounds Properties menu illustrated in Figure 5.15.

Sounds are stored in .WAV files in the main Windows subdirectory. Many different sound editors are available that allow you to create or modify sounds to fully customize your Windows environment.

FIGURE 5.14 Regional Settings Properties menu

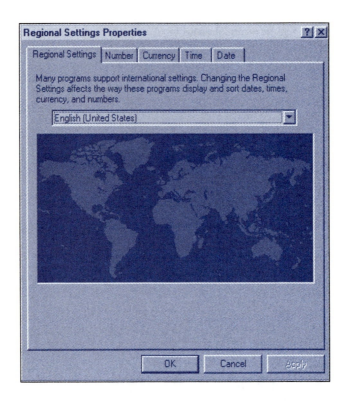

FIGURE 5.15 Sounds Properties menu

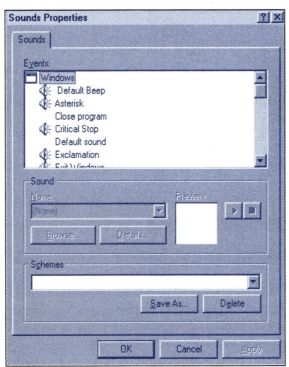

SYSTEM

The System Properties menu is the central location for all system-related information. The initial properties window for Windows 98 and Windows NT is shown in Figure 5.16. The Windows 95 menu is virtually the same as the Windows 98 menu. The operating system version, processor type, and amount of RAM are some of the system properties displayed.

The other submenus (Device Manager, Hardware Profiles, and Performance, to name a few) provide detailed access to the inner workings of the installed hardware as

FIGURE 5.16 (a) Windows 98 System Properties menu, and (b) Windows NT System Properties menu

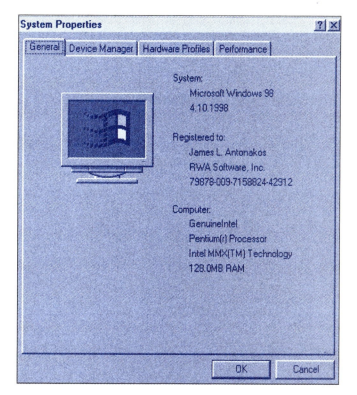

(a)

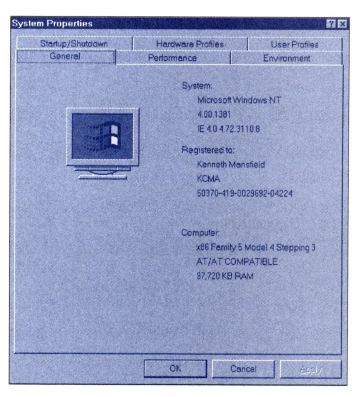

(b)

well as critical Windows variables, such as virtual memory settings and file system configuration.

There are many icons yet to be examined. You are encouraged to examine their properties on your own.

When troubleshooting on a Windows 95 or 98 computer, the System Properties menu can be used to investigate and diagnose many types of problems. The Device Manager submenu provides a single location to examine all the hardware components installed on a system. After installing a new piece of hardware, it is good practice to view the new hardware properties and make note of any configuration parameters.

On some occasions, by viewing these screens, problems may be discovered. This is indicated in Figure 5.17. When reviewing the System devices category, the exclamation point next to the Plug and Play BIOS item indicates a problem exists.

By double-clicking the Plug and Play BIOS item, the associated Properties menu is displayed, as shown in Figure 5.18. The device status portion of the menu indicates that the device drivers for the Plug and Play BIOS have not been installed correctly and then goes on to suggest that the user click on the Drivers submenu tab to change the drivers. When the problem is resolved, the error indication will be removed from the Device Manager display.

FIGURE 5.17   System Properties reports an error

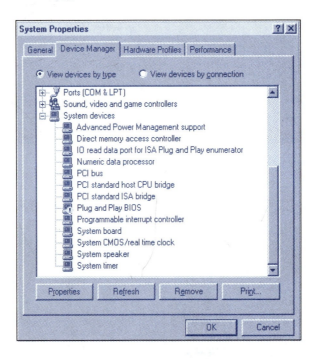

FIGURE 5.18   A course of action is suggested

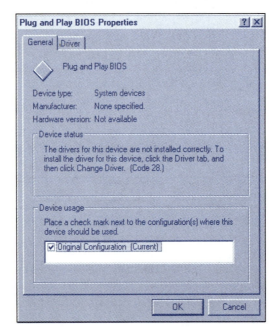

It pays to become familiar with the System Properties menu when troubleshooting problems using Windows 95 or 98.

In Windows NT, hardware errors are written to a log file, which can be examined for details concerning the nature of the problem.

This self-test is designed to help you check your understanding of the background information presented in this exercise.

True/False

Answer *true* or *false*.

1. The Control Panel is used to control hardware only. *False*
2. Only one language can be used with the keyboard. *True*
3. Windows takes advantage of MMX technology. *True*
4. Each user of a computer can maintain his or her own desktop. *True*
5. The Printers window shows all active printers. *True*

Multiple Choice

Select the best answer.

6. Passwords are stored in
   a. PWL files.
   b. PAS files.
   c. DAT files.
7. Changing the way numbers are displayed is performed in
   a. Multimedia Properties.
   b. Display Properties.
   c. Regional Settings.
8. Control Panel icons
   a. Are identical in all versions of Windows.
   b. Are only found in Windows NT.
   c. Establish control over almost all hardware and software properties.
9. The Systems Properties menu shows
   a. Information about the system printers.
   b. The version of the operating system.
   c. Information about system performance.
   d. All of the above.
10. The Accessibility options are used to
    a. Network several Windows machines together.
    b. Customize how the system and user interact.
    c. Control how the hard drive is accessed.

Completion

Fill in the blank or blanks with the best answers.

11. Windows can automatically play _____ video files.
12. To connect to a remote computer, we may use the _____ or _____.
13. The Network menu is used to change the _____ for the network interface card.
14. The _____ _____ submenu is used to display information about all the hardware components installed in a Windows 95 or 98 system.
15. Sounds are stored in _____ files.

1. Go through each of the Control Panel operations covered in this exercise. Be sure to click on every tab to see the full extent of control you have over the operating system and the computer.

2. Briefly examine the remaining operations provided by Control Panel. Determine what ODBC stands for. Find out how to send a fax.
3. Make a Windows startup disk using Add/Remove Software. Windows NT calls this disk an emergency repair disk.

1. How must Windows be configured to allow each user to maintain his or her own desktop settings?
2. Why is it necessary to have different dialing locations maintained inside of Dial-Up Networking?
3. How does Windows indicate problems with hardware settings?

Under the supervision of your instructor,

1. Identify several of the many useful icons in the Control Panel and explain their functions.
2. Use the Control Panel to modify system properties such as time/date, sounds, the display format, and keyboard/mouse functionality.
3. Show how printers and modems are configured.

6 Windows Explorer

Joe Tekk was busy using Windows Explorer to organize his hard drive. He moved directories, renamed them, created a folder of his most often used shortcuts, and copied several files from his hard drive to a floppy disk. Then he installed an application from a shared CD-ROM on RWA Software's Pentium II file server.

Don, Joe's manager, was watching. "Joe, do you ever close Explorer?" Don asked.

Joe laughed. "Sometimes, Don, but usually I leave it minimized on the taskbar, just in case I need it for something."

PERFORMANCE OBJECTIVES

Upon completion of this exercise, you will be able to

1. Explain the basic features of Windows Explorer.
2. Show how to start an application.
3. Create a shortcut, move, find, and delete files and folders.
4. Map a network drive.

BACKGROUND INFORMATION

Windows Explorer can be thought of as the *command post* of Windows, performing duties similar to the Program Manager and File Manager utilities in Windows 3.x.

Figure 6.1 shows a typical Explorer display of the folders on drive C:. Although there are indeed similarities between Windows 95 Explorer and Windows 98 Explorer (shown, respectively, in Figures 6.1(a) and 6.1(b) for comparison), Windows 98 Explorer offers additional features because Internet Explorer 4.0 has been integrated into the desktop. For example, Web pages can be viewed in the rightmost panel, without having to first open a browser. We will examine the new Windows 98 Explorer features at the end of this exercise. Bear in mind that much of what follows regarding Windows 95 Explorer operation also applies to Windows 98 Explorer. Furthermore, Windows NT Explorer is practically identical to Windows 95 Explorer, although you may install Internet Explorer 4.0 and gain the enhancements for your Windows NT environment. For now, let us take a detailed look at Windows 95 Explorer.

FIGURE 6.1   (a) Windows 95 Explorer window, and (b) Windows 98 Explorer Window

(a)

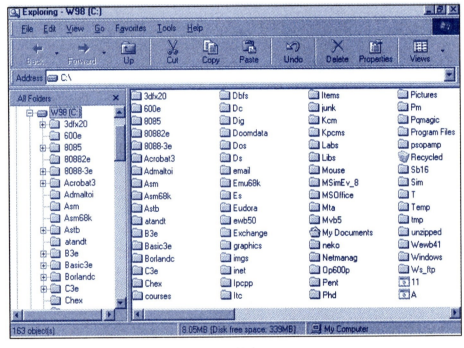

(b)

The Windows 95 Explorer window can be broken down into four areas: the pull-down menus, the folder display window, the file display window, and the status bar. The pull-down menus provide a way to access the features of Explorer, to configure it, and do many other useful things, such as map network drives. The folder display window allows the selection of any resource on the computer. This includes all the drives (floppy, hard, CD-ROM, network), special folders (Control Panel, Printers, Dial-Up Networking), and other system and user items.

The file display window shows a list of the folders and files in the currently selected location. For example, in Figure 6.1(a), the file display window shows the contents of the root directory of drive C:. Note the drive label "W95" at the top of the window. The status bar, located at the bottom of the Explorer window, shows the number of objects in the file display window as well as the amount of disk space used by the objects and the disk free space. Single-clicking on an item in the folder display window will show the contents of the item in the file display window.

FIGURE 6.2   Explorer display with graphical toolbar

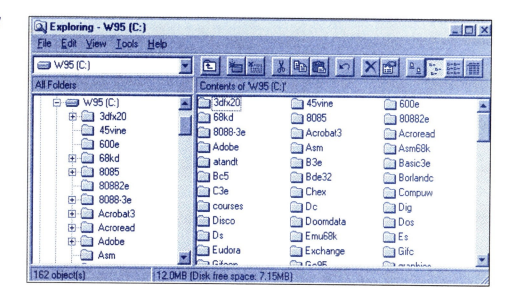

FIGURE 6.3   Explorer toolbar functions

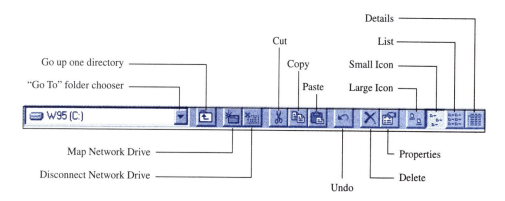

CHANGING THE VIEW

The Windows 95 Explorer interface can be customized in many ways. First, let's add a graphical toolbar to the Explorer window. This is accomplished by selecting the Toolbar option in the View menu. The new Explorer display window should look similar to Figure 6.2. Many of the pull-down menu items are represented in the graphical toolbar. Holding the mouse still over a toolbar button will produce a small pop-up description window, such as "Delete" or "Map Network Drive." The function of each button is shown in Figure 6.3.

Another way to change Explorer's view is to adjust the method used to show files and folders. These methods are as follows:

- By large icon
- By small icon
- As a list
- With details

Figure 6.4 shows the large icon format in the file display window. The files displayed in the window scroll vertically, and are listed in alphabetical order with folders first.

Figure 6.5 shows an example of small icon format. This display mode also scrolls vertically.

When files and folders are displayed using the list option, the display looks similar to that shown in Figure 6.6. Notice that the list scrolls *horizontally,* not vertically, and uses the small icon format.

FIGURE 6.4 Explorer display window showing large icons

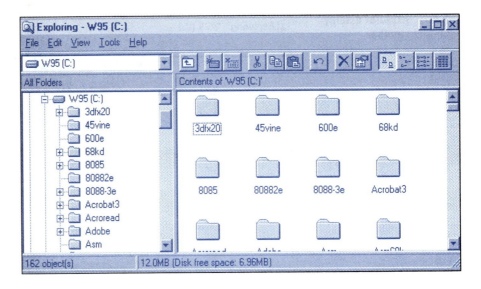

FIGURE 6.5 Explorer display window showing small icons

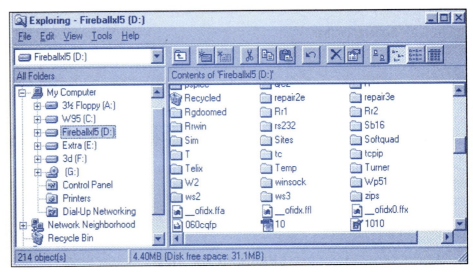

FIGURE 6.6 Explorer display window showing icon list

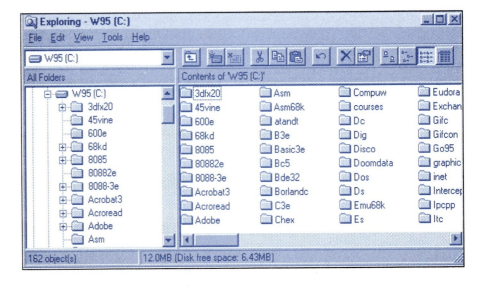

It may be necessary to view the details of each file. This includes the type of file or folder, its size, and last modification date. Figures 6.7 and 6.8 show examples of this display mode.

The arrangement of files and folders in the file display window is flexible (by name, date, size, and so on) and can be limited to certain types if desired. The Options selection in the View menu is used to adjust these features.

FIGURE 6.7 Explorer display window showing folder details

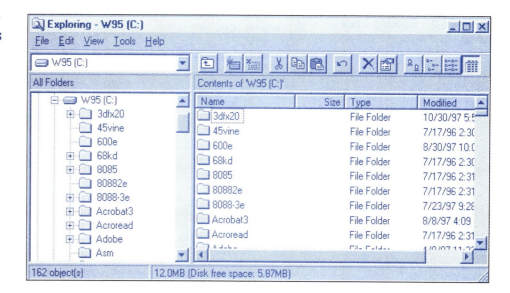

FIGURE 6.8 Explorer display window showing file details

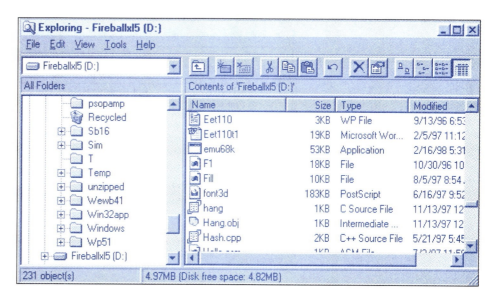

CREATING NEW FOLDERS

To help organize files, it is often necessary to create folders. By selecting "File, New, Folder," Windows 95 automatically creates the new folder in the current directory (the directory shown in the file display window). You are prompted to enter a name for the new folder.

The destination of the new folder is important. For example, if you want the folder to be in the root directory of drive E:, you must make this directory the current directory. This is easily done by left-clicking on the icon for drive E: in the folder display window. Figure 6.9 shows a new folder being created. Windows 95 automatically names the folder New Folder and gives you the opportunity to rename it by typing in a new name. Figure 6.10 shows the new folder renamed Ken's Stuff.

To create a new folder inside an existing folder, left-click on the existing folder to select it and then create the folder as you normally would.

DELETING A FOLDER

Occasionally it becomes necessary to delete the contents of a folder. This is typically a result of the hard disk running out of free space. By selecting the folder, and then selecting "Delete," Windows 95 will prepare to delete the contents of the folder. First, as Figure 6.11

FIGURE 6.9 Creating a new folder

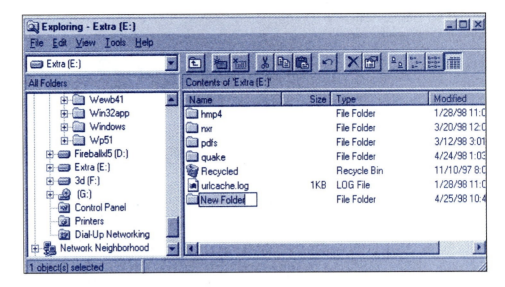

FIGURE 6.10 Naming the new folder

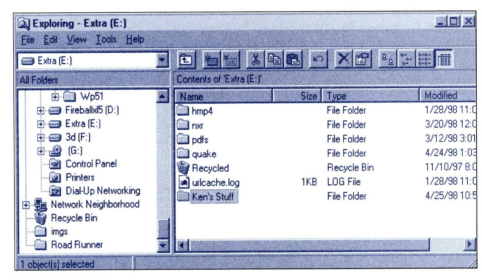

FIGURE 6.11 Confirmation window

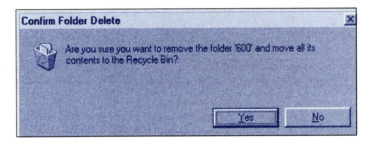

illustrates, Windows 95 will issue a confirmation message to make sure you really want to move the folder to the Recycle Bin. This is because you might have selected the wrong menu item or pressed a wrong button by accident. Or, being human, you may simply have changed your mind.

Windows 95 makes doubly sure that a file you have chosen for deletion should really be deleted. Figure 6.12 shows the warning window that is displayed when a program is deleted. Other, selected file types, such as drivers or fonts, may require confirmation as well. In addition, Windows 95 usually knows when a file you try to delete is in use by an application (or itself), and may disallow the delete operation.

While files are being moved into the Recycle Bin, Windows 95 displays an animation of trash being thrown into a wastebasket. This is shown in Figure 6.13.

The entire folder can be recovered from the Recycle Bin at a later time, if necessary, by clicking File . . . Restore inside the Recycle Bin.

FIGURE 6.12 Confirming deletion of a program

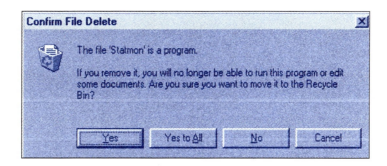

FIGURE 6.13 Deleting the files in a folder

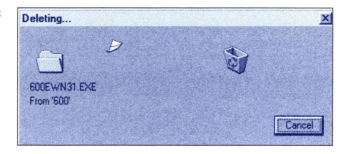

CREATING SHORTCUTS

A shortcut is an icon that you can double-click to start an application, rather than navigating to the application using Windows Explorer and double-clicking it there. For ease of use, shortcuts to often-used applications are usually placed on the desktop, so you can instantly access the applications without opening Explorer or using the Run or Programs menu.

To create a shortcut, use Explorer to navigate to the folder where the application is stored. Select the application by left-clicking it. Then choose File, Create Shortcut to create the shortcut icon, which is placed into the current folder. The name of the shortcut is automatically "Shortcut to . . . ," although you may rename it if you want. Drag and drop the shortcut onto the desktop for easy access to it.

By right-clicking the shortcut icon and selecting Properties from the menu that appears, you can examine or set its properties, such as the type of window it starts the application in (minimized, maximized), its attributes (hidden, read-only), and the working directory. Figure 6.14 shows the initial Properties display for a shortcut to an MS-DOS application called ASM. The Program tab displays the properties shown in Figure 6.15. Note that the shortcut can be started with the same command line parameters you would use while in DOS, just by entering them in the Cmd line box.

For appearance, you may want to change the icon used by a shortcut. Left-clicking the Change Icon button brings up the window shown in Figure 6.16. The new icon can be chosen from the group provided, or you can use Browse to select an icon from a different location.

The Font and Screen tabs control the appearance of the window the shortcut application runs in, and the Memory tab allows you to adjust the way memory is allocated to the application. Last, the Misc tab provides control over other important properties, such as the ability to use a screen saver with the application, and how the application may be terminated.

Note: If you right-click a non-DOS application's shortcut icon, such as Word, and then select Properties, you get a screen with only two tabs (General and Shortcut)—not six tabs as in Figure 6.14. This is further proof that menus are context sensitive.

CHECKING/SETTING PROPERTIES

All the items in the Folder and File display windows can be examined in great detail by looking at the item properties. For example, by right-clicking a disk drive and selecting Properties from the menu, the hard disk Properties window is displayed, as illustrated in Figure 6.17.

FIGURE 6.14   Initial shortcut properties screen

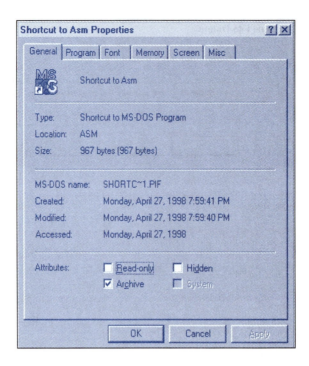

FIGURE 6.15   Program properties for shortcut

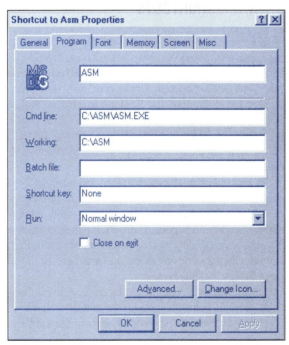

FIGURE 6.16   Choosing a shortcut icon

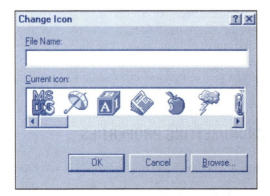

FIGURE 6.17   Disk Properties
window

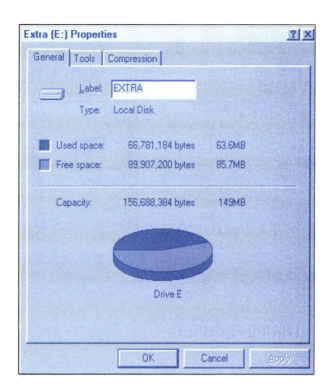

FIGURE 6.18   A Folder
Properties menu

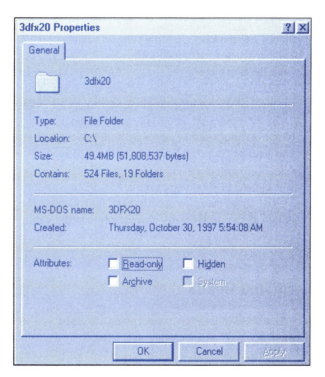

From the disk Properties menu, any of the hard disk properties can be modified or disk tools run. Similarly, if we right-click on a folder and select Properties from the menu, folder properties are displayed. Examine Figure 6.18 to view the properties of a folder. The folder properties include items such as the file name, creation date, file location, size of all the files stored inside the folder, and file attributes. The file attributes can be changed as necessary.

Many property windows have an Apply button, which causes the changes made in the Properties window to take effect. Properties are typically set or cleared by clicking check boxes or radio buttons, or by entering data into a text box.

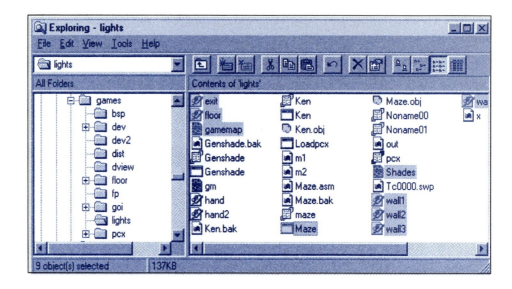

FIGURE 6.19 Selecting multiple files

EDITING FEATURES

Explorer provides four essential editing features: Cut, Copy, Paste, and Undo. These operations are available in the Edit pull-down menu and as buttons on the toolbar. To cut a file or folder, left-click it and then select Edit, Cut, or click the Cut button. Cutting is typically used when you want to move a file or folder to another location. Unlike Delete, nothing is placed into the Recycle Bin when you cut an object. To move the object that has been cut to another location, navigate to the destination, and click the Paste button (or choose Paste from the Edit menu).

Clicking the Copy button (or selecting Copy from the Edit menu) instead of Cut leaves the original file or folder in place, storing a copy when Paste is clicked.

To select more than one file from a folder, or multiple folders, press and hold the Ctrl (Control) button on the keyboard, and left-click on each file or folder you want to select. If you change your mind, simply left-click anywhere in a blank portion of the file display window to deselect everything.

Figure 6.19 shows an example of selecting multiple files in a folder.

The Undo button is used to back up and erase the results of the last operation. For example, if you accidentally move a folder to the wrong location, Undo will move it back for you. You may need to refresh the display to see the results.

FINDING THINGS

Windows Explorer has many options to assist you in finding a file or set of files on your machine. You can search subfolders, search by the date of creation or last access, or search for files of a certain size. You can even search for files containing specific text strings.

Clicking on the Find option under the Tools menu allows you to choose "Files or Folders" or "Computer" as the search location. Choosing "Files or Folders" produces the search window shown in Figure 6.20. The file ASM.C on drive C: is the subject of the search. You may enter any legal file name in the Named box, including file names with wild card characters. Clicking on "Find Now" begins the search, which can be stopped at any time by clicking "Stop." Figure 6.21 shows the results of the search. As indicated, two copies of ASM.C were found, each in a different directory on drive C:.

Clicking the Date Modified tab brings up the search options shown in Figure 6.22. Notice that you can search for a recently created or modified file by selecting a *time frame* for the search.

FIGURE 6.20 Finding a file

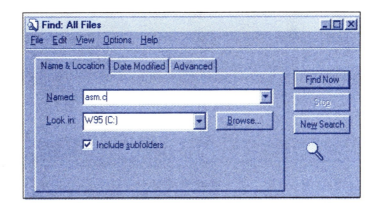

FIGURE 6.21 Search results

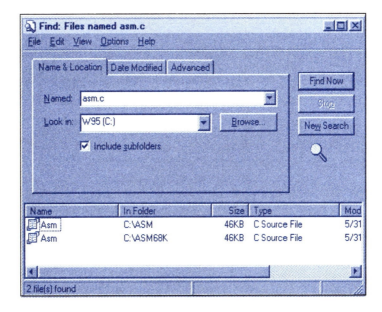

FIGURE 6.22 Using the date to limit the search

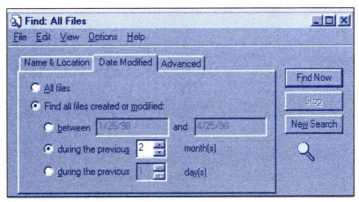

Advanced search options, illustrated in Figure 6.23, include searching files for a string of text (the word "microprocessor" in Figure 6.23), searching for files of a particular size, or searching files associated with a specific application (illustrated in Figure 6.24).

Combinations of each search option may also be used to further restrict the scope of the search.

The second option on the "Tools . . . Find . . ." menu is "Computer." Instead of searching for a file or folder, you can search the network your computer is connected to (even if you are using Dial-Up Networking) for a specific machine. In Figure 6.25 a machine called "Waveguide" is found using this search method.

Altogether there are many ways to search for an object using Windows.

FIGURE 6.23 Searching for text

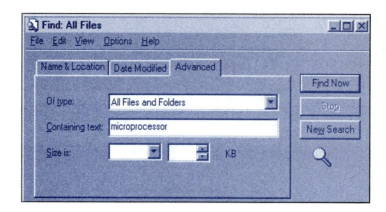

FIGURE 6.24 Selecting a file type to search for

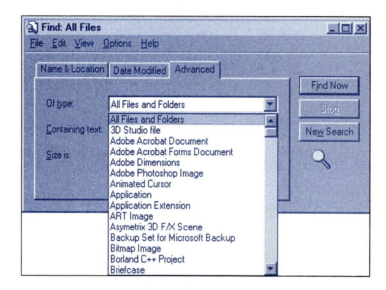

FIGURE 6.25 Searching for a computer on a network

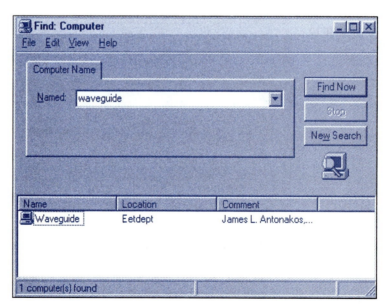

WORKING WITH NETWORK DRIVES

If you have a connection to a network (dial-up PPP or network interface card), you can use Explorer to *map* a network drive to your machine. This is done by selecting Map Network Drive on the Tools menu. Figure 6.26 shows the menu window used to map a network drive.

FIGURE 6.26 Mapping a
network drive

FIGURE 6.27 Supplying a
network password

FIGURE 6.28 Contents of
network drive H:

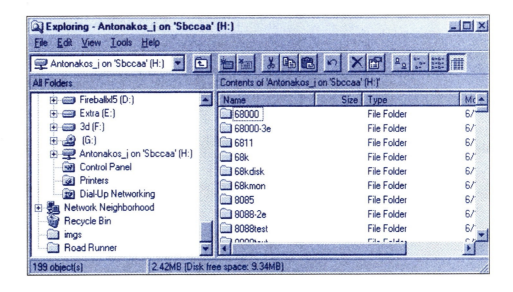

The computer automatically picks the first free drive letter (you can pick a different one) and requires a path to the network drive. In Figure 6.26 the path is \\SBCCAA\ ANTONAKOS_J. The general format is \\machine-name\user-name.

Access to the network drive may require a password, as indicated in Figure 6.27. If an invalid password is entered, the drive is not mapped.

If the drive is successfully mapped, it will show up in Explorer's folder display window. Figure 6.28 shows the contents of the mapped drive. Note that drive H: has a different icon from the other hard drives.

When you have finished using the network drive, you can disconnect it (via the Tools menu). This is illustrated in Figure 6.29.

USING GO TO

When the toolbar is turned on, the Go To folder chooser (above the folder display window) is available; this provides access to all components of the computer. Figure 6.30 shows many of the typical items found in the Go To list.

FIGURE 6.29 Disconnecting a network drive

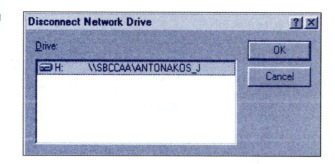

FIGURE 6.30 Using the Go To menu to select an item

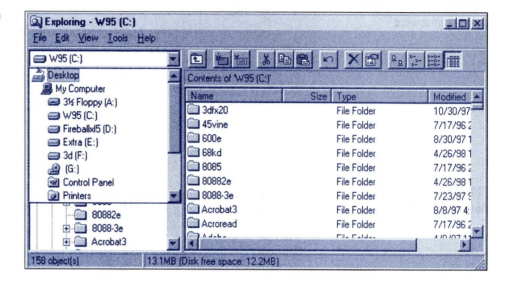

FIGURE 6.31 More Go To items

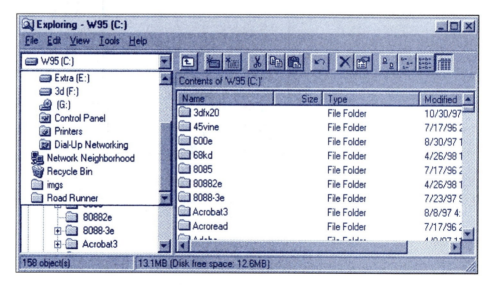

Notice that the first item in the list is the Desktop. Built-in desktop components such as My Computer, Network Neighborhood, and the Recycle Bin are in the Go To list, as well as other desktop items added by the user. This is illustrated in Figure 6.31, which shows the remaining items in the Go To list. The imgs and Road Runner folders, which reside on the desktop, are part of the Go To list as well.

Left-clicking on an item (such as My Computer) in the Go To list displays its contents in the file display window. This is illustrated in Figure 6.32.

So almost anywhere you want to go on your machine is only a click away with the Go To chooser.

FIGURE 6.32 The contents of
My Computer

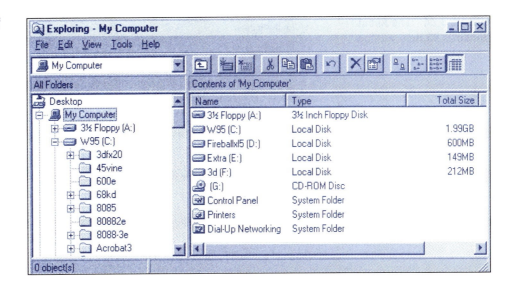

FIGURE 6.33 Initial Help
screen

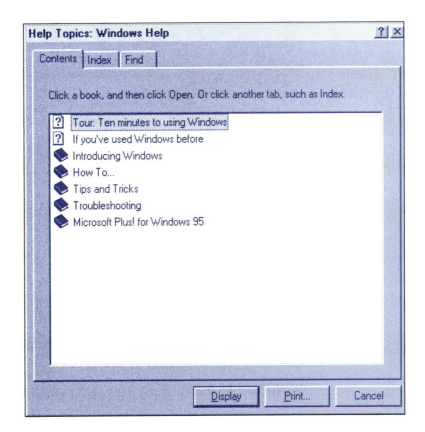

HELP

Help on almost any Windows 95 topic is provided through the Help menu (previously examined in Exercise 2). This is the same Help that is found on the Start menu. The initial Help screen is shown in Figure 6.33.

In addition, selecting the About Windows 95 item on the Help menu brings up the information window shown in Figure 6.34.

The Help menu completes a set of Explorer menus that provide a large amount of control over the Windows 95 environment. Spend some time getting familiar with all the features of Explorer.

FIGURE 6.34   The About Windows 95 Help window

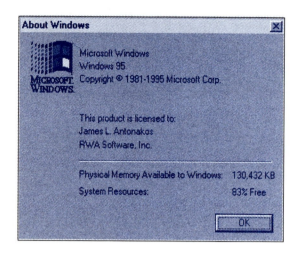

FIGURE 6.35   Windows 98 Explorer showing a Web page

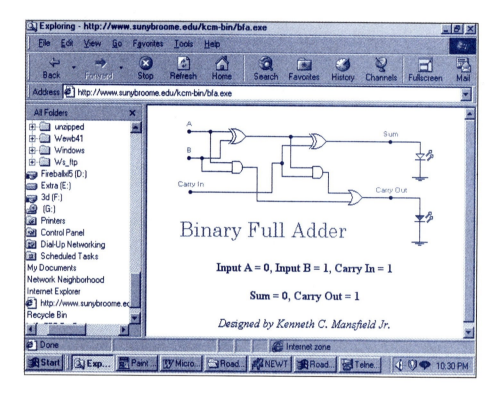

WINDOWS 98 EXPLORER

As previously mentioned, many of the features found in Windows 95 Explorer are also found in Windows 98 Explorer. You can still map drives, navigate around directories, and change views. However, many new features are provided to improve the Windows 98 experience. One enhancement is illustrated in Figure 6.35, where a Web page is being viewed in the rightmost panel. Compare this with Figure 6.1(b), which displays files and folders in the same panel.

In general, the features of Windows 98 Explorer are provided by Internet Explorer 4.0. The toolbar is context dependent with larger buttons that are more visible and convenient. The Address Bar eliminates the need for the Go To option in Windows 95 Explorer. The path to a directory folder or file, or the WWW address of a Web page, can be entered in the Address Bar (with Explorer 4.0 automatically completing the entry if it has been entered before). Furthermore, you may easily add a file, folder, or Web page to your list of favorite locations (with Web pages automatically updated in a background process as frequently as desired).

The context-sensitive aspects of Explorer 4.0 can be seen by comparing the graphical toolbars in Figure 6.35 and 6.1(b). In Figure 6.35 the buttons provide options normally available when browsing the Web, such as Stop, Refresh, and Home. The graphical toolbar in Figure 6.1(b) provides the typical editing functions necessary when working with folders and files, such as Cut, Copy, and Paste.

WINDOWS NT

The Windows NT 4.0 operating system comes standard with the same interface as the Windows 95 operating system. All the discussion about Windows 95 Explorer earlier in this exercise is applicable.

Fortunately, the newer features available in Windows 98 Explorer can also be added to Windows NT 4.0 by upgrading to Windows Explorer 4.0. The upgrade can be downloaded directly from Microsoft or when installing other Microsoft software (such as Visual C++ 6.0). In either case, it is easy to get the new software and take advantage of the new features.

TROUBLESHOOTING TECHNIQUES

While the graphical toolbar contains buttons for many useful operations, it is important to remember that some operations are missing. For example, you may get frustrated searching the toolbar for the Create Folder button, only to find that there is no button for the operation, which is available only as a pull-down menu item. It pays to periodically reexamine the meaning of each toolbar button. You may find that you've been using the Cut and Paste operations from the pull-down menu quite frequently, instead of their built-in toolbar buttons.

SELF-TEST

This self-test is designed to help you check your understanding of the background information presented in this exercise.

True/False

Answer *true* or *false*.

1. The Windows Explorer window contains three separate areas. *False Four*
2. The graphical toolbar contains a limited set of the available features from the pull-down *True* menus.
3. Using the List option scrolls the files vertically. *False Horzontaly*
4. Deleting a file consists of moving the file contents into the Recycle Bin. *False*
5. New folders are always created in the currently selected drive's root directory. *True*

Multiple Choice

Select the best answer.

6. The file display shows
   a. The number of objects in the window and the amount of disk space used.
   b. The graphical toolbar.
   c. A list of files and folders in the currently selected location.
   d. Help topics.
7. New folders are created in
   a. The root directory on the C: drive.
   b. The Windows subdirectory on the D: drive.
   c. The currently selected drive's root directory.
   d. The currently selected location.
8. When deleting a folder, Explorer
   a. Moves the folder into the Recycle Bin immediately.
   b. Prompts two times to be sure all files are to be deleted.
   c. Prompts one time to be sure all files are to be deleted.
   d. Prompts multiple times to be sure all files are to be deleted.

9. Explorer performs four essential editing features
    a. Cut, Copy, Paste, Redo.
    b. Cut, Edit, Move, Rename.
    c. Cut, Paste, Undo, Redo.
    d. Cut, Copy, Paste, Undo.
10. When mapping a network drive
    a. The computer automatically picks the last free drive letter.
    b. The disk shows up in the Explorer display just like the local drives.
    c. Use of the Go To menu item is required to identify the specific network resources.
    d. None of the above.

Completion

Fill in the blank or blanks with the best answers.

11. The list of items on the desktop contains the built-in desktop components _My computer_ _Network_ , and _Recycle bin_
12. _Right_ -clicking on an item displays its contents in the file display window.
13. Many of the _controle_ menu items are located in the graphical toolbar.
14. _File_ properties include file name, creation date, file location, and total size.
15. The _Begin_ _Search_ button begins the search when finding files.

FAMILIARIZATION ACTIVITY

1. Practice using all the display modes of Explorer to determine which one you prefer.
2. Use Explorer to map a network drive. What kinds of software are available on the network drive?
3. Practice creating and renaming files and folders, moving them to different locations, copying them, and navigating through the drives on your system.
4. Create shortcuts to your frequently used applications.

QUESTIONS/ACTIVITIES

1. What is the value of a mapped network drive?
2. Make a list of ten things you normally do while using your computer (for example, rename or copy files, run applications). Explain how these chores can be performed by Windows Explorer (if possible).

REVIEW QUIZ

Under the supervision of your instructor,

1. Explain the basic features of Windows Explorer.
2. Show how to start an application.
3. Create a shortcut, move, find, and delete files and folders.
4. Map a network drive.

7

Managing Printers

<table>
<tr><td>

INTRODUCTION

</td><td>

Joe Tekk lugged a heavy laser printer into his office and set it down on his desk. He plugged the power cable in, loaded paper into the paper tray, and connected the printer to his Windows 98 machine. The printer was an old HP LaserJet Series II that Joe had bought at a hamfest for $25.

Joe ran the Add Printer Wizard and loaded the drivers for the laser printer. However, when he printed a test page, nothing happened. After several minutes of troubleshooting, Joe found the problem: a bent pin on the printer connector. He straightened the pin with a set of needlenose pliers and tried the test page again.

His $25 laser printer worked just fine.

</td></tr>
</table>

PERFORMANCE OBJECTIVES

Upon completion of this exercise, you will be able to

1. Check the properties of any available printers.
2. Install a new printer.
3. Connect to a network printer.

BACKGROUND INFORMATION

Windows has many built-in features to assist you in using your printer, from installing it to sharing the printer on a network. Figure 7.1 shows the starting point for many printer operations, the Printers folder. You can open the Printers folder by left-clicking Start, Settings, and Printer. Or, from inside the Control Panel, you can double-click the Printers folder. The printers installed on the system appear in the Printers folder as separate printer icons. Figure 7.1 shows just one installed printer, the Okidata OL-400e.

PRINTER PROPERTIES

Printer properties can be examined by right-clicking the desired Printer icon. Doing so opens up a context-sensitive menu that provides some basic printer control (pause/purge print jobs, set default printer) and a properties section. The Properties window for the Okidata laser printer is shown in Figure 7.2(a).

FIGURE 7.1 Printers folder

FIGURE 7.2 (a) Initial Windows 95/98 printer Properties window, and (b) Windows NT printer Properties

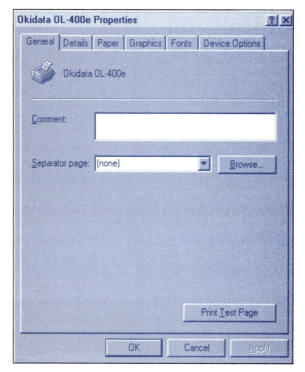

(a)

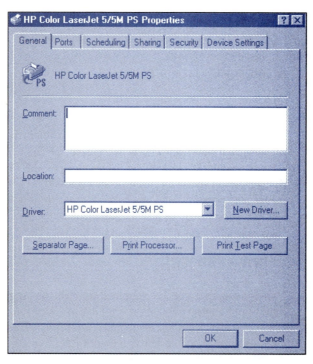

(b)

FIGURE 7.3 Built-in printer
help

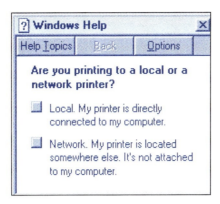

Figure 7.2(a) illustrates the Windows 95/98 Properties window for the Okidata printer. Note that the format of the tabs displayed along the top of the Properties window may contain a Sharing tab if printer sharing is enabled. Figure 7.2(b) shows the Windows NT properties menu. Although much of the same information is available on Windows NT, it is located on different tabs. You are encouraged to familiarize yourself with these screens.

If the printer has been installed correctly, left-clicking the Print Test Page button will cause the printer to print a test page. The test page contains a graphical Windows logo and information about the printer and its various drivers. A dialog box appears asking whether the test page printed correctly. If the answer is no, Windows starts a printer help session. Figure 7.3 shows the initial Help window.

Windows will ask several printer-related questions to help determine why the printer is not working. The causes are different for network printers, so Windows provides two different troubleshooting paths (network vs. local).

In Windows 95/98 the Details tab brings up the window illustrated in Figure 7.4(a). Many of the software and hardware settings are accessed from this window. Notice that LPT1 is set up as an ECP Printer Port. ECP stands for extended capabilities port, a parallel port standard that allows 8-bit bidirectional data flow. This allows ECP to support other peripherals over the parallel printer port, such as external CD-ROM drives or scanners.

The Ports tab shown in Figure 7.4(b) shows a comparable Windows NT menu that indicates an HP Color LaserJet printer is attached on LPT2. The Enable printer pooling

FIGURE 7.4 (a) Windows
95/98 Printer details menu

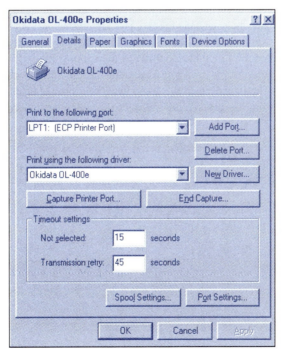

(a)

(continued on the next page)

FIGURE 7.4 *(continued)*
(b) Windows NT printer Ports
menu

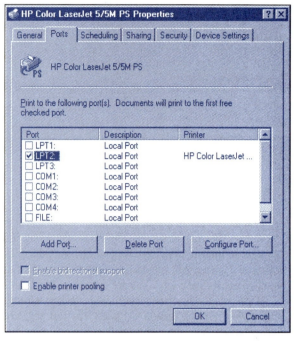

(b)

feature in Windows NT allows for several printers to work together to handle a heavy print workload. Print jobs in the queue are assigned to available printers in the pool so that each of the printers in the pool can be used to service many different print jobs simultaneously.

Left-clicking the Spool Settings button (or the Scheduling tab in Windows NT) allows you to check/set the spool settings for the printer. These settings are indicated in Figure 7.5. Print spooling is a technique that speeds up the time required by an application to send data to the printer. The hard disk is used as a temporary holding place for the print job. The application prints everything very quickly to a spooling file on the hard drive. Windows then prints the spooled file in the background, while the user continues working on other things. The application does not have to wait for each page to be printed before returning control back to the user. This process is diagrammed in Figure 7.6.

Notice that in Figure 7.5(b), other useful information is available on the Windows NT Scheduling menu in addition to the spool settings. The times that the printer is available, priority for the print jobs, and other various settings can be modified as required.

The Paper tab in the Windows 95/98 printer Properties window brings up the window shown in Figure 7.7(a). Here the size and orientation of the paper are selected, as well as

FIGURE 7.5 (a) Windows 95/98 Spool Settings window

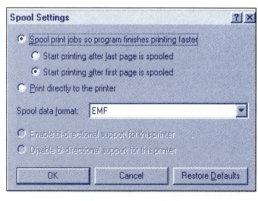

(a)

FIGURE 7.5 *(continued)*
(b) Windows NT printer
scheduling menu

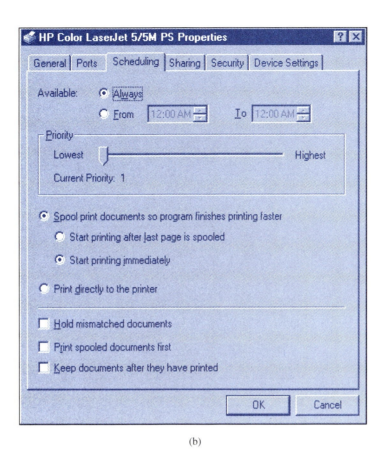

(b)

FIGURE 7.6 Operation of a
print spooler

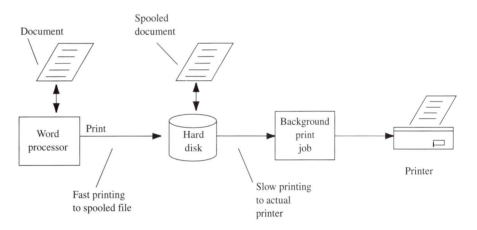

the default number of copies. Normally, pages are printed in Portrait mode. Selecting Landscape turns the printed document 90 degrees, which is useful for printing wide documents such as spreadsheets. In Figure 7.7(b) the Device Settings tab shows the paper settings as well as the Additional Postscript Memory setting for the HP Color LaserJet printer running under Windows NT.

The Graphics settings menu illustrated in Figure 7.8 allows the resolution, *dithering,* and intensity of the printer to be adjusted. Dithering is a method used to represent a particular color by using one or more colors that are similar. This is a necessary step when printing graphical documents that may contain many more colors (or shades of gray) than the printer supports.

You should experiment with the graphics settings until you find the right combination for your printer. Note that the printer properties window does not contain a Graphics tab in Windows NT.

FIGURE 7.7 (a) Windows 95/98 Paper settings menu, and (b) Windows NT printer Device Settings menu

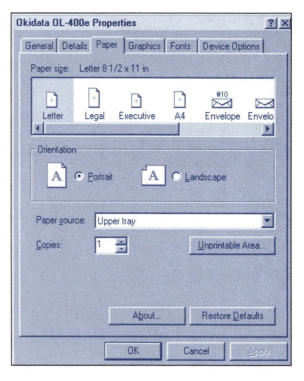

(a)

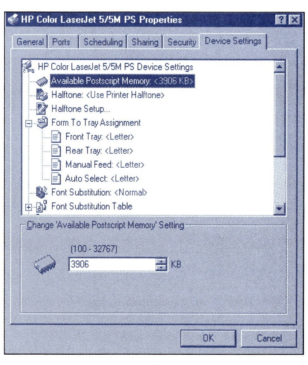

(b)

ADDING A NEW PRINTER

To add a new printer, double-click the Add Printer icon in the Printers folder shown in Figure 7.1. This will start up the Add Printer Wizard, an automated process that guides you through the installation process.

The first choice you must make is shown in Figure 7.9. A local printer is local to your machine. Only your machine can print to your printer, even if your computer is networked. A network printer can be printed to by anyone on the network who has made a connection to

FIGURE 7.8   Windows 95/98 Graphics settings menu

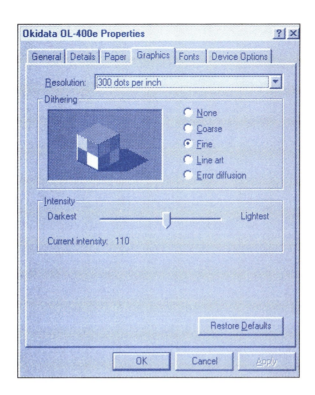

FIGURE 7.9   Choosing local/ network printing

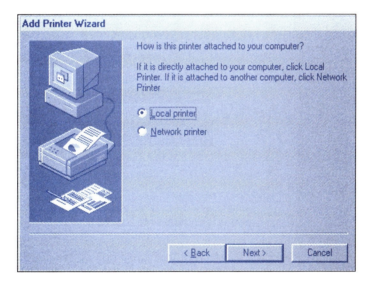

that printer. A network printer is also a local printer to the machine that hosts it. If you are installing a network printer, the next window will look like that shown in Figure 7.10. The printer being mapped is an HP LaserJet II (named "hplaserii" on the network) connected to the machine "deepspace." You can also browse the Network Neighborhood to select a network printer. DOS accessibility to the network printer is controlled from this window as well.

Next, the manufacturer and model of your printer must be chosen. Windows has a large database of printers to choose from. Figure 7.11 shows the initial set of choices. If your printer is not on the list, you must insert a disk with the appropriate drivers (usually supplied by the printer manufacturer).

Once the printer has been selected, the last step is to name it (as in the network printer "hplaserii").

If only one printer is installed, it is automatically the default printer for Windows. For two or more printers (including network printers), one must be set as the default. This can be done by right-clicking on the Printers icon and selecting Set As Default. You can also access printer properties and change the default printer from inside the printer status window, using the Printer pull-down menu.

FIGURE 7.10   Mapping a
network printer

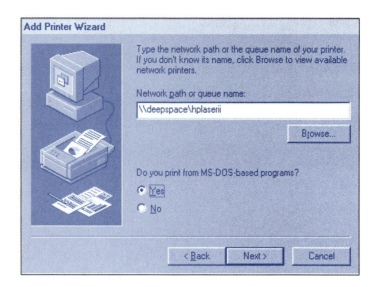

FIGURE 7.11   Choosing a
printer manufacturer/model

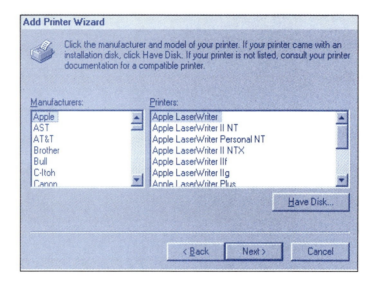

CHECKING THE PRINTER STATUS

To check the status of the current printer, double-click the Printer icon in the system settings area of the taskbar (next to the time display) or double-click the desired Printer icon in the Printers folder. The printer status window is shown in Figure 7.12. The printer status indicates that a document is being printed by the printer.

If Windows detects a problem with the printer (out of paper, offline, powered off), it will report the error without having to bring up the status window. Figure 7.13 shows a typical printer error window.

If a document cannot be printed for some reason before the system is shut down, Windows will save the print job and try to print it the next time the system is booted.

PAUSING A PRINT JOB

To pause a print job (possibly to add more paper), select Pause Printing from the Printer or Document pull-down menus in the printer status window. This will shut off the flow of new data to the printer. If several pages have already been sent, they will be printed before the pause takes effect.

FIGURE 7.12 Printer status window

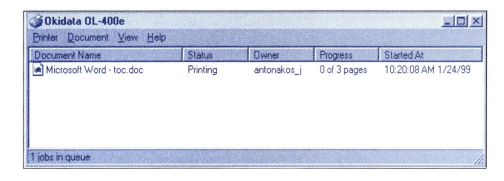

FIGURE 7.13 Printer error window

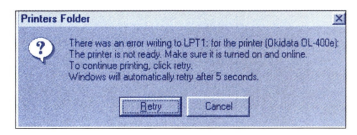

RESUMING A PRINT JOB

To resume a paused print job, select Pause Printing again (it should have a black check mark next to it when paused).

DELETING A PRINT JOB

Individual print jobs can be deleted by selecting them with a single left-click and then choosing Cancel Printing from the Document pull-down menu. *All* the jobs in the printer can be deleted at the same time by selecting Purge Print Jobs from the Printer pull-down menu.

NETWORK PRINTING

To make the printer on your machine a network printer, you need to double-click the Network icon in Control Panel and then left-click the File and Print Sharing button. This opens up the window shown in Figure 7.14. The second box must be checked to allow network access to your printer.

After a network printer connection has been established, you may use it like an ordinary local printer. Windows communicates with the network printer's host machine using NetBEUI. What this means is that jobs sent to a network printer are sent in small bursts (packets) and typically require additional time to print due to the network overhead. In a busy environment, such as an office or college laboratory, printer packets compete with all the other data flying around on the network, and thus take longer to transmit than data traveling over a simple parallel connection between the computer and the printer.

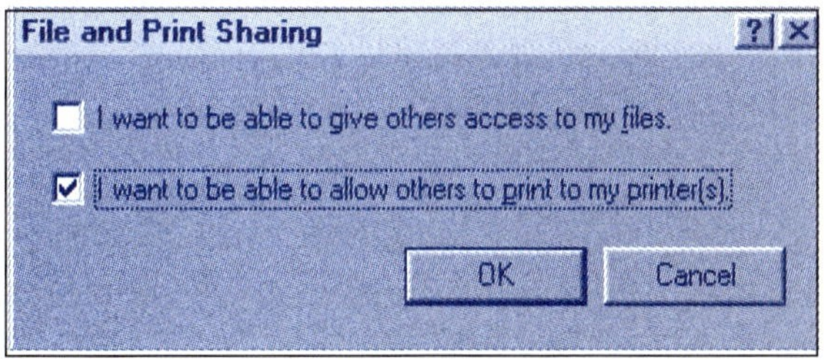

FIGURE 7.14 Giving network access to your printer

111

FIGURE 7.15   Windows NT
printer Security menu

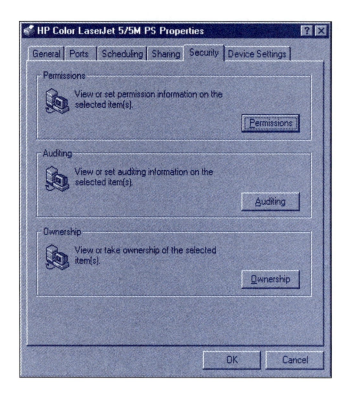

WINDOWS NT PRINTER SECURITY

The Security tab in Windows NT printer properties offers several options not available in Windows 95/98. Figure 7.15 shows the window that allows access to view or change the Permissions, Auditing, and Ownership properties of a Windows NT printer. The system user can assign the printer permissions so that individuals, groups, or everyone in the entire domain may access a particular printer. The type of access (No Access, Print, Manage Documents, or Full Control) is specified for each user or group.

Windows NT printer auditing can be configured to retain certain events. These events (Print, Full Control, Delete, Change Permission, and Take Ownership) are grouped into two categories: success and failure. In general, Auditing involves tracking certain users or examining the data to determine a trend such as paper usage, job length, or who prints the most.

The ownership of a printer can be changed by selecting the Ownership menu button. The owner of a printer can modify permission settings and grant permission to other users. The type of access to the printer determines if ownership of a printer can be taken.

By setting the Permissions, Auditing, and Ownership options appropriately, the system administrator can manage every aspect of the printing process for every computer user in the Windows NT domain. You are encouraged to examine each of these menus in detail to determine what types of settings are currently selected on your computer.

TROUBLESHOOTING TECHNIQUES

You may run into a situation in which you need new drivers for your printer, but you do not have a disk from the manufacturer. One of the easiest solutions is to search the Web for the printer manufacturer and look for a driver download page for your printer. Figure 7.16 shows a Netscape window of a portion of Okidata's driver page. The printer drivers are contained in self-extracting executables. Download the one you need and run it. Be sure to read the README file or other preliminary information. Some printer manufacturers require the old drivers to be completely removed before beginning a new installation or update.

FIGURE 7.16   Downloading a printer driver

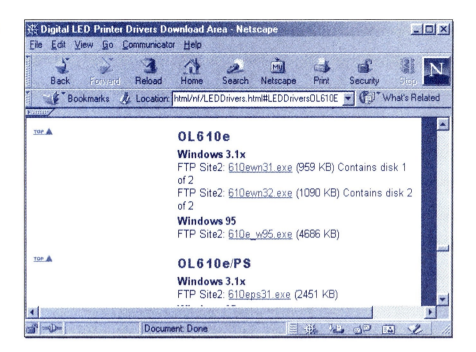

This self-test is designed to help you check your understanding of the background information presented in this exercise.

True/False

Answer *true* or *false*.

1. Only one printer may be installed at a time.
2. Print spooling refers to the coiling of the printer cable around a spool.
3. One printer driver will work for many different printers.
4. Print jobs can be paused.
5. ECP stands for easily connected printer.
6. Windows NT has no security advantages over Windows 95/98.

Multiple Choice

Select the best answer.

7. A print job is deleted by
   a. Dragging it to the Recycle Bin.
   b. Selecting it and choosing Cancel Printing.
   c. Turning the printer off.
8. When Purge Print Jobs is selected,
   a. The current job is purged.
   b. Only network printer jobs are purged.
   c. All print jobs are purged.
9. A network print job typically takes
   a. Less time than a local print job.
   b. The same time as a local print job.
   c. More time than a local print job.
10. When you pause a print job,
    a. Paper immediately stops coming out of the printer.
    b. The current page finishes printing.
    c. All complete pages sent to the printer finish printing.

11. In the network printer path \\waveguide\riko, the name of the printer is
   a. Waveguide.
   b. Riko.
   c. Neither.
12. The Windows NT printer Security menu shows
   a. Administrator, Access, and Permission.
   b. Owner, Permissions, and Manager.
   c. Permissions, Auditing, and Ownership.

Completion

Fill in the blank or blanks with the best answers.

13. A technique that attempts to match colors and gray levels during printing is called
   _____.
14. Data is sent to network printers using the _____ protocol.
15. The two types of printer connections are network and _____.
16. Saving a print job on the hard drive and printing it in the background is called
   _____.
17. One printer is always set as the _____ printer.
18. Windows NT printer auditing groups events into two categories: _____ and
   _____.

FAMILIARIZATION ACTIVITY

1. Examine the properties of all the printers connected to your computer, including any network printers.
2. Install a new local printer.
3. Install a new network printer.
4. Connect to a network printer.
5. Send a ten-page document containing text and graphics to a network printer. Keep track of how long it takes to finish printing. Repeat the print job three more times and compare the timing results.
6. Download a new printer driver from the Web and install it.

QUESTIONS/ACTIVITIES

1. Make a list of all the printers in the lab. How many of them can be found in the installation list of manufacturers and models (Figure 7.11)?
2. Go to a local computer store and check the prices of various printers. What is available for less than $200?

REVIEW QUIZ

Under the supervision of your instructor,

1. Check the properties of any available printers.
2. Install a new printer.
3. Connect to a network printer.

8 Accessories

INTRODUCTION

Joe Tekk stopped in his office for a few minutes on the way to a meeting. He sat down at his computer and checked his schedule with the Calendar utility, opened WordPad to write a quick note to a friend, and examined a graphical display of network traffic using System Monitor. Then his telephone rang. It was Jeff Page, wondering what a $550 motherboard would cost with a 17% discount. Joe brought up the Calculator utility, performed the calculation, and gave Jeff the answer.

As he closed the Calculator window, Joe wondered what he would do without all the utilities provided by Windows.

PERFORMANCE OBJECTIVES

Upon completion of this exercise, you will be able to

1. Describe the basic contents of the Accessories folder.
2. Demonstrate the operation of several utilities, such as the Calculator, Notepad, and Paint.
3. Explain the importance of the System Tools and Administrative Tools folders.

BACKGROUND INFORMATION

Windows provides a large number of accessory applications designed to help perform all the small chores we might require when using a computer to simplify our lives. Figure 8.1 shows the Accessories menu, which is full of useful applications. Notice that there are several folders in the submenu as well.

Notice the similarities between the Windows 95/98 Accessories menu in Figure 8.1(a) and the Windows NT Accessories menu in Figure 8.1(b). Many of the discussions that follow apply to Accessories applications found in both operating systems.

In this exercise we will take a brief look at many of the accessory applications. You are encouraged to spend additional time learning how to use any applications that are useful to you or recommended by your instructor.

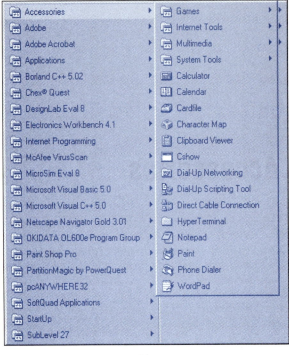

(a)

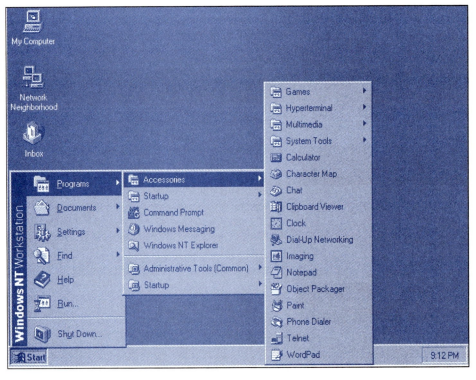

(b)

GAMES

Windows comes equipped with several games, just as Windows 3.x did. These games can be useful tools when introducing a new user to the Windows environment. The Solitaire game shown in Figure 8.2 can be used to introduce a user to the mouse. Experienced Windows users rely on the mouse regularly, so why not get used to using the mouse in a fun way?

FIGURE 8.2 Solitaire game window

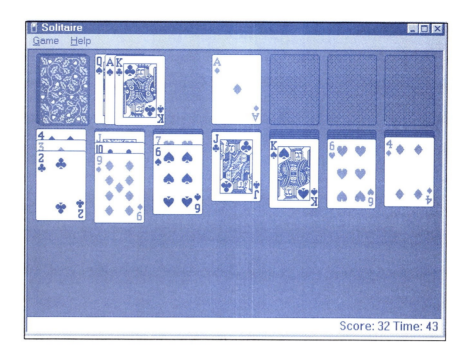

FIGURE 8.3 Internet Explorer window

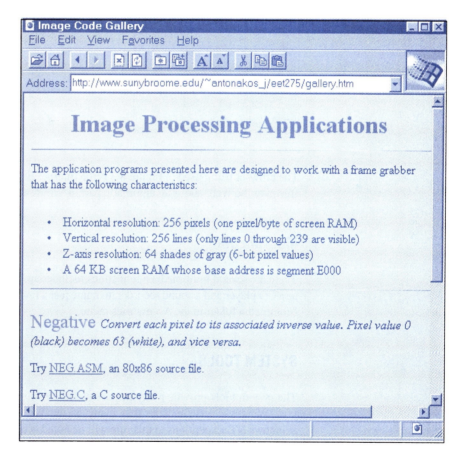

INTERNET TOOLS

The Internet (covered in depth in Exercise 13) is a worldwide collection of networks that all communicate with each other, sharing files and data. One very popular form of posting and sharing information is through a *Web page*. A Web page is written in a language called HTML (hypertext markup language) and is viewed using a *browser*, an application capable of interpreting HTML and displaying a graphical page layout, as illustrated in Figure 8.3.

FIGURE 8.4 Multimedia applications

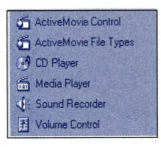

FIGURE 8.5 System Tools menu

The browser being used in Figure 8.3 is the Internet Explorer, which can be downloaded for free over the Web, and is an automatic part of Windows.

MULTIMEDIA

There are several multimedia applications included in Windows. They provide features such as a CD player and a sound recorder, to name just a few. Figure 8.4 shows the contents of the Multimedia folder in the Accessories menu.

SYSTEM TOOLS

The system tools provide many important features for controlling, monitoring, and tuning your Windows system. Note that many of the Windows 95/98 tools are also found in Windows NT. Exceptions to this rule will be noted accordingly. Figure 8.5 shows the selections available in the System Tools menu. Let us look at several of these applications.

Windows 95/98 Disk Defragmenter

When a hard drive is used for a long period of time, many files are created, modified, and deleted. The nature of the file allocation mechanism eventually causes the files of the hard drive (or even a floppy) to become *fragmented*. A fragmented file is spread out over many different areas of the disk, rather than being stored in one big block. This fragmentation increases the time required to read or write to the file, and can lower performance significantly if a large portion of the disk becomes fragmented.

FIGURE 8.6   Selecting a drive
to defragment

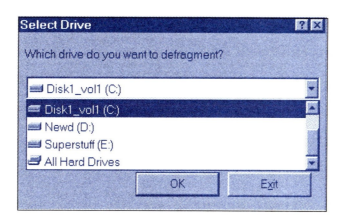

FIGURE 8.7   Defragmentation
in progress

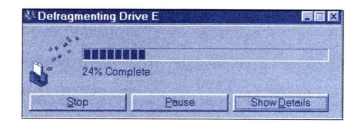

Fortunately, for Windows 95/98 users, the Disk Defragmenter application is included in the system tools to automatically defragment a disk drive. The files are read one by one and written back unfragmented (space is made available before writing the file back). Figure 8.6 shows the initial window for the Disk Defragmenter. Any or all of the disk drives may be selected. Depending on the size of the drive and the amount of fragmentation, the defragmentation process can be quite time-consuming. A simplified status window reports the current progress, as shown in Figure 8.7. Clicking Show Details brings up a multicolored graphical display of the drive being defragmented, with different colors used to identify files being moved, files that cannot be moved, and free space. Stopping the process does not destroy any data on the drive.

Windows NT does not come with a defragmenting program of its own, but you may install a third-party product for disk maintenance, such as Diskeeper, from Executive Software (www.execsoft.com). Diskeeper works with NTFS, the native file system structure used by Windows NT.

The old-fashioned method of defragmentation was to back up the hard disk to tape, format the hard disk, and restore from the backup. Cautious users may still prefer this method, the reward being a *permanent* copy of the hard drive data.

Resource Meter

Three important resources used by Windows 95/98 are system, user, and GDI. The amount of each resource available determines how well Windows 95/98 can handle a new event, such as the launch of a new application. The Resource Meter, shown in Figure 8.8, displays the free percentage of each resource in a bar-graph display. The display is updated as resources are allocated and deallocated.

ScanDisk

Occasionally, a file or directory may develop a problem that prevents Windows 95/98 from reading or writing the file, or accessing the directory. This could be the result of having shut the computer down improperly, or having an application run amok and cause some damage. Or you may accidentally have bumped the computer, causing a head crash in the hard drive.

No matter what the cause, it is always a good idea to run the ScanDisk utility to check the integrity of a disk. Figure 8.9(a) shows the initial ScanDisk window. ScanDisk checks the organization of the FAT (file allocation table), the directory areas, and every file on the

119

**FIGURE 8.8   Windows 95/98
Resource Meter display**

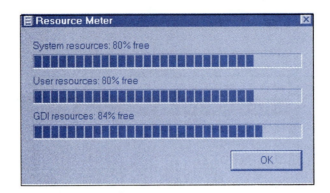

**FIGURE 8.9   (a) ScanDisk
control window, and (b) Sample
Chkdsk execution in
Windows NT**

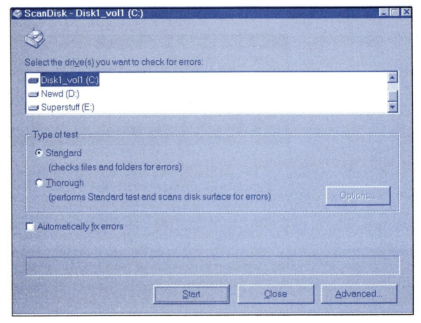

(a)

(b)

disk. The entire surface of the disk can be searched for bad sectors as well. Windows 95 version B and Windows 98 automatically run ScanDisk if they detect that the computer was not properly shut down the last time it was used.

In Windows NT Chkdsk, not ScanDisk, is used to check the structure of a disk drive. Figure 8.9(b) shows the results of running Chkdsk on a Windows NT disk.

FIGURE 8.10 System Information window

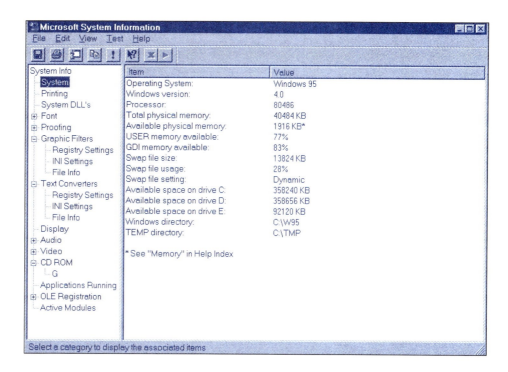

FIGURE 8.11 Monitoring system resources

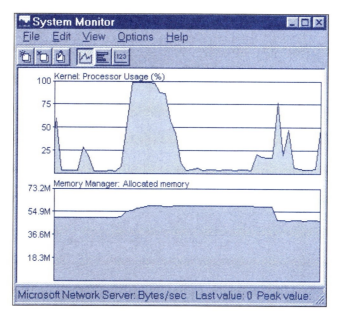

System Information

For a quick look at the information particular to your computer, use the System Information utility. Figure 8.10 shows the first screen of information, which gives a good summary of the pertinent data regarding the overall system. It would be worthwhile to spend some time looking at the different items available in the System Information window. You might be surprised at what you find, such as applications running in the background that you did not even know were there.

System Monitor

To really monitor the performance of your computer, use the System Monitor utility. With this tool, you can display the *history* of usage for one or more resources. As Figure 8.11 shows, the percentage of processor usage and the amount of allocated memory are being tracked by the System Monitor. The Edit menu allows you to add items to track, choosing from four different areas: file system, kernel, memory manager, and network.

FIGURE 8.12   Standard Calculator window

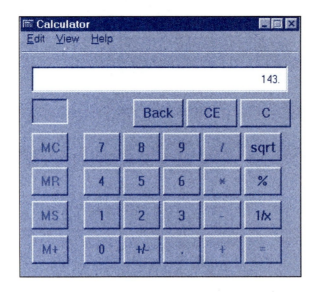

FIGURE 8.13   Scientific Calculator window

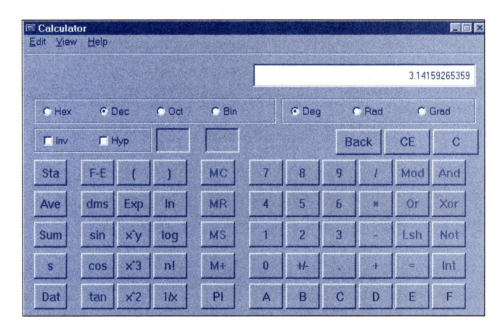

Compression Agent and DriveSpace

There are other tools in the System Tools menu that you may want to examine. One of these is Compression Agent. The Compression Agent utility is used to compress files on a *DriveSpace* compressed drive. DriveSpace is the technology used by Windows 95/98 to compress files on a hard drive or floppy disk, to increase the amount of storage capacity. There is a slight overhead required for decompression when a file is accessed, which can lower performance. Only compress your drive if you've run out of space and cannot add a new drive to your system.

CALCULATOR

If you need to do a few quick calculations, there is no need to go looking for your calculator. Windows has a built-in calculator that performs basic math functions in *standard mode* (shown in Figure 8.12), or in *scientific mode,* where the calculator has many additional features, such as the use of scientific numbers, conversion between different bases, and several transcendental functions. Figure 8.13 illustrates the scientific calculator.

FIGURE 8.14 Friday's
schedule

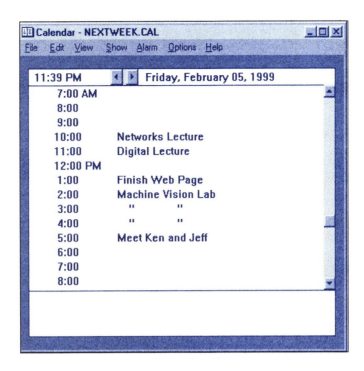

FIGURE 8.15 Sample Cardfile
contents

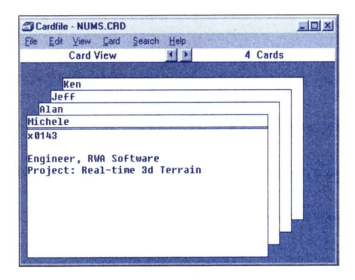

CALENDAR

The Calendar application program is used to maintain a listing of important events or appointments. Each day is displayed showing time increments along the left-hand margin and can be modified very easily by simply typing information at the appropriate time. The calendar can also be set to generate an alarm up to 10 minutes before each appointment or event. Figure 8.14 shows a typical day in the calendar application.

When many computers are networked together, electronic calendars are used to simplify the scheduling process for groups of people. This may be one reason why use of electronic calendars is becoming very widespread.

CARDFILE

The Cardfile utility is an electronic form of the popular 3-by-5-inch index card. You can label a card with a heading, put other information on the body of the card, and save groups of cards in their own file. Figure 8.15 shows a sample set of cards.

FIGURE 8.16   Character Map window

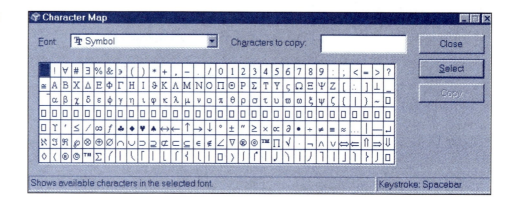

FIGURE 8.17   An image saved on the Clipboard

CHARACTER MAP

The Character Map tool provides a graphical display of each character available in a particular font. If you require a special character to include in a report or project, you can use Character Map to browse the installed fonts on your system and find the character. Figure 8.16 shows the contents of the Symbol font.

CLIPBOARD VIEWER

Whenever you cut or copy an object, or press the Print Screen button, data is placed on the Windows Clipboard. The Clipboard is a temporary holding place for information that you may want to exchange between applications. For example, when developing a Web page, you may use the Clipboard to cut and paste text and images from several different applications into your Web design tool. Figure 8.17 shows the Clipboard Viewer window, which contains a small graphic image placed on the Clipboard by a paint program.

MISSING SHORTCUTS

Occasionally you may come across an icon that does not have an application associated with it anymore (the application was deleted or moved to another folder or drive). In this case Windows will display a small searchlight while it attempts to locate the missing application. Figure 8.18 demonstrates this process.

A new application can be assigned by using the Browse feature. If there is no need for the shortcut any longer, it can be deleted by navigating to the Accessories folder using Windows Explorer and then right-clicking on the shortcut and choosing the Delete operation.

FIGURE 8.18   Missing Shortcut window

FIGURE 8.19   Dial-Up Networking window

FIGURE 8.20   Initial Direct Cable Connection window

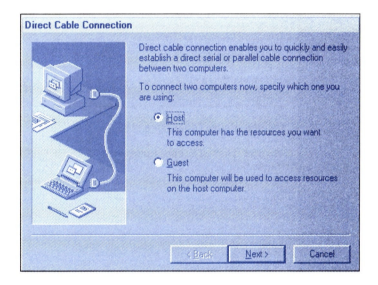

DIAL-UP NETWORKING

Dial-Up Networking (covered in detail in Exercise 11), allows your computer to use its modem to connect to a computer network (using a special protocol called PPP, for point-to-point protocol). Figure 8.19 shows the initial dial-up window. Double-clicking the My Office icon automatically dials the network mainframe computer and establishes the connection. Once connected, your computer can share files as if it were part of the actual network.

DIRECT CABLE CONNECTION

Two Windows machines can communicate with each other directly by using a direct cable connection between the serial or parallel ports of each machine. The initial window (shown in Figure 8.20) allows the user to specify Host or Guest mode. The type of connection (parallel vs. serial) and port are specified next. More information about using a direct cable connection is presented in Exercise 11.

FIGURE 8.21 Using HyperTerminal to communicate with a mainframe

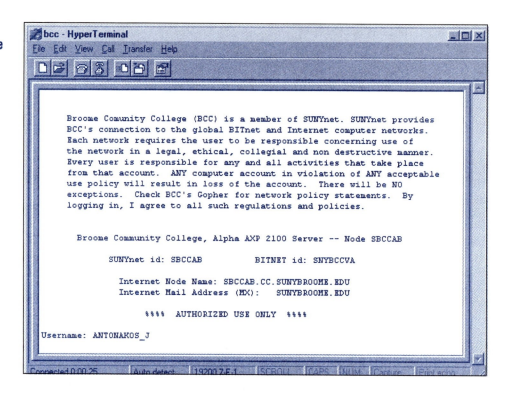

FIGURE 8.22 Notepad window

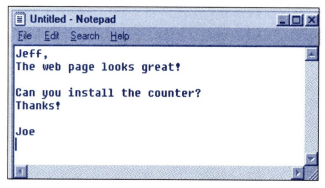

HYPERTERMINAL

If you do not own a modem communications program, Windows comes equipped with the HyperTerminal application, which emulates an ASCII data terminal and controls the modem at the same time. Figure 8.21 shows the start of a HyperTerminal session, with the user ANTONAKOS_J logging into the Broome Community College Alpha server. HyperTerminal will capture its screen to a file, transfer files between computers, and emulate several different types of terminals.

NOTEPAD

The Notepad is a screen-based text editor. You can paste information into the notepad from the Clipboard, or enter it directly from the keyboard. Figure 8.22 shows a quick note being entered. Notepad is useful for small text files (it can search for text, print, and perform simple editing chores), basically replacing the EDIT utility from DOS.

PAINT

The Paint utility allows you to create and edit graphical files (which may include text, graphics, or images). Figure 8.23 shows a simple schematic being edited. Paint saves files

FIGURE 8.23   Using Paint to create a graphics file

FIGURE 8.23   Using Paint to create a graphics file

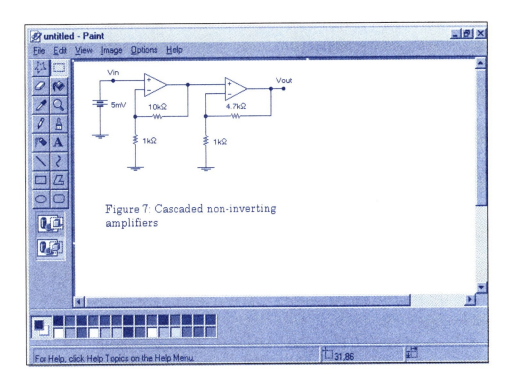

FIGURE 8.24   Phone Dialer window

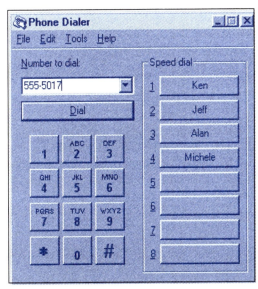

in one format: bitmap. Bitmap files are native to the Windows operating system and have a special binary format that makes them compatible with the desktop display.

PHONE DIALER

Dialing a touch-tone number is as easy as a single click when you use Phone Dialer. Numbers may be saved for speed dialing, or entered from a graphical keypad (or manually from the keyboard). The Phone Dialer window is shown in Figure 8.24. After the number has been dialed (via the internal modem), Phone Dialer instructs you to lift the handset and begin speaking.

WORDPAD

WordPad has many of the features of a full-fledged word-processing application. These include insertion of objects, expanded editing functions, and more control over the printer.

FIGURE 8.25   Editing a
document using WordPad

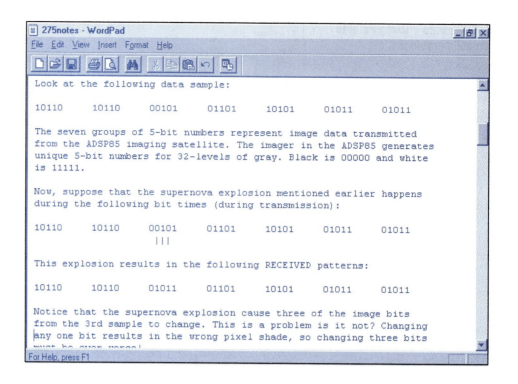

Figure 8.25 shows a sample document being edited in WordPad. Together, WordPad and Paint provide you with a significant amount of text and graphical file processing, all included automatically.

WINDOWS NT ADMINISTRATIVE TOOLS

Windows NT provides additional tools for system administration. The very names of the tools suggest a different atmosphere from the Windows 95/98 environment. In fact, Windows NT allows users to have their own accounts on the system, with varying degrees of privilege and control over the system. Figure 8.26 shows the Administrative Tools

**FIGURE 8.26   Windows NT
Administrative Tools menu**

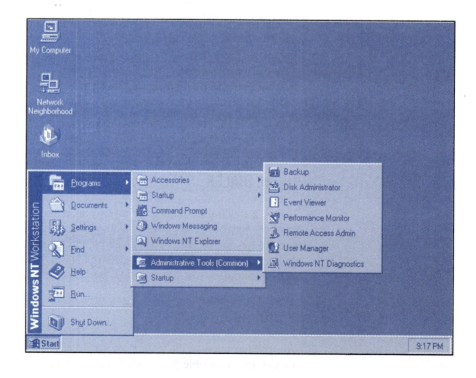

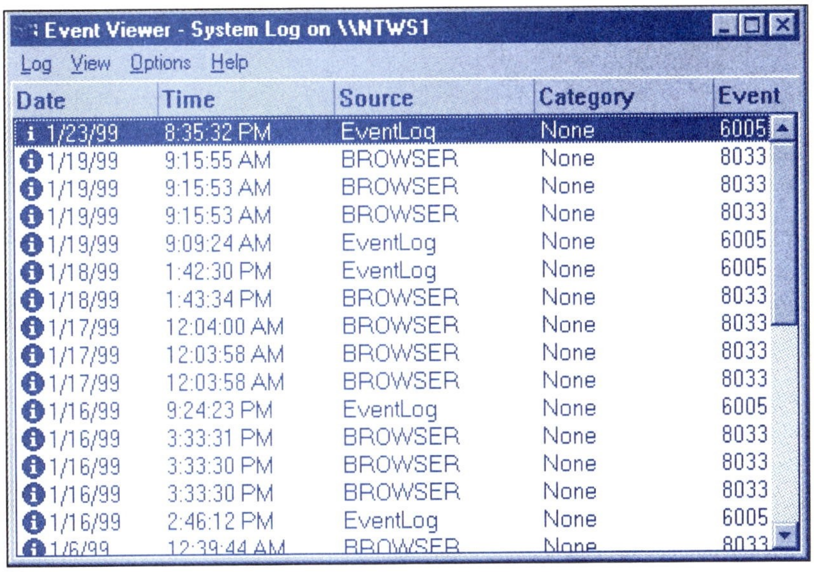

FIGURE 8.27   Event Viewer display

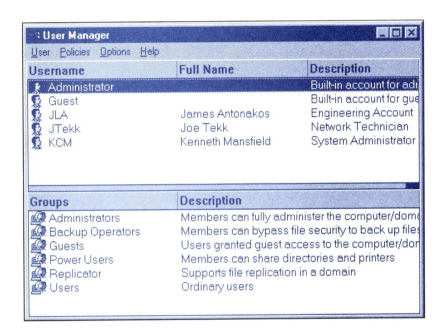

FIGURE 8.28   User Manager display

(Common) menu. The applications listed in the menu are designed to monitor and configure many different aspects of the Windows NT operating system. Let us briefly examine three of these applications. You are encouraged to spend some time exploring these tools in the Familiarization Activity section.

Event Viewer

The Event Viewer tool, shown in Figure 8.27, keeps a running log of various system events. The historical information in the log may prove helpful when troubleshooting problems.

User Manager

The User Manager tool allows user accounts to be created, modified, and deleted (if your own account has sufficient privilege). Figure 8.28 illustrates the user base of a small system. Users who are not displayed in the list cannot log on to the Windows NT system. When Windows NT is first installed, an Administrator account is created automatically that has all available privileges, including the ability to begin creating other accounts on the new system.

129

FIGURE 8.29 Windows NT Memory Diagnostics display

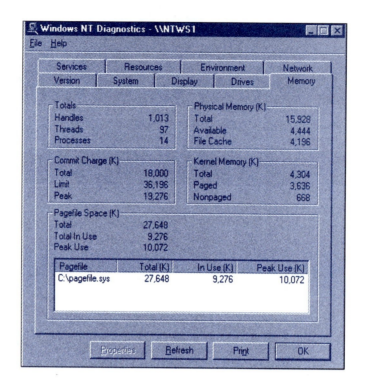

Windows NT Diagnostics

Windows NT keeps track of a significant amount of system parameters. Viewing these parameters and statistics is useful when fine-tuning the operating system for peak performance or locating a bottleneck. This activity is typically done by an experienced system administrator. The presentation of the system data is interesting in its own right, so a tour of the individual diagnostic tabs would be worthwhile. For example, the Memory diagnostic tab display shown in Figure 8.29 uses nomenclature that is not familiar to the typical system user.

TROUBLESHOOTING TECHNIQUES

It is easy to forget that the Accessories folder contains so many useful utilities. It would be good to create shortcuts to the utilities you use the most, placing them on the desktop for quick access.

You may want to periodically review the contents of the Accessories folder. You may re-encounter a utility that you had forgotten about and now could use.

SELF-TEST

This self-test is designed to help you check your understanding of the background information presented in this exercise.

True/False

Answer *true* or *false*.

1. Games have no value for the beginning user. False
2. HTML is a program that executes when browsing the Web. True
3. The Character Map tool provides directions on where to download fonts on the Internet. False
4. A modem is required for Dial-Up Networking. True
5. A direct cable connection can be used to connect two Windows machines. True
6. HyperTerminal is the only utility that can dial a number with the modem. False
7. Any user can create, modify, and delete accounts in Windows NT. True False

130

Multiple Choice

Select the best answer.

8. A list of names might be stored using the
   a. Notepad.
   b. Cardfile. *(circled)*
   c. Calendar.
   d. Clipboard Viewer.
9. An image can be transferred between applications using the
   a. Notepad.
   b. Cardfile.
   c. Calendar.
   d. Clipboard Viewer. *(circled)*
10. The time/date of an important meeting can be stored using the
    a. Notepad.
    b. Cardfile.
    c. Calendar.
    d. Clipboard Viewer.
11. Windows NT systems events are viewed with the
    a. Notepad.
    b. Event Viewer.
    c. Calendar. *(circled)*
    d. Clipboard Viewer.

Matching

Match each utility on the right with each item on the left.

12. Used to create an image. C
13. Connects with a mainframe. D
14. Used to create a text document. B
15. Performs numeric operations. A

a. Calculator
b. Notepad
c. Paint
d. HyperTerminal

Completion

Fill in the blank or blanks with the best answers.

16. HTML stands for _Hypertex markup Language_
17. A disk that has each file stored in different locations, rather than one long block, is said to be _Fragmented File_
18. The _scan Disk_ utility checks a disk for errors (such as bad FAT entries).
19. Two editors capable of creating a text file are Notepad and _Clipboard_
20. Windows NT administrative accounts have all available _____.

FAMILIARIZATION ACTIVITY

1. Use Netscape or Internet Explorer to browse the Web. Try www.windows95.com or www.winfiles.com, two very good sites for Windows users.
2. Check the amount of fragmentation for each drive on your lab system. If you have a floppy disk, check it also.
3. If approved by your instructor, defragment each drive. Make note of how long it takes to do a single drive and the size of the drive.
4. Determine the amount of available resources (system, user, GDI).
5. Use the System Monitor to watch at least four parameters (processor usage, etc.) for 10 minutes. Perform other work on the system as you watch the display. What do you notice?
6. Use the calculator to determine how long, in minutes, a photon of light takes to travel from the Sun to Earth. The distance is 93 million miles and the speed of light is 186,000 miles/second.
7. Make a cardfile of your favorite television shows. Explain how you organized them.

131

8. Open the Clipboard Viewer. Is there anything on it? If so, where do you think the data came from?

9. Use Notepad to write a short letter requesting a sample catalog from a computer manufacturer. Print out the final version.

10. Open the letter from step 9 using WordPad. Use the text editing features to make some words bold, others italic, and others a different size. Center and right justify some of the text. Print out the final version.

11. Explore each of the Windows NT Administrative Tools on the Administrative Tools display menu.

QUESTIONS/ACTIVITIES

1. What other utilities might be useful in the Windows environment?

2. Search the Web for the utilities you listed in question 1. With approval, download one of your choices and install it. How well does it meet your expectations?

REVIEW QUIZ

Under the supervision of your instructor,

1. Describe the basic contents of the Accessories folder.

2. Demonstrate the operation of several utilities, such as the Calculator, Notepad, and Paint.

3. Explain the importance of the System Tools and Administrative Tools folders.

9 Installing New Software

INTRODUCTION

Joe Tekk was very excited. He was downloading a demo version of a game over the Internet. He had considered buying the game, but he did not really know if it was worth the money.

Joe finished downloading a 12MB self-expanding .EXE file that contained all the installation files in a compressed format. He planned to install the software on his D: drive, since the C: drive was quite low in disk space.

However, as he ran the installation procedure, it failed, stating the disk was out of room. Joe noticed the files were being extracted to the TEMP directory on the C: drive during the installation process.

Joe ended up having to reclaim some space on the C: drive so he could install the program. As it turned out, the 12MB of compressed files grew to more than 40MB during the installation process. When the C: drive finally contained enough free space, the installation was successful.

Joe had never run into this problem before, but it was one he would not soon forget, since he had wasted quite a bit of time.

PERFORMANCE OBJECTIVES

Upon completion of this exercise, you will be able to

1. Determine which Windows applications are currently installed.
2. Locate the updates for the Windows operating system on the Microsoft Web site.
3. Discuss the general steps required to perform an application installation and deinstallation.

BACKGROUND INFORMATION

The process of installing software involves moving files from a floppy disk or CD-ROM to a suitable hard disk location (local or network). Every single file located on a computer disk has been installed, one way or another. The files actually consist of executable images, data files, initialization files, dynamic link libraries, and other custom files necessary for the computer or application to run. The files are placed in a directory structure determined by an application's installation program.

What types of software do we install? The operating system itself, plus every single application. Note that many applications register themselves with the operating system during the installation process. Specific application settings are stored in the Registry.

EXISTING WINDOWS 3.X SOFTWARE

Most data files and application programs are upward compatible. This means that Windows 3.x applications can be installed on Windows 95, 98, and NT. Unfortunately, Windows 95, 98, and NT applications cannot be installed on a Windows 3.x computer. When a Windows 3.x computer is upgraded to Windows 95 or 98, all installed applications are also upgraded, by copying information from the old .INI files to the new system Registry. There is no upgrade path to Windows NT from any of the other Windows operating systems.

INSTALLING NEW SOFTWARE FROM THE WINDOWS CD-ROM

The Windows operating system can be configured in many different ways. There are likely to be many files that were not installed during the initial Windows installation. To see the current system configuration, double-click the Add/Remove Programs icon in the system Control Panel. Then select the Windows Setup tab as shown in Figure 9.1. Just by looking at the window, it is apparent what components are installed.

Check boxes along the left margin of the component window identify three situations. First, if a box is checked and the inside of the box is white, all components of the category have been installed. If a box is checked and the inside of the box is gray, only some of the components from that particular category are installed. The Details button is used to show the specific status of each component. If the Windows operating system components are changed, the computer will request the Windows CD-ROM or floppy installation media to copy the additional files.

The Windows 95/98 files are stored in .CAB files on the CD-ROM drive. These files contain the Windows operating system components in compressed form for distribution. Each file is extracted from the .CAB file using a special procedure built into the system.

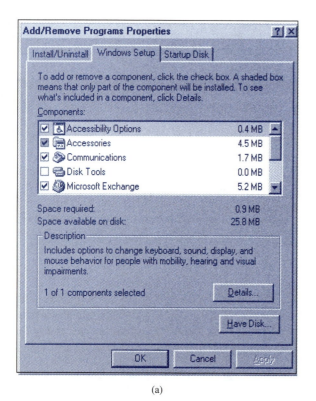

(a)

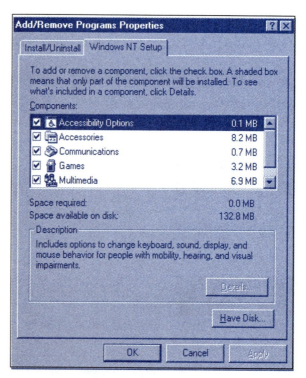

(b)

FIGURE 9.1   (a) Windows 95/98 Setup menu, and (b) Windows NT Setup menu

GETTING THE LATEST UPDATES FROM MICROSOFT

As improvements are made in the Windows operating system, they are posted on the World Wide Web. Figure 9.2 shows the downloads page for support drivers, patches, and service packs from Microsoft. As you can see, each category contains specific applications such as Word, Exchange, and the different Microsoft operating systems.

INSTALLING MATHPRO EXPLORER

Application software is usually installed with a custom software installation wizard. Let us examine an installation process under Windows. The program being installed is MathPro Explorer. The first step in performing a software installation is identifying which installation program to run. Figure 9.3 shows the Run window with the path specified to a setup file. The setup file is located in the root directory of the MathPro CD-ROM. The CD-ROM must be placed into the CD-ROM drive. When the OK button is selected, the system begins the installation process. Figure 9.4 shows the initial MathPro installation screen. This information is usually displayed by the application each time it is run. Each window has option buttons that allow the program user to move back and forward through the screens presented during the installation.

During an installation, the user is given choices about how to install or reconfigure the application software. A complete installation usually installs all components of an application, and a custom installation may allow for one or more components of a product to be installed individually.

FIGURE 9.2   Operating system update from the Web

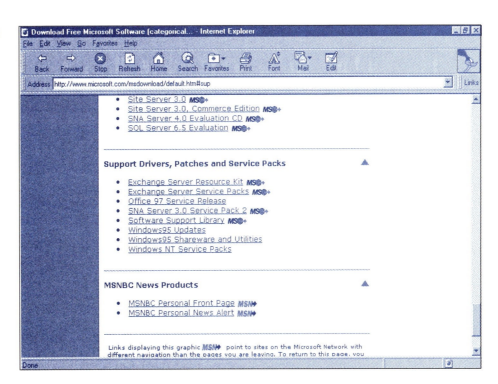

FIGURE 9.3   Location and name of the installation file

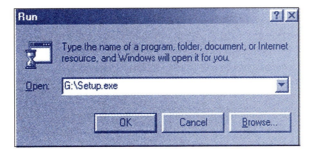

FIGURE 9.4 Initial MathPro installation screen

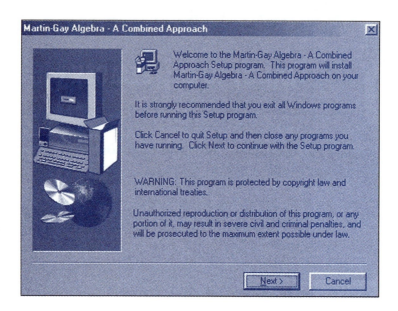

FIGURE 9.5 Choosing a Destination Location

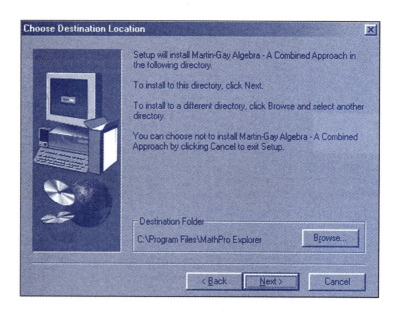

Some applications will search for a previous installation of the application to identify which particular components are installed. When the user is prompted to confirm the program location, it will display the current installation locations. Figure 9.5 shows the default destination folder for MathPro. Clicking the Browse button allows you to select a different storage location. The directory will be created when the Next button is clicked.

The installation process continues with a question asking about how the new application software is to be configured, such as what files may exist from a previous installation, as shown in Figure 9.6 and the name of the program folder, as shown in Figure 9.7. These options are usually set to a default answer. When in doubt, use the default responses.

Eventually, the Setup Wizard begins the process of copying files from the installation media and the selected installation location. The progress meter shown in Figure 9.8 shows a percentage of the how close the installation is from being complete and the current file being processed.

When the installation process completes, it may require the system to be rebooted. This allows for new drivers to be loaded during system initialization. Figure 9.9 shows the final message for successfully completing the MathPro installation.

FIGURE 9.6   Installation question

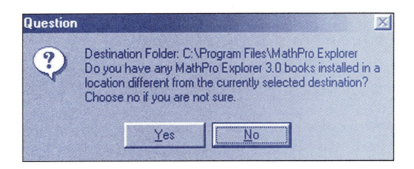

Question

? Destination Folder: C:\Program Files\MathPro Explorer
Do you have any MathPro Explorer 3.0 books installed in a location different from the currently selected destination? Choose no if you are not sure.

[Yes]   [No]

FIGURE 9.7   Selecting a program folder

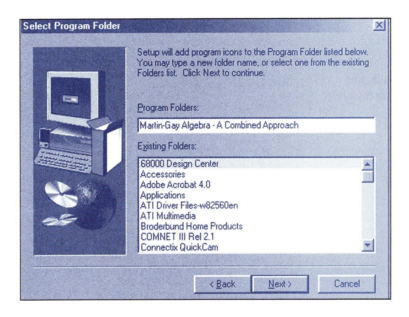

Select Program Folder

Setup will add program icons to the Program Folder listed below. You may type a new folder name, or select one from the existing Folders list. Click Next to continue.

Program Folders:

Martin-Gay Algebra - A Combined Approach

Existing Folders:

68000 Design Center
Accessories
Adobe Acrobat 4.0
Applications
ATI Driver Files-w82560en
ATI Multimedia
Broderbund Home Products
COMNET III Rel 2.1
Connectix QuickCam

[< Back]   [Next >]   [Cancel]

FIGURE 9.8   Files being copied to the hard disk

Copying movies (please wait, this may take a while)
c:\...\mathpro explorer\books\mgcombo\media\mpv124a.mov

25 %

[Cancel]

FIGURE 9.9   Setup Complete screen

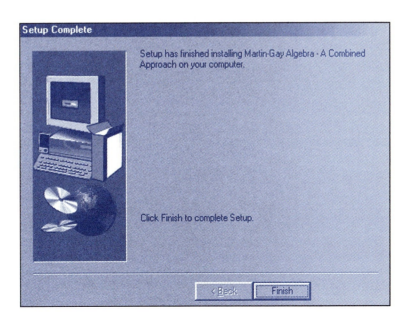

Setup Complete

Setup has finished installing Martin-Gay Algebra - A Combined Approach on your computer.

Click Finish to complete Setup.

[< Back]   [Finish]

REMOVING APPLICATION SOFTWARE

Many situations may require a software application to be removed from a computer. Because an installation program may install application programs in many different directories on a disk, an application cannot be deleted by simply deleting the files in the installation directory. This action deletes many files and will remove the ability to run the program but leaves behind a trail of other files, which remain on the hard disk for no purpose.

To solve this dilemma, newer Windows applications provide an uninstall feature. Using the Add/Remove Programs icon in the Control Panel, the Install/Uninstall tab shows all installed products, as shown in Figure 9.10. These are the applications that can be uninstalled. To uninstall any product, simply select it from the list and press the Add/Remove button. Confirmation prompts will double-check to make sure the choice to remove a product is correct. With positive acknowledgment, the item is removed from the computer. Figure 9.11

FIGURE 9.10   Application selection window for Install/ Uninstall feature

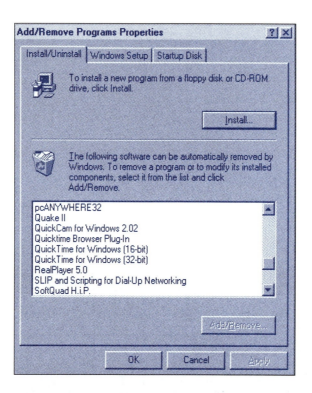

FIGURE 9.11   Progress display during removal of software

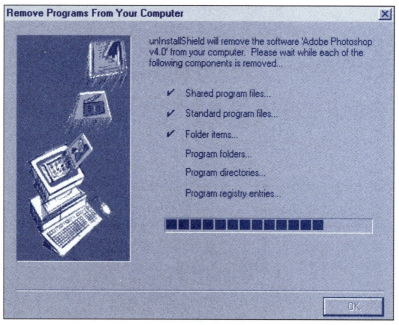

shows the current step being performed by the uninstall process and a progress display indicating the status of the procedure. Exercise extreme caution when uninstalling software. Unless the files have been backed up, it may be impossible to retrieve deleted data files.

TROUBLESHOOTING TECHNIQUES

When a software installation fails, it can usually be attributed to a lack of available resources on the computer, such as disk space or memory. Refer to the product literature to determine the application requirements and be sure your system meets the requirements before you begin an installation.

Be cautious when trying to make room for new applications. Do not delete directories manually or run an uninstall program without making quite sure that any files that may be deleted can be replaced if necessary. Remember, if you delete the wrong files, your system can be rendered unusable.

SELF-TEST

This self-test is designed to help you check your understanding of the background information presented in this exercise.

True/False

Answer *true* or *false*.

1. All the optional Windows operating system components are installed when Windows is installed. *False*
2. Due to software licensing issues, program updates cannot be made available over the Web. *False*
3. The Windows operating system is stored in compressed form on the Windows installation CD-ROM.
4. Every installation must do the same thing, regardless of the application being installed.
5. A checked box with a white background on the Windows Setup menu indicates only a portion of a component is installed.
6. Windows 3.x may be upgraded to Windows NT.

Multiple Choice

Select the best answer.

7. Windows updates can be retrieved from the
    a. World Wide Web.
    b. Electronic mail attachments.
    c. Newest version of the Windows CD-ROM.
8. When installing software, the setup program
    a. Must reboot the system before the installation process can proceed.
    b. Prepares the files on the CD-ROM for transfer.
    c. Copies files from the CD-ROM to a suitable installation location.
9. A .CAB file contains
    a. Compressed Windows installation files.
    b. Security information and passwords.
    c. Application program data files.
10. Windows 95 applications are
    a. Not compatible with new Windows operating systems.
    b. Not compatible with old Windows 3.x operating systems.
    c. Not a factor as far as the operating system is concerned.
11. When a program is uninstalled, the application files are
    a. Deleted by the uninstall process.
    b. Moved to the Recycle Bin by the uninstall process.
    c. Moved into the TEMP directory.

Completion

Fill in the blank or blanks with the best answers.

12. The _____ _____ on the Windows Setup menu show which components are installed.

13. Files are copied to a hard drive during a(n) _____ process.

14. _____ compatibility means Windows 3.x software can be installed on a Windows 95, 98, or NT system.

15. The system Registry contains information about each registered _____ _____.

16. Access to the Add/Remove Programs icon is found on the _____ _____.

FAMILIARIZATION ACTIVITY

1. Determine the status of each component in the Windows Setup window. Which ones are completely installed, partially installed, or not installed at all?
2. Determine what application programs have been installed.
3. Visit the Microsoft Web site to examine support information and data.
4. Install MathPro Explorer (or any other approved Application).

QUESTIONS/ACTIVITIES

1. What are .CAB files?
2. What actions does a Setup Wizard program perform?
3. What actions are performed by an uninstall program?
4. Was anything else installed during the MathPro installation besides the MathPro software?

REVIEW QUIZ

Under the supervision of your instructor,

1. Determine which Windows applications are currently installed.
2. Locate the updates for the Windows operating system on the Microsoft Web site.
3. Discuss the general steps required to perform an application installation and deinstallation.

10

What Is a Network?

INTRODUCTION

Joe Tekk was visiting his friend Julie Plume, an instructor at a local community college. Julie was interested in setting up a network in her classroom.

"Joe," she began, "I need to know a number of things. How much will it all cost? Where do I buy everything? Who can set it up for me?"

Joe laughed. "Hold on, Julie, one thing at a time. The cost depends on how many computers you want to network, the type of network used, and who you buy your equipment from. I have a number of networking catalogs you can look at, and you can also browse the Web for networking products."

Joe looked around the room. There were 14 computers, two laser printers, and a color scanner. "You could probably buy a 16-port Ethernet hub that would take care of this entire room. One network interface card for each PC, some UTP cable, and that's about it. Probably a few hundred dollars will do it. I could set it up with you some afternoon."

Julie had more questions. "Will I need to buy software?"

"Most of the stuff you'll want to do, such as network printing and sharing files, is already built into Windows. You may need to purchase special network versions of some of your software."

"Just one more question, Joe," Julie said. "How does it all work?"

PERFORMANCE OBJECTIVES

Upon completion of this exercise, you will be able to

1. Sketch and discuss the different types of network topologies and their advantages and disadvantages.
2. Sketch and explain examples of digital data encoding.
3. Discuss the OSI reference model.
4. Explain the basic operation of Ethernet and token-ring networks.

BACKGROUND INFORMATION

A computer network is a collection of computers and devices connected so that they can share information. Such networks are called local area networks or LANs (networks in office buildings or on college campuses) and wide area networks or WANs (networks for very large geographical areas). Computer networks are becoming increasingly popular. With the **Internet** spanning the globe, and the **information superhighway** on the drawing boards, the exchange of information among computer users is increasing every day. In this exercise we will examine the basic operation of a computer network, how it is connected, how it transmits information, and what is required to connect a computer to a network. This exercise lays the foundation for the remaining exercises in this unit.

COMPUTER NETWORK TOPOLOGY

Topology has to do with the way things are connected. The topology of a computer network is the way the individual computers or devices (called *nodes*) are connected. Figure 10.1 shows some common topologies.

Figure 10.1(a) illustrates a *fully connected network*. This kind of network is the most expensive to build, because every node must be connected to every other node in the network. The five-node network pictured requires 10 connections. A 20-node network would require 190 connections. The advantage of the fully connected network is that data need only traverse a single link to get from any node to any other node.

Figure 10.1(b) shows the *star network*. Note that one node in the network is a centralized communications point. This makes the star connection inexpensive to build, since a minimum number of communication links are needed (always one less than the number of nodes). However, if the center node fails, the entire network shuts down. This does not happen in the fully connected network.

The *bus network* is shown in Figure 10.1(c). All nodes in the bus network are connected to the same communication link. One popular bus network is **Ethernet,** which we will be covering shortly. The communication link in an Ethernet network is often a coaxial cable connected to each node through a T-connector. The bus network is inexpensive to build, and it is easy to add a new node to the network just by tapping into the communication link. One thing to consider in the bus network is the maximum distance between two nodes, because this affects the time required to send data between the nodes at each end of the link.

The last topology is the *ring network,* shown in Figure 10.1(d). This connection scheme puts the nodes into a circular communication path. Unlike the other topologies, the communication links between nodes may be one-way links in a ring network. Thus, as in the bus network, the maximum communication time depends on how many nodes there are in the network.

FIGURE 10.1   Topologies for a five-node network

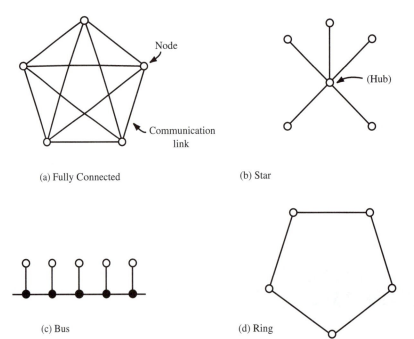

(a) Fully Connected

(b) Star

(c) Bus

(d) Ring

REPRESENTING DIGITAL DATA

The information exchanged between computers in a network is of necessity digital, the only form of data with which a computer can work. However, the actual way in which the digital data is represented varies. Figure 10.2 shows some of the more common methods used to represent digital data.

When an analog medium is used to transmit digital data (such as through the telephone system with a modem), the digital data may be represented by various forms of a *carrier-modulated* signal. Two forms of carrier modulation are amplitude modulation and frequency-shift keying. In amplitude modulation, the digital data controls the presence of a fixed-frequency carrier signal. In frequency-shift keying, the 0s and 1s are assigned two different frequencies, resulting in a shift in carrier frequency when the data changes from 0 to 1 or from 1 to 0. A third method is called *phase-shift keying,* where the digital data controls the phase shift of the carrier signal.

When a digital medium is used to transmit digital data (between COM1 of two PCs, for example), some form of digital waveform is used to represent the data. A digital waveform is a waveform that contains only two different voltages. Inside the computer, these two voltages are usually 0 volts and 5 volts. Outside the computer, plus and minus 12 volts are often used for digital waveforms. Refer again to Figure 10.2. The nonreturn to zero (NRZ) technique simply uses a positive voltage to represent a 0 and a negative voltage to represent a 1. The signal *never* returns to zero.

Another popular method is Manchester encoding. In this technique, phase transitions are used to represent the digital data. A one-to-zero transition is used for 0s and a zero-to-one transition is used for 1s. Thus, each bit being transmitted causes a transition in the Manchester waveform. This is not the case for the NRZ waveform, which may have long periods between transitions. The result is that Manchester encoding includes both data *and* a clock signal, which is helpful in extracting the original data in the receiver.

COMMUNICATION PROTOCOLS

Just throwing 1s and 0s onto a communication link is not enough to establish coherent communication between two nodes in a network. Both nodes must agree in advance on what the format of the information will look like. This format is called a *protocol* and is firmly defined. Figure 10.3 shows one of the accepted standards governing the use of protocols in computer networks. The Open Systems Interconnection (OSI) reference model defines seven layers required to establish reliable communication between two nodes. Different protocols are used between layers to handle such things as error recovery and information routing between nodes. A handy way to remember the names of each layer is contained in a simple statement: All Packets Should Take New Data Paths.

FIGURE 10.2   Methods of representing digital data

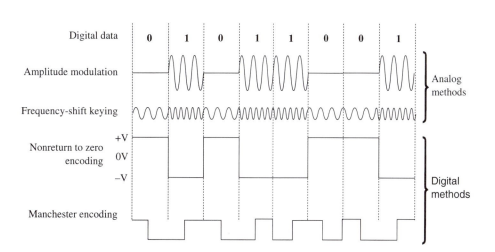

FIGURE 10.3 OSI reference model

Layer	Function
7	Application
6	Presentation
5	Session
4	Transport
3	Network
2	Data Link
1	Physical

Not all of the seven layers are always used in a computer network. For example, Ethernet uses only the first two layers. The OSI reference model is really just a guide to establishing standards for network communications.

The Physical Layer

The Physical layer (layer 1) controls how the digital information is transmitted between nodes. In this layer, the encoding technique, the type of connector used, and the data rate, all of which are *physical* properties, are established.

The Data-Link Layer

The Data-Link layer (layer 2) takes care of error detection and correction.

The Network Layer

The Network layer (layer 3) is responsible for assembling blocks of data into **packets.** A typical data packet consists of control and addressing information, followed by the actual data.

The Transport Layer

The Transport layer (layer 4) is the first layer that is not concerned with how the data actually gets from node to node. Instead, the Transport layer assumes that the physical data is error-free, and concentrates on providing correct communication between nodes from a *logical* perspective. For example, the Transport layer guarantees that a large block of data transmitted in smaller chunks is reassembled in the proper order when received.

The Session Layer

The Session layer (layer 5) handles communication between processes running on two different nodes. For example, two mail programs running on different nodes must communicate with each other.

The Presentation Layer

The Presentation layer (layer 6) deals with matters such as text compression and encryption.

The Application Layer

The Application layer (layer 7) is where the actual user program executes.

ETHERNET

One of the most popular communication networks in use is Ethernet. Ethernet was developed jointly by Digital Equipment Corporation, Intel, and Xerox in 1980. Ethernet is referred to as a *baseband system,* which means that a single digital signal is transmitted. Contrast this with a *broadband system,* which uses multiple channels of data.

Ethernet transmits data at the rate of 10 million bits per second (which translates to 1.25 million bytes per second). This corresponds to a bit time of 100 ns. Manchester encoding is used for the digital data. New 100Mbit and 1000Mbit Ethernet is already being used.

FIGURE 10.4   Eleven-user
Ethernet LAN

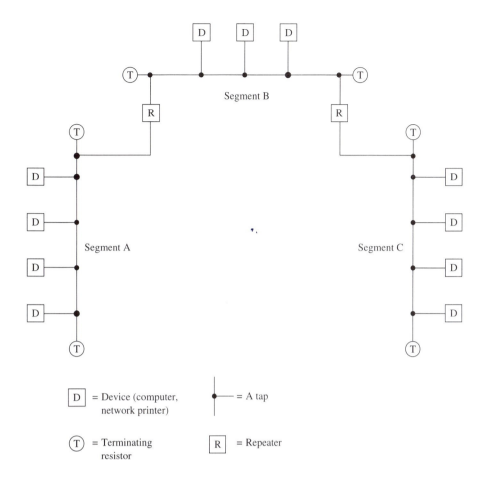

Each device connected to the Ethernet must contain a **transceiver** that provides the electronic connection between the device and the coaxial cable commonly used to connect nodes. Figure 10.4 shows a typical Ethernet installation. The 11 devices on the Ethernet are grouped into three *segments.* Each segment consists of a coaxial cable with a *tap* for each device. It is very important to correctly terminate both ends of the coaxial cable in each segment; otherwise, signal reflections will distort the information on the network and result in poor communications. Segments are connected to each other through the use of *repeaters,* which allow two-way communication between segments.

Each Ethernet device has its own unique binary address. The Ethernet card in each device waits to see its own address on the coaxial cable before actually paying attention to the data being transmitted. Thus, when one device transmits data to another, every device listens. This is called *broadcasting,* much like the operation of a radio. However, Ethernet contains special hardware that detects when two or more devices attempt to transmit data at the same time (called a *collision*). When a collision occurs, all devices that are transmitting stop and wait a random period of time before transmitting the same data again. The random waiting period is designed to help reduce multiple collisions. This procedure represents a protocol called Carrier Sense Multiple Access with Collision Detection (CSMA/CD).

The format in which Ethernet transmits data is called a *frame.* Figure 10.5 details the individual components of the Ethernet frame. Recall that the physical and data-link layers are responsible for handling data at this level. Note that the length of the data section is limited to

FIGURE 10.5   Ethernet frame
format

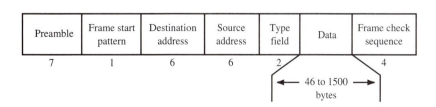

145

a range of 46 to 1500 bytes, which means that frame lengths are also limited in range. Because of the 10Mbit/second data rate and the format of the Ethernet frame, the lengths of the various segments making up an Ethernet LAN are limited to 500 meters. This guarantees that a collision can be detected no matter which two nodes on a segment are active.

TOKEN-RING NETWORKS

Token-ring networks are not as popular as Ethernet but have their own advantages. The high collision rate of an Ethernet system with a lot of communication taking place is eliminated in a token-ring setup.

The basic operation of a token-ring network involves the use of a special token (just another binary pattern) that circulates between nodes in the ring. When a node receives the token, it simply transmits it to the next node if there is nothing else to transmit. But if a node has its own frame of data to transmit, it holds onto the token and transmits the frame instead. Token-ring frames are similar to Ethernet frames in that both contain source and destination addresses. Each node that receives the frame checks the frame's destination address with its own address. If they match, the node captures the frame data and then retransmits the frame to the next node. If the addresses do not match, the frame is simply retransmitted.

When the node that originated the frame receives its own frame again (a complete trip through the ring), it transmits the original token again. Thus, even with no data being transmitted between nodes, the token is still being circulated.

Unfortunately, only one node's frame can circulate at any one time. Other nodes waiting to send their own frames must wait until they receive the token, which tends to reduce the amount of data that can be transmitted over a period of time. However, this is a small price to pay for the elimination of collisions.

TROUBLESHOOTING TECHNIQUES

Troubleshooting a network problem can take many forms. Before the network is even installed, decisions must be made about it that will affect the way it is troubleshot in the future. For example, Ethernet and token-ring networks use different data encoding schemes and connections, as well as different support software. Each has its own set of peculiar problems and solutions.

Troubleshooting a network may take you down a hardware path (bad crimps on the cable connectors causing intermittent errors), a software path (the machine does not have its network addresses set up correctly), or both. There may even be nothing wrong with the network, the failure coming from the application using the network. So, a good deal of trial and error may be required to determine the exact nature of the problem. In the remaining exercises, many of these troubleshooting scenarios will be discussed.

SELF-TEST

This self-test is designed to help you check your understanding of the background information presented in this exercise.

True/False

Answer *true* or *false*.

1. The Internet is a computer network. True
2. Both analog and digital media can be used to transmit digital data. True
3. All seven layers of the OSI reference model are always used for communication in a network. True False
4. Ethernet uses collision detection to handle transmission errors. False

Multiple Choice

Select the best answer.

5. The term LAN stands for
   a. Logical access node.
   b. Local area network.
   c. Large access network.
6. Phase transitions for each bit are used in
   a. Amplitude modulation.
   b. Carrier modulation.
   c. Manchester encoding.
   d. NRZ encoding.
7. Ethernet transmits data in
   a. Continuous streams of 0s and 1s.
   b. Frames.
   c. Blocks of 256 bytes.

Matching

Match a description of the topology property on the right with each item on the left.

D 8. Fully connected   a. The whole network shuts down when the central node fails.
C 9. Star   b. All nodes connect to the same communication link.
B 10. Bus   c. Communication links may be one-way.
A 11. Ring   d. Most expensive to build.

Completion

Fill in the blank or blanks with the best answers.

12. The _____ of a network concerns how the nodes are connected.
13. A(n) _____ _____ network provides the fastest communication be-tween any two nodes.
14. Using two different frequencies to represent digital data is called _____ _____.
15. The layer responsible for error detection and recovery is the _____ layer.
16. Collisions are eliminated in the _____ network.

QUESTIONS/ACTIVITIES

1. Visit the computer center of your school. Find out who the network administrator is and discuss the overall structure of the school's network with him or her.
2. Visit a local computer store and find out how much it would cost to set up a 16-user LAN.

REVIEW QUIZ

Under the supervision of your instructor,

1. Sketch and discuss the different types of network topologies and their advantages and disadvantages.
2. Sketch and explain examples of digital data encoding.
3. Discuss the OSI reference model.
4. Explain the basic operation of Ethernet and token-ring networks.

11 Networking with Windows

INTRODUCTION

"That's the computer right there."

Those were the first words Joe Tekk heard when he entered a high school laboratory maintained under contract by RWA Software. "Pardon me?" Joe asked.

The laboratory technician was a senior, ready to graduate in a few months, with little patience for computers that did not work.

"It's that one right there. It won't connect to the network." He pointed at the computer until Joe got to it. Joe walked around to the back of the computer, pulled the T-connector off the back of the network card, and looked at it closely.

"Here's your problem," he said, to the surprise of the student. "The metal pin is missing from the center of the connector."

The student looked at the connector and then back at Joe. "How did you know to look for that?"

"I always pull the connector out first. I've seen this happen before. Now, it's just a habit."

PERFORMANCE OBJECTIVES

Upon completion of this exercise, you will be able to

1. Identify hard disk resources available on a network computer.
2. Identify printer resources available on a network computer.
3. Create a Dial-Up Networking connection.

BACKGROUND INFORMATION

Windows offers many different ways to connect your machine to one or more computers and plenty of applications to assist you with your networking needs. In this exercise we will examine the basics of networking in Windows.

MICROSOFT NETWORKING

Although Windows supports many different types of common networking protocols the backbone of its network operations is **NetBEUI** (NetBIOS Extended User Interface), a specialized Microsoft protocol used in Windows for Workgroups, Windows 95/98, and Windows NT. NetBEUI allows small (up to 200 nodes) networks of users to share resources (files and printers).

THE NETWORK NEIGHBORHOOD

The Network Neighborhood is a hierarchical collection of the machines capable of communicating with each other over a Windows network. Note that systems running Windows for Workgroups have the ability to connect to the network as well.

Figure 11.1 shows a typical Network Neighborhood. The three small PC icons named At213_tower, Nomad, and Waveguide all represent different machines connected to the network. Each machine is also a member of a *workgroup,* or *domain* of computers that share a common set of properties.

Double-clicking on Waveguide brings up the items being shared by Waveguide. As indicated in Figure 11.2, Waveguide is sharing two folders: pcx and pub.

The Network Neighborhood gives you a way to graphically navigate to shared resources (files, CD-ROM drives, printers).

NETWORK PRINTING

A network printer is a printer that a user has decided to share. For the user's machine it is a local printer. But other users on the network can map to the network printer and use it as if it were their own printer. Figure 11.3 shows a shared printer offered by a computer named Nomad. Nomad is offering an hp 890c.

It is necessary to install the printer on your machine before you can begin using it over the network.

FIGURE 11.1   Network Neighborhood window

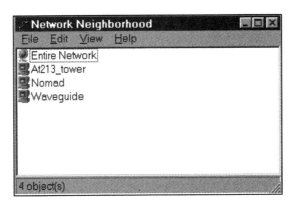

FIGURE 11.2   Items shared by Waveguide

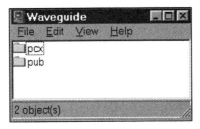

FIGURE 11.3   A printer shared by Nomad

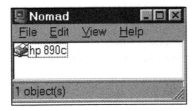

FIGURE 11.4   Indicating a shared drive

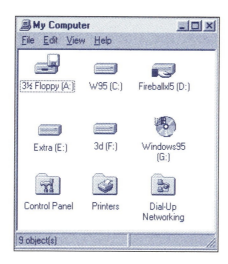

FIGURE 11.5   Sharing Properties window for drive D:

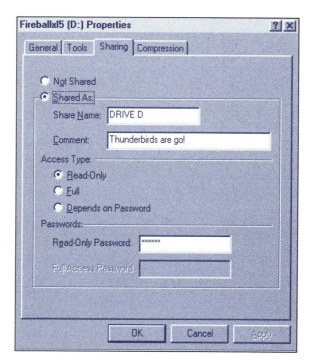

SHARING FILES OVER A NETWORK

A computer can share its disks with the network and allow remote users to map them for use as an available drive on a remote computer. The first time a disk is shared and a connection is established, it is necessary to provide a password to gain access to the data. The password is typically provided by the network administrator. This password is usually stored in the password file for subsequent access to the disk if it is reconnected after a reboot. Figure 11.4 shows the contents of My Computer. The small hand holding drive D: (Fireballxl5) indicates the drive is shared.

The user sharing the drive controls the access others will have to it over the network. Figure 11.5 shows the sharing properties for drive D: (right-click on the drive icon and select Properties). Clearly, the user has a good deal of control over how sharing takes place.

DIAL-UP NETWORKING

Dial-Up Networking is designed to provide reliable data connections using a modem and a telephone line. Figure 11.6 shows two icons in the Dial-Up Networking folder (found in

151

FIGURE 11.6 Dial-Up Networking icons

FIGURE 11.7 Make New Connection window

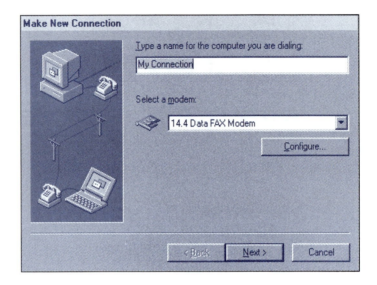

FIGURE 11.8 Information required to access host

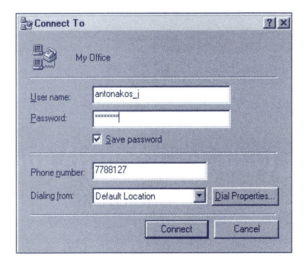

Accessories on the Start menu). Double-clicking the Make New Connection icon will start the process to make a new connection as shown in Figure 11.7. The name of the connection and the modem for the connection are specified.

It is also necessary to provide an area code and telephone number during the configuration process. This number must be for a machine capable of supporting a PPP (point-to-point protocol) connection.

Once the connection has been created, it is activated by double-clicking it. To connect to a remote host, it is necessary to supply a user name and a password. This can be done automatically by the Dial-Up Networking software. Figure 11.8 shows the connection window for the My Office icon.

FIGURE 11.9 Active Dial-Up
Networking connection

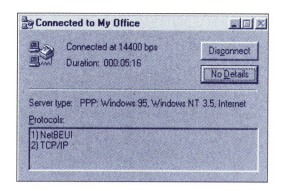

After the information has been entered, the Connect button is used to start up a connection. When the connection has been established, Windows displays a status window showing the current duration of the connection and the active protocols. Figure 11.9 shows the status for the My Office connection. Left-clicking the Disconnect button shuts the connection down and hangs up the modem.

CONNECTING TO THE INTERNET

Besides a modem or a network interface card and the associated software, one more piece is needed to complete the networking picture: the *Internet service provider* (ISP). An ISP is any facility that contains its own direct connection to the Internet. For example, many schools and businesses now have their own dedicated high-speed connection (typically a T1 line, which provides data transfers of more than 1.5 million bits/second).

Many users sign up with a company (such as AOL or MSN) and then dial in to these companies' computers, which themselves provide the Internet connection. The company is the ISP in this case.

Even the local cable company is an ISP now, offering high-speed cable modems that use unassigned television channels for Internet data. The cable modem is many times faster than the fastest telephone modems on the market.

Once you have an ISP, the rest is up to you. You may design your own Web page (many ISPs host Web pages for their customers), use e-mail, browse the Web, Telnet to your school's mainframe and work on an assignment, or download a cool game from an ftp site. Exercise 13 provides additional details about the Internet.

TROUBLESHOOTING TECHNIQUES

Troubleshooting a network connection requires familiarity with several levels of operation. At the hardware level, the physical connection (parallel cable, modem, network interface card) must be working properly. A noisy phone line, the wrong interrupt selected during setup for the network interface card, incompatible parallel ports, and many other types of hardware glitches can prevent a good network connection.

At the software level there are two areas of concern: the network operating system software and the application software. For example, if Internet Explorer will not open any Web pages, is the cause of the problem Internet Explorer or the underlying TCP/IP protocol software?

Even with all of the built-in functions Windows automatically performs, there is still a need for human intervention to get things up and running in the world of networking.

SELF-TEST

This self-test is designed to help you check your understanding of the background information presented in this exercise.

True/False

Answer *true* or *false*.

1. NetBEUI is a protocol only used by Windows NT.
2. Dial-up connections work with ordinary dial-up phone numbers.
3. The cable company is an example of an ISP.
4. Network printers can be used as soon as you map them.
5. Anyone who wants to can delete all the files on a shared drive.

Multiple Choice

Select the best answer.

6. NetBEUI is a protocol used
   a. Only for network printers.
   b. Only for file sharing.
   c. For sharing files and printers.
7. A workgroup is a set of users that
   a. Share common properties.
   b. Use the same printer.
   c. Work as a team on projects.
8. The Network Neighborhood shows
   a. The networked computers within 20 meters of your machine.
   b. Every computer on the entire network.
   c. Machines sharing resources.
9. What is required for Dial-Up Networking?
   a. A modem.
   b. A network interface card.
   c. A direct cable connection.
10. The Network Neighborhood shows
    a. Every computer on the network at the same time.
    b. Hierarchical groups of networked computers.
    c. All the computers on the Internet.

Completion

Fill in the blank or blanks with the best answers.

11. NetBEUI stands for NetBIOS _____ _____ _____.
12. Another term for workgroup is _____.
13. Dial-Up Networking is accessed via the _____ folder in the Start menu.
14. The Dial-Up Networking connection uses the _____ protocol.
15. ISP stands for _____ _____ _____.

Do one or both of the following activities:

Activity 1

1. Set up a new modem connection in Dial-Up Networking. Your instructor will supply the Dial-Up Networking number and other parameters.
2. Begin a Dial-Up Networking session. How long does it take to connect?
3. Use the Network Neighborhood to view the machines reachable over your connection.
4. Try to find three machines sharing files or printers.

Activity 2

1. Establish a network connection using the network interface card. This may be automatically done at boot time.
2. If possible, determine the number of machines on your Windows network by counting the icons in the Network Neighborhood windows.

1. Go to a local business that advertises on the Web. Ask them to describe their network connection. Do they have Dial-Up service? What is the cost? Who maintains their systems?
2. Search the Web for information on how satellites are used in Internet connections.

Under the supervision of your instructor,

1. Identify hard disk resources available on a network computer.
2. Identify printer resources available on a network computer.
3. Create a Dial-Up Networking connection.

12

Computer Security and Viruses

INTRODUCTION

Joe Tekk was puzzled. First, the time displayed on his Windows 98 taskbar was incorrect. Next, for no apparent reason, some applications would launch with a single-click instead of a double-click.

Then the printer began working erratically. Joe checked the cable, which was fine. He connected the printer to a different computer, where it also worked fine.

The problem must be inside my computer, he thought to himself. Deciding to look inside, he shut Windows 98 down and powered off. He took the case off and gave the motherboard, plug-in cards, and connecting cables a good visual. Seeing nothing out of the ordinary, he powered his machine back on and waited while it booted up.

A warning tone from his computer took Joe by surprise. One look at the screen told him what was wrong. His computer had a virus.

PERFORMANCE OBJECTIVES

Upon completion of this exercise, you will be able to

1. Explain the purpose of a firewall.
2. Describe the operation of a typical file-infecting virus.
3. Discuss how viruses are transmitted between computers and files.
4. Use MSAV to scan a system for viruses.

BACKGROUND INFORMATION

It may be surprising to learn that your Windows computer is vulnerable to two major types of threats: those involving intrusion into your networked computer and those associated with software viruses. In this exercise we will examine both kinds of threats, and what can be done to combat them.

COMPUTER SECURITY

The threat of intrusion is never ending. Malicious programs relentlessly search the Internet for vulnerable machines. The operation of the networking layer of Windows makes it possible for other networked machines to discover information about your machine (what files and printers are being shared, the machine name and characteristics, and important network addresses). Once a vulnerable computer is discovered, it may have its files deleted or altered or have its resources stolen (a large unwanted print job sent to the printer, hundreds of programs launched to eat up CPU time and memory). In addition, a virus or other type of malicious program may be secretly loaded into the computer, making it an unwitting accomplice in a future attack on another machine, or on itself. Visit www.grc.com to find out how vulnerable your own computer is to threats from the network.

FIREWALLS

To protect your computer from attacks, it is necessary to utilize a *firewall*, a program designed to limit network access to and from your computer. For example, a malicious program might be scanning the network for all machines having an open TCP/IP port 139. Without a firewall, your machine will send a reply to the port request, identifying itself as an open (and therefore vulnerable) machine. Firewall client software is available in many forms. Three places to look first are:

- www.zonelabs.com   Look for ZoneAlarm.
- www.networkice.com   Look for BlackICE.
- www.hallogram.com   Look for Intruder Alert.

These firewall clients, once installed, load at boot time and protect your computer from unwanted network accesses.

DATA ENCRYPTION

Another measure designed to help thwart intrusion into private information is **data encryption**. Data encryption involves modifying the data to conceal its contents without the use of a proper code or password. Data may be encrypted during transmission on a network to prevent a network sniffer program from being able to determine the contents of the packets. For example, consider this simple message and its encrypted counterpart:

```
The cat jumped over the lazy dog.
Gur png whzcrq bire gur ynml qbt.
```

If may be difficult to decrypt the second message depending on the encryption method used. One popular encryption program is called PGP (Pretty Good Privacy) and is available for free for personal use. You are encouraged to visit www.pgp.com for more information.

In the remaining sections, we will examine the characteristics of the most pervasive type of threat: the software virus.

There are literally thousands of programs available for the personal computer that provide meaningful and constructive service to the user. Unfortunately, there is a growing group of destructive programs, called *computer viruses,* as well. These virus programs are written by clever but dangerous programmers with the intent of doing some kind of damage to the computers of others. This damage can be as simple as a message on the display that reads "You are infected!" or as devastating as a destructive hard drive format. In the remainder of this exercise, we will examine the operation of a computer virus, its method of infection and replication, its classification, and methods of preventing and eliminating virus infections.

OPERATION OF A TYPICAL FILE-INFECTING VIRUS

A virus is a computer program designed to place a copy of itself into another program. Programs are stored on floppy and hard disks as .COM and .EXE files. A virus program intercepts the .COM or .EXE file as DOS begins to load it into memory for execution. The virus checks the .COM or .EXE file to see if it already contains an infection. If the file is not infected, the virus inserts a copy of itself into the .COM or .EXE file and makes whatever other changes are necessary to the file so that the virus, and not the original .COM or .EXE program, executes first the next time the file is executed.

If the file is already infected, the virus does nothing, and allows the program to load normally. This is a clever way of avoiding detection, and it makes no sense to reinfect an already-infected file. This process is illustrated in Figure 12.1. Notice that control in the computer switches from DOS to the virus program, and back to DOS. This implies that the virus is already in memory, watching what is going on in the computer. How did the virus get there in the first place? The answer is given in the next section.

GETTING THE FIRST INFECTION

Virus infections come from a limited number of sources. First, you may get an infection by copying or running a program from someone else's floppy disk (or a swapped hard drive). If you copy an infected program (e.g., a game program, a popular hiding place for viruses), you must run the copied program to activate the virus. If you run an infected program, the virus code gets control first, and loads itself into memory to reinfect more programs later. The virus is active until you turn your computer off. It is not good enough to just reboot your machine using Ctrl-Alt-Del when you think you have an infection, because some viruses take over the reboot code and keep themselves in memory during a warm boot.

FIGURE 12.1 Typical virus infection sequence

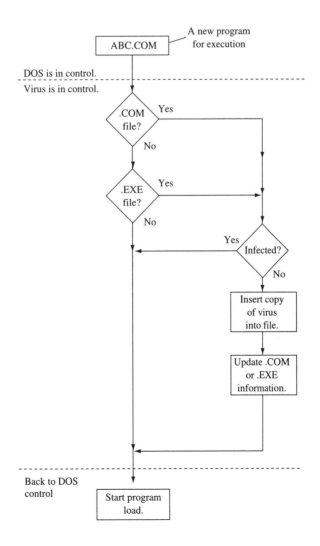

159

**FIGURE 12.2 How a virus
spreads**

Floppy with
infected file

STARGATE.EXE

(a) User puts friend's floppy into disk drive.

Copy STARGATE.EXE C:\Games

(b) User copies the infected program.

CD\Games
STARGATE

(c) User runs the infected game program.
The virus installs itself into memory.

CHESS

(d) User gets tired of game and starts
a new one.

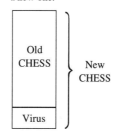

Old
CHESS   New
         CHESS

Virus

(e) Virus infects the CHESS program prior
to execution.

The second way to get an infection is to download an infected program from a computer bulletin board or network site. Most system operators in control of these types of installations already scan their software to help guarantee it is virus-free.

It is rare to find a virus hiding in a newly purchased software product or a box of preformatted disks, but anything is possible. It only takes one infected program to begin the spread of a virus, as you can see in Figure 12.2. The user in this example has unknowingly infected two programs. What happens the next time the STARGATE or CHESS program is executed? The virus will be back in business, resident in memory, waiting to infect any other programs that are run that day. So the user will spread the virus, once again unknowingly, to more and more files.

THE ANATOMY OF A VIRUS

For a self-replicating virus to be able to survive, it must be capable of the following:

• Operating as a memory-resident program
• Interfacing with the disk I/O routines
• Duplicating itself

You may think that the code for a virus that does all of this might be substantial. But the whole trick to writing a virus is to make it as small as possible, because large chunks of

FIGURE 12.3 Hooking an interrupt

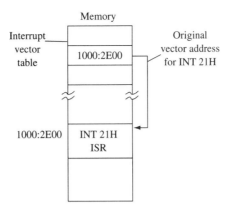

(a) Original vector

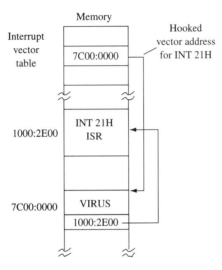

(b) Hooked vector. Virus gets control first.

virus code might be more easily spotted. Many self-replicating viruses are written with fewer than 1000 bytes of machine code, but are still capable of mass destruction of system files.

For the virus to install itself as a memory-resident program, it requires a method of getting control of the system. This is usually done through the use of an interrupt *hook*. BIOS and DOS initialize the interrupt vector table at boot time with the addresses for each interrupt service routine. All the virus has to do is pick an interrupt frequently used by BIOS and DOS (e.g., the keyboard interrupt or INT 21H) and make the interrupt vector point to itself, instead of to the BIOS or DOS service routine. To give the impression that everything is fine, the virus always completes its job by running the original interrupt service routine. This concept of hooking an interrupt is illustrated in Figure 12.3. As you can see, the virus makes a copy of the original INT 21H service routine address, and uses it upon exit.

The process of hooking the interrupt and installation as a memory-resident program is called *initialization*.

When a virus detects that a file should be infected, it does the following:

1. Positions the file pointer at the end of the file. This is accomplished with calls to BIOS or DOS file I/O routines.
2. Writes a copy of itself (directly from memory) to the end of the file being infected.
3. Modifies the initial code of a .COM file (usually a JMP instruction), or the header information in an .EXE file, so that the virus code is executed first whenever the file loads.

4. Tells DOS to update the file information. This is done with another call to BIOS/DOS file routines.
5. Resumes the process of loading the file into memory, so that the user thinks everything is normal.

The duplication (or replication) phase of the virus continues either forever or until the virus has determined that it has reached a preset number of duplications. In the latter case, the virus takes whatever destructive action it was designed to take. Some viruses merely slow down the system response time when they are activated. For example, files that get sent to the printer take longer to print, the mouse gets sluggish, the display scrolls at a slower rate—anything annoying to the user. Other viruses are more troublesome, scrambling the internal key codes so that the keyboard becomes impossible to use (and Ctrl-Alt-Del does not work anymore). A destructive virus might swap numbers around in a spreadsheet file, encrypt a file so that it is impossible to decipher, format a few random tracks on a floppy disk, or delete important system files such as AUTOEXEC.BAT, CONFIG.SYS, and COMMAND.COM. Some viruses are mutated into *strains* or *variants* by other programmers who make a few changes in the virus code so that it does slightly different things when activated. File-infecting viruses are the most plentiful of the known viruses.

BOOT SECTOR VIRUSES

The viruses discussed so far infect actual files residing on floppy or hard disks. A **boot sector virus** is a virus that takes over the boot sector of a floppy or hard drive. The boot sector is the *first* piece of code read in from the disk and is in control of what happens next during the boot process. Boot sector viruses are very sophisticated and require good programming skills, which is why there are fewer boot sector viruses than file viruses.

The operation of a typical boot sector virus is detailed in Figure 12.4. The boot sector virus moves the original boot sector to a new sector on the disk and marks the new sector as bad so that DOS does not try to access it. The portion of the boot sector virus code that does not fit into the boot sector is stored in sectors marked bad also, to help avoid detection.

Any system booted from an infected disk begins running the virus immediately. Once again, the virus operates transparently to the user as a memory-resident program, possibly infecting the boot sector of any disk placed in the floppy drives.

WORMS

A **worm** is essentially a virus that does not replicate. When a worm-infected program is loaded into memory, the worm gets control first and does its damage (possibly transparently). Since it does not try to replicate, the worm is more difficult to discover by the user.

TROJAN HORSES

A **Trojan horse** is a virus disguised as a normal program. For example, a chess program with sophisticated 3-D graphics and sound could be a Trojan horse program containing a file-infecting virus. So, while the user is having fun playing chess, the virus is secretly running, too, examining directories on the hard drive for files to infect. The hard drive activity seems normal to the user because of the 3-D effects on the screen. But the effects are a mask for the file infection happening under the user's nose.

MACRO VIRUSES

Relative newcomers to the world of viruses are the *macro viruses*. Typically, a macro virus is written in a scripting language, such as Visual BASIC, and contains commands that can be processed from within a word-processing or spreadsheet application. These commands,

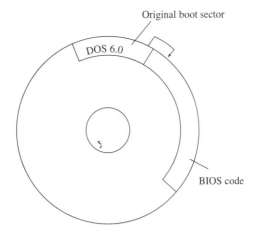

FIGURE 12.4 Typical boot sector virus operation

Original boot sector

DOS 6.0

BIOS code

(a) Original floppy

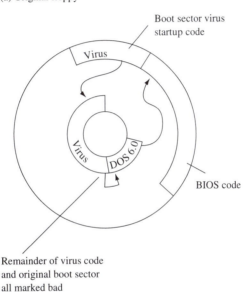

Boot sector virus
startup code

Virus

Virus

DOS 6.0

BIOS code

Remainder of virus code
and original boot sector
all marked bad

(b) Infected floppy

stored as a macro for the particular application, can be very destructive, for instance by deleting files or randomly rearranging data within a spreadsheet when the macro is executed. Putting a simple macro virus together does not require the programming skills of an assembly language or C/C++ programmer, making them easier to write. Word-processing and spreadsheet documents should be included in the list of files routinely scanned by your virus scanner.

VIRUS DETECTION

Fortunately, there are methods that can easily detect the presence of a virus hiding in *any* file. One method involves the use of a file **checksum.** A checksum is a numerical sum based on every byte contained in a file. For example, an 8-bit checksum is made by adding all the bytes of the file, ignoring any carries out of the most significant bit. A file infected by a virus will practically always have a different checksum, thereby making detection a simple matter of comparing the current checksum of a file with a previous one.

A second method of virus detection involves searching a file for a known virus **signature.** A virus signature is an encoded string of characters that represents a portion of the actual virus code for a specific virus. A file is scanned for an entire collection of virus

163

signatures to help guarantee that it is free of infection. The only problem with this technique is that a brand-new virus does not have a signature that the virus detection software will recognize, and it will escape detection.

There are many good virus scanning programs on the market. Many are also available as shareware, or through the computer centers at many colleges, which have an interest in keeping the number of virus infections on students' disks to a minimum. Students migrate from machine to machine on a large campus and will rapidly spread a virus before it is detected.

When DOS 6.0 was released, a virus scanning program called MSAV.EXE (Microsoft AntiVirus) was included; it runs under DOS and is very easy to use. MSAV contains a detailed list of all the viruses it scans for and is capable of removing a virus from an infected file. When MSAV is first started, it scans the current drive for all possible directories. A menu allows the user to choose one of the following operations:

1. Detect
2. Detect and clean
3. Select new drive
4. Options
5. Exit

Selections are made by using the arrows keys followed by ENTER, by typing in specific letters, or by using the mouse. MSAV also contains an information menu, activated when F9 is pressed, which allows the user to view details of a specific virus chosen from MSAV's internal list of viruses.

VIRUS DETECTION IN WINDOWS

A number of companies make virus detection products for Windows. One popular virus scanner package is McAfee VirusScan. VirusScan's VShield runs at boot time, installing itself so that it can watch everything that is going on. VirusScan is used to scan entire drives for infected files. VShield scans files when they are accessed. When a floppy disk is inserted into the drive, it is automatically scanned for viruses. VirusScan is also able to automatically update its virus information database over the Web (for registered users). A screen shot of VirusScan's control window is shown in Figure 12.5. As shown, the hard drive C: has been scanned and no viruses were found. Several configuration options allow you to control what types of files are scanned and what to do when a virus is found. Figures 12.6 and 12.7 illustrate VShields Detection and Action option menus, respectively.

FIGURE 12.5   McAfee VirusScan control window

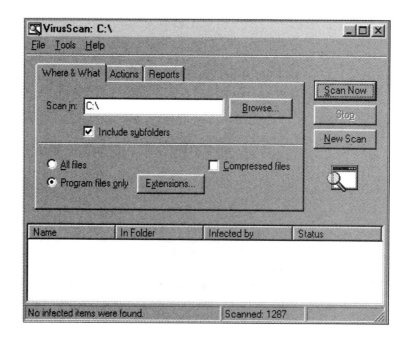

FIGURE 12.6   Detection menu

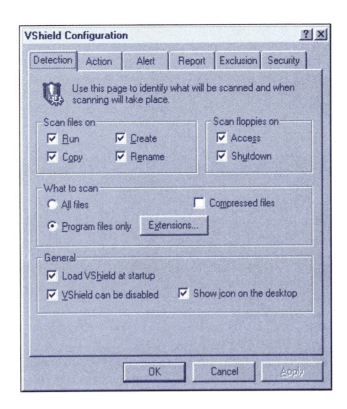

FIGURE 12.7   Action menu

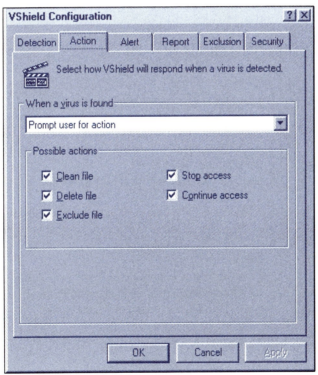

It is well worth the investment (money and scanning time) to use a good virus scanner. You will appreciate its value the first time it finds a virus.

VIRUS PREVENTION

The simple rules that follow should help eliminate the threat of infection.

• Never share your floppy disks with another person.
• Make sure your floppy disks are write-protected.

- Never copy or run software from another person's floppy before it has been scanned for viruses.
- Never execute downloaded software (from a computer bulletin board or network site) before it has been scanned for viruses.
- Never run a program from your disk on another person's computer.
- Run a virus scanner periodically on your computer to ensure that there are no spreading infections.

Keeping your disks and software to yourself is the best method of avoiding a virus infection.

TROUBLESHOOTING TECHNIQUES

Four simple words say it all: When in doubt, scan.

SELF-TEST

This self-test is designed to help you check your understanding of the background information presented in this exercise.

True/False

Answer *true* or *false*.

1. Only .COM files can be infected.
2. File infections come only from other infected files.
3. The hard drive is safe from infection.
4. Viruses are easy to find because they require huge amounts of code and cannot easily hide in a file.

Multiple Choice

Select the best answer.

5. In order to reproduce, a virus must be able to
   a. Install itself as a memory-resident program.
   b. Use BIOS/DOS disk I/O routines.
   c. Duplicate itself from its own image in memory.
   d. All of the above.
6. The best way to stop a virus from spreading is to
   a. Reboot the computer.
   b. Run as many programs as possible to tire the virus out.
   c. Eliminate the virus with a virus scanning program.
   d. Format the hard drive.
7. Viruses are written
   a. By beginning programmers.
   b. With the intent of doing damage to PCs.
   c. For amusement at computer trade shows.

Completion

Fill in the blanks with the best answers.

8. A virus disguised as a normal program is called a(n) _____ _____.
9. Viruses are detected by comparing _____ or _____.
10. A virus may also hide inside the _____ _____ of a floppy or hard disk.

FAMILIARIZATION ACTIVITY

1. Start up the MSAV program from hard drive C:.
2. When MSAV finishes checking directories, choose Detect from the menu. Detect will scan system RAM for memory-resident viruses and then begin scanning the hard drive. The current file (and its associated directory path) is displayed in the upper left corner

of the display. Warning! If MSAV finds a virus, the speaker will emit a short tone, and a message will be displayed such as:

Virus Friday 13th was found in: SHIP.COM

MSAV will give you the option of cleaning the file, continuing, or stopping the scan. Consult with your instructor if this happens.

You may also get a *verify* error for a particular file. This indicates that MSAV found a different time/date stamp on the file, or that the file's length or checksum has changed. MSAV allows you to update the file's information, or ignore the error and resume scanning. Consult with your instructor if this happens (in case you have discovered a brand-new virus).

3. Use the Select option to choose a new drive. Select the drive appropriate for the floppy disk you are going to scan.
4. Use Detect to scan your floppy disk. Once again, consult with your instructor if a virus is detected.
5. Press the F9 button to get into MSAV's information menu. An alphabetical list of viruses should appear, beginning with the As.
6. Enter the three characters "F," "R," and "I." MSAV should update the virus list so that the selected virus begins with the name "Friday." Use the arrow keys to select the Friday the 13th virus.
7. Press Tab until the Info box is highlighted. Then press ENTER. You should get a screen of information on the virus that shows what types of files are infected, how long the virus code is, and what the side effects of the virus are.
8. Exit MSAV.
9. Restart MSAV with the following command line:

```
C> MSAV  C:\DOS
```

MSAV will now scan only the \DOS directory, not the entire hard drive. Individual files can also be scanned in this way by using their full path names after MSAV, as in:

```
C> MSAV C:\DOS\PRINT.EXE
```

QUESTIONS/ACTIVITIES

1. Download a firewall client and observe its operation for a lab period. What, if any, type of attacks were discovered?
2. Use MSAV to make a list of 10 boot sector viruses. Compare their relative sizes and infection information. Include the Pakistani Brain, Michelangelo, and Alameda viruses in your list.
3. Repeat step 2 for the file-infecting viruses. Include the Friday the 13th and Columbus Day viruses in your list.

REVIEW QUIZ

Under the supervision of your instructor,

1. Explain the characteristics of a firewall.
2. Describe the operation of a typical file-infecting virus.
3. Discuss how viruses are transmitted between computers and files.
4. Use MSAV to scan a system for viruses.

13 Internet Technology

INTRODUCTION

It was 2:45 A.M. Joe Tekk was awake, sitting in his darkened living room. The only light in the room was coming from the monitor of his computer. Joe was exhausted, but didn't want to stop browsing the Web. He had stumbled onto a Web page containing links to computer graphics, game design, and protected-mode programming. For three hours, Joe had been going back and forth from one page to another, adding some links to his bookmarks and ignoring others. When he finally decided to quit, it was not because of lack of interest, but simply time to go to sleep.

"From now on, I'm only browsing for 30 minutes," Joe vowed to himself. But he knew he would have another late-night browsing session that would last much longer. It was too much fun having so much information available instantly.

PERFORMANCE OBJECTIVES

Upon completion of this exercise, you will be able to

1. Describe the basic organization of the Internet.
2. Explain the purpose of a browser and its relationship to HTML.
3. Discuss the usefulness of CGI applications.

BACKGROUND INFORMATION

The Internet started as a small network of computers connecting a few large mainframe computers. It has grown to become the largest computer network in the world, connecting virtually all types of computers. The Internet offers a method to achieve *universal service,* or a connection to virtually any computer, anywhere in the world, at any time. This concept is similar to the use of a telephone, which provides a voice connection anywhere at any time. The Internet provides a way to connect all types of computers together regardless of their manufacturer, size, and resources. The *one* requirement is a connection to the network. Figure 13.1 shows how several networks are connected together.

The type of connection to the Internet can take many different forms, such as a simple modem connection, a cable modem connection, a T1 line, a T3 line, or a frame relay connection.

FIGURE 13.1   Concept of Internet connections

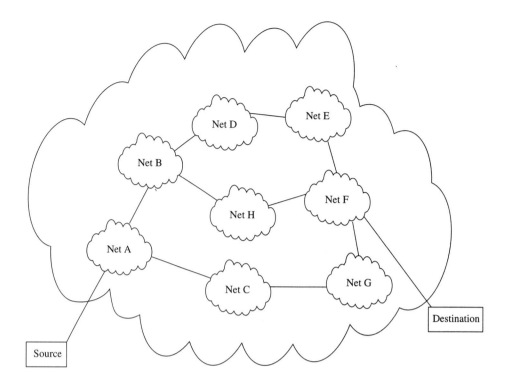

TABLE 13.1   Organization of the Internet

Domain Type	Organization Type
edu	Educational institution
com	Commercial organization
gov	Government
mil	Military
net	Network providers and support
org	Other organizations not listed above
country code	A country code, for example, .us for United States, .ca for Canada

THE ORGANIZATION OF THE INTERNET

The current version of the Internet (V4) is organized into several categories, as shown in Table 13.1. The name of an Internet host shows the category to which it is assigned. For example, the rwa.com domain is the name of a company, and the bcc.edu domain is an educational institution. Each domain is registered on the appropriate root server. For example, the domain rwa.com is known by the com root server. Then, within each domain, a locally administered Domain Name Server allows for each host to be configured.

WORLD WIDE WEB

The World Wide Web, or WWW as it is commonly referred to, is actually the Hypertext Transport Protocol (HTTP) in use on the Internet. The HTTP protocol allows for hypermedia information to be exchanged, such as text, video, audio, animation, Java applets, images, and more. The hypertext markup language, or HTML, is used to determine how the hypermedia information is to be displayed on a WWW browser screen.

The WWW browser is used to navigate the Internet by selecting *links* on any WWW page or by specifying a Universal Resource Locator, or URL, to point to a specific *page* of information. There are many different WWW browsers. The two most popular are Microsoft Internet Explorer and Netscape Navigator, shown in Figures 13.2 and 13.3,

FIGURE 13.2 Sample home page displayed using Internet Explorer

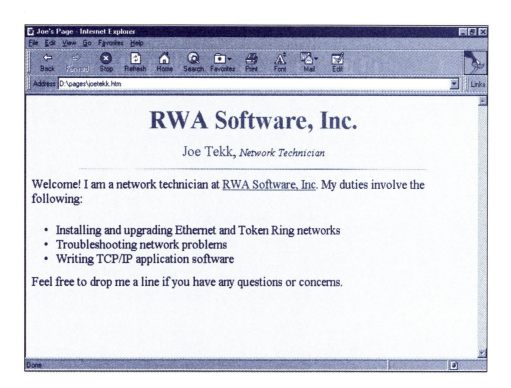

FIGURE 13.3 Sample home page displayed using Netscape Navigator

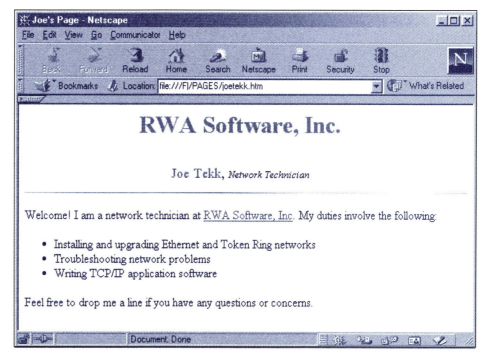

respectively. Both of these browsers are available free over the Internet, and contain familiar pull-down menus and graphical toolbars to access the most commonly used functions such as forward, backward, stop, print, and reload.

Note the differences in page layout between Figures 13.2 and 13.3. Individuals who design Web pages must take into account the different requirements of each browser so that the page looks acceptable in both browsers.

HTML

HTML stands for hypertext markup language. HTML is the core component of the information that composes a Web page. The HTML *source* code for a Web page has an overall

171

syntax and structure that contains formatting commands (called *tags*) understood by a Web browser. Here is a sample HTML source. The actual Web page for this HTML code was shown in Figure 13.3.

```
<HTML>

<HEAD>
<TITLE>Joe's Page</TITLE>
</HEAD>

<BODY BGCOLOR="#FFFF80">

<P ALIGN="CENTER">
<B><FONT SIZE="+3" COLOR="#FF0000">RWA Software, Inc.</FONT></B>
</P>

<P ALIGN="CENTER">
<FONT SIZE="+1"><FONT COLOR="#008000">Joe Tekk</FONT>,
<FONT SIZE="-1"><I>Network Technician</I></FONT></FONT>
</P>

<P ALIGN="CENTER">
<IMG SRC="bar.gif" ALT="Color Bar">
</P>

<P ALIGN="LEFT">
Welcome! I am a network technician at
<A HREF="http://www.rwasoftware.com">RWA Software, Inc</A>. My duties
 involve the following:
</P>

<UL>
<LI>Installing and upgrading Ethernet and Token Ring networks</LI>
<LI>Troubleshooting network problems</LI>
<LI>Writing TCP/IP application software </LI>
</UL>

<P ALIGN="LEFT">
Feel free to drop me a line if you have any questions or
concerns.
</P>

</BODY>
</HTML>
```

The HTML source consists of many different tags that instruct the browser what to do when preparing the graphical Web page. Table 13.2 shows some of the more common tags. The main portion of the Web page is contained between the BODY tags. Note that BGCOLOR= "#FFFF80" sets the background color of the Web page. The six-digit hexadecimal number contains three pairs of values for the red, green, and blue color levels desired.

Pay attention to the tags used in the HTML source and what actually appears on the Web page in the browser (Figure 13.3). The browser ignores whitespace (multiple blanks between words or lines of text) when it processes the HTML source. For example, the anchor for the RWA Software link begins on its own line in the source, but the actual link for the anchor is displayed on the same line as the text that comes before and after it.

Many people use HTML editors, such as HoTMetaL or Front Page, to create and maintain their Web pages. Options to display the page in HTML format, or in WYSIWYG (what you see is what you get), are usually available, along with sample pages, image editing, and conversion tools that convert many different file types (such as a Word document) into

TABLE 13.2   Assorted HTML tags

Tag	Meaning
<P>	Begin paragraph
</P>	End paragraph
<B>	Bold
<I>	Italics
<IMG SRC...>	Image source
<UL>	Unordered list
<LI>	List item
<TABLE>	Table
<TR>	Table row
<TD>	Table data
<A>	Anchor

FIGURE 13.4   HoTMetaL PRO with sample page

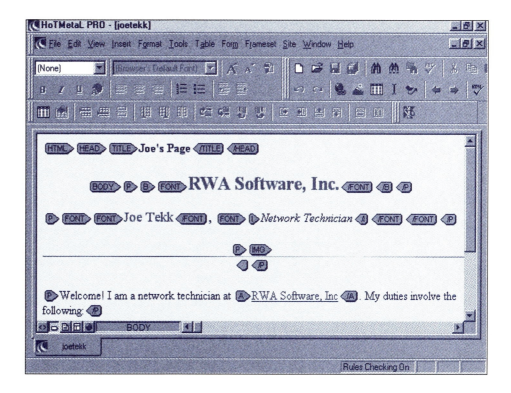

HTML. Demo versions of these HTML editors, and others, can be downloaded from the Web. Figure 13.4 shows HoTMetaL's graphical page editor with Joe Tekk's page loaded.

WWW pages are classified into three categories: static, dynamic, and active. The easiest to make are static and involve only HTML code. The page content is determined by what is contained in the HTML code. Dynamic WWW pages contain a combination of HTML code and a "call" to a server using a Common Gateway Interface application, or a CGI application. In this scenario, information supplied by the user into an HTML form is transferred back to a host computer for processing. The host computer then returns a dynamic customized WWW page. Active pages contain a combination of HTML code and applets. Therefore, the WWW page is not completely specified during the HTML coding process. Instead, using a Java applet, it is specified while being displayed by the WWW browser.

CGI

The Common Gateway Interface (CGI) is a software interface that allows a small amount of interactive processing to take place with information provided on a Web page. For example, consider the Web page shown in Figure 13.5. The Web page contains a FORM

173

FIGURE 13.5   Web page with FORM element

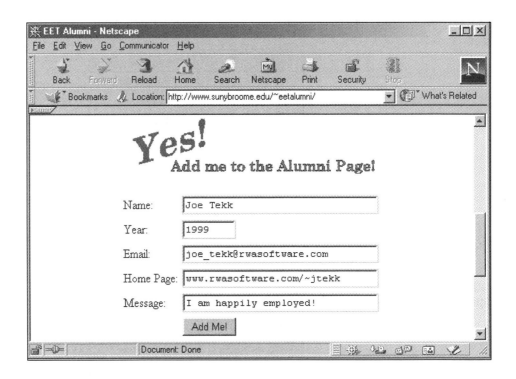

element, which itself can contain many different types of inputs, such as text boxes, radio buttons, lists with scroll bars, and other types of buttons and elements. The user browsing the page enters his or her information and then clicks the Add Me! button. This begins the following chain of events:

1. The form data entered by the user is placed into a message.
2. The browser POSTs the form data (sends message to CGI server).
3. The CGI server application processes the form data.
4. The CGI server application sends the results back to the CGI client (Netscape or Internet Explorer).

Let us take a closer look. The HTML code for the alumni page form looks like this:

```
<FORM ACTION="/htbin/cgi-mailto/eetalumni" METHOD="POST">
<P ALIGN="CENTER"><IMG SRC="yes.gif"></P>
<CENTER>
<TABLE WIDTH="50%" ALIGN="CENTER">
<TR><TD>Name:</TD>
<TD><INPUT TYPE="TEXT" NAME="name" SIZE="32"></TD></TR>
<TR><TD>Year:</TD>
<TD><INPUT TYPE="TEXT" NAME="year" SIZE="8"></TD></TR>
<TR><TD>Email:</TD>
<TD><INPUT TYPE="TEXT" NAME="email" SIZE="32"></TD></TR>
<TR><TD>Home Page:</TD>
<TD><INPUT TYPE="TEXT" NAME="home" SIZE="32"></TD></TR>
<TR><TD>Message:</TD>
<TD><INPUT TYPE="TEXT" NAME="msg" SIZE="32"></TD></TR>
<TR><TD></TD>
<TD><INPUT TYPE="SUBMIT" VALUE="Add Me!"></TD></TR>
</TABLE>
</CENTER>
</FORM>
```

The first line of the form element specifies POST as the method used to send the form data out for processing. The CGI application that will receive the POSTed form data is

174

the cgi-mailto program in the *htbin* directory. Specifically, cgi-mailto processes the form data and sends an e-mail message to the *eetalumni* account. The e-mail message looks like this:

```
From:    SBCCVA::WWWSERVER
To:      eetalumni
CC:
Subj:

REMOTE_ADDRESS: 204.210.159.19
name: Joe Tekk
year: 1998
email: joe_tekk@rwa.software.com
home: www.rwasoftware.com/~jtekk
msg: I am happily employed!
```

Note that the identifiers (name, year, email, home, and msg) match the names used to identify the text input elements in the form.

Instead of e-mailing the form data, another CGI application might create a Web page on-the-fly containing custom information based on the form data submitted. CGI applications are written in C/C++, Visual BASIC, Java, Perl, and many other languages. The Web is full of sample forms and CGI applications available for download and inclusion in your own Web pages.

JAVA

The Java programming language is the method used to create active WWW pages using Java applets. An active WWW page is specified by the Java applet when the WWW page is displayed rather than during the HTML coding process. A Java applet is actually a program transferred from an Internet host to the WWW browser. The WWW browser executes the Java applet code on a Java virtual machine (which is built into the WWW browser). The Java language can be characterized by the following nonexhaustive list:

- General purpose
- High level
- Object oriented
- Dynamic
- Concurrent

Java consists of a programming language, a run-time environment, and a class library. The Java programming language resembles C++ and can be used to create conventional computer applications or applets. Only an applet is used to create an active WWW page. The run-time environment provides the facilities to execute an application or applet. The class library contains prewritten code that can simply be included in the application or applet. Table 13.3 shows the Java class library functional areas.

TABLE 13.3   Java class library categories

Class	Description
Graphics	Abstract window tool kit (AWT).
Network I/O	Socket level connnections.
File I/O	Local and remote file access.
Event capture	User actions (mouse, keyboard, etc.).
Run-time system calls	Access to built-in functions.
Exception handling	Method to handle any type of error condition.
Server interaction	Built-in code to interact with a server.

The following Java program is used to switch from one image to a second image (and back) whenever the mouse moves over the Java applet window. Furthermore, a mouse click while the mouse is over the applet window causes a new page to load.

```
import java.awt.Graphics;
import java.awt.Image;
import java.awt.Color;
import java.awt.Event;
import java.net.URL;
import java.net.MalformedURLException;

public class myswitch extends java.applet.Applet implements Runnable
{
    Image swoffpic;
    Image swonpic;
    Image currentimg;
    Thread runner;

public void start()
{
    if (runner == null)
    {
        runner = new Thread(this);
        runner.start();
    }
}

public void stop() {
    if (runner != null)
    {
        runner.stop();
        runner = null;
    }
}

public void run()
{
    swoffpic = getImage(getCodeBase(), "swoff.gif");
    swonpic = getImage(getCodeBase(), "swon.gif");
    currentimg = swoffpic;
    setBackground(Color.red);
    repaint();
}

public void paint(Graphics g)
{
    g.drawImage(currentimg, 8, 8, this);
}

public boolean mouseEnter(Event evt, int x, int y)
{
    currentimg = swonpic;
    repaint();
    return(true);
}

public boolean mouseExit(Event evt, int x, int y)
{
    currentimg = swoffpic;
```

```
        repaint();
        return(true);
    }

    public boolean mouseDown(Event evt, int x, int y)
    {
        URL destURL = null;
        String url = "http://www.sunybroome.edu/~eet_dept";

        try
        {
            destURL = new URL(url);
        }
        catch(MalformedURLException e)
        {
            System.out.println("Bad destination URL: " + destURL);
        }
        if (destURL != null)
            getAppletContext().showDocument(destURL);
        return(true);
    }


}
```

Programming in Java, like any other language, requires practice and skill. With its popularity still increasing, now would be a good time to experiment with Java yourself by downloading the free Java compiler and writing some applets.

RELATED SITES

Here are a number of service, reference, and technology-based sites that may be of interest:

- www.prenhall.com   Engineering and technology textbooks
- www.yahoo.com   Search engine
- www.internic.net   Internet authority
- www.intel.com   Intel Corporation
- www.microsoft.com   Microsoft Corporation
- www.sunybroome.edu/~mansfield_k   Author's home page
- www.sunybroome.edu/~antonakos_j   Author's home page
- www.netscape.com   Netscape corporation

The Internet is full of information about every aspect of the Web page development process. Many people put a tremendous amount of information on their own Web pages. You are encouraged to learn more about Web pages and Web programming.

TROUBLESHOOTING TECHNIQUES

The Internet and the World Wide Web are not the same thing. The Internet is a physical collection of networked computers. The World Wide Web is a logical collection of information contained on many of the computers comprising the Internet. To download a file from a Web page, the two computers (client machine running a browser and server machine hosting the Web page) must exchange the file data, along with other control information. If the download speed is slow, what could be the cause? A short list identifies many suspects:

- Noise in the communication channel forces retransmission of many packets.
- The path through the Internet introduces delay.
- The server is sending data at a limited rate.
- The Internet service provider has limited bandwidth.

So, before buying a new modem or upgrading your network, determine where the bottleneck is. The Internet gets more popular every day. New home pages are added, additional files are placed on FTP sites for downloading, news and entertainment services are coming online and broadcasting digitally, and more and more machines are being connected. The 10- and 100Mbit Ethernet technology is already hard-pressed to keep up with the Internet traffic. Gigabit networking is coming, but will only provide a short respite from the ever-increasing demands of global information exchange.

SELF-TEST

This self-test is designed to help you check your understanding of the background information presented in this exercise.

True/False

Answer *true* or *false*.

1. The hypertext markup language is used to encode GIF images.
2. CGI stands for Common Gateway Interchange.
3. HTML contains formatting commands called tags.
4. The main portion of a Web page is contained between the HEADER tags.
5. Java is a Web browser produced by Microsoft Corporation.

Multiple Choice

Select the best answer.

6. CGI applications can be written using
   a. Perl, Java, C, C++, and Visual BASIC.
   b. Only the Javascript language.
   c. An HTML editor.
7. The three different categories of Web pages are
   a. Large, medium, and small.
   b. Active, passive, and neutral.
   c. Static, dynamic, and active.
8. When the network is slow,
   a. Turn off all power to the computer and perform a reset.
   b. Try to determine where the bottleneck is located.
   c. Immediately upgrade to the newest, most expensive hardware available.
9. CGI applications use FORMs to
   a. Receive the input required for processing.
   b. Post the data to an e-mail application.
   c. Send information to the browser display.
10. The concept of universal service and the Internet involves
    a. Being able to connect to a universal router on the Internet.
    b. Being able to exchange information between computers at any time or place.
    c. Allowing all users to access the universal Internet database.

Completion

Fill in the blank or blanks with the best answers.

11. The same _____ code is displayed differently using different Internet browsers.
12. A CGI application provides the ability to create _____ _____ on-the-fly.
13. Java is used to create _____ Web pages.
14. _____ information includes text, video, audio, Java applets, and images.
15. The _____ protocol is used to exchange hypermedia information.

FAMILIARIZATION ACTIVITY

WWW

1. Examine each of the pull-down menu items in Netscape Navigator and Internet Explorer.
2. Read the online help to learn about browser features.

3. Identify similarities between the two browsers discussed in this exercise.
4. Identify differences between the two browsers discussed in this exercise.

CGI

1. Search the Web to locate information about Perl.
2. Locate a source for Perl, available free of charge.
3. Download Perl.
4. Install Perl.
5. Run some of the sample Perl scripts.

Java

1. Search the Web to locate information about the Java language.
2. Locate the source of a Java compiler available free of charge.
3. Download the Java compiler.
4. Install the Java compiler.
5. Compile some of the sample Java applets.
6. Execute a sample Java applet.

QUESTIONS/ACTIVITIES

1. Determine how to clear the browser's cache memory.
2. Determine the current allocation settings for the browser cache.
3. Search the Web to locate some useful resources related to active Web page development.

REVIEW QUIZ

Under the supervision of your instructor,

1. Describe the basic organization of the Internet.
2. Explain the purpose of a browser and its relationship to HTML.
3. Discuss the usefulness of CGI applications.

UNIT  **Productivity Software**

14 Word Processors

Joe Tekk finished installing the latest version of Microsoft Word in the RWA Software corporate training center. Joe thought about how many people would be using the computers during the next few days. A corporate training session, a tour for a local scout troop, and a group of high school students coming in to write their resumes would all be running the new version of Word.

Joe also thought about the company-wide documentation project that had been going on for several months. All the company documents were being entered into Word and then exported in HTML for use on RWA's internal network. Joe never realized how much time was spent doing word processing.

PERFORMANCE OBJECTIVES

Upon completion of this exercise, you will be able to:

1. Enter and leave Microsoft Word.
2. Resize and move the Word window.
3. Create a new document.
4. Save a document.
5. Open and modify existing documents.
6. Spell check your document.
7. Use bold and underline format tools.
8. Print documents.
9. Use online help.

BACKGROUND INFORMATION

Using Microsoft Word, all types of word-processing chores can be accomplished. Aside from the basic editing features necessary to create letters, memos, mailing labels, reports, and essays, Word provides several advanced features, such as mail merge, table editing, drawing tools, word art, and WWW page editing, to name just a few. Word provides a graphical WYSIWYG (what-you-see-is-what-you-get) window that is powerful and easy to learn. Figure 14.1 shows a typical Word editing window.

As you can see from Figure 14.1, some of the text displayed on the screen is bold, other text is justified, there are bulleted lists, and the text is displayed using different font sizes.

FIGURE 14.1   Microsoft Word
editing window

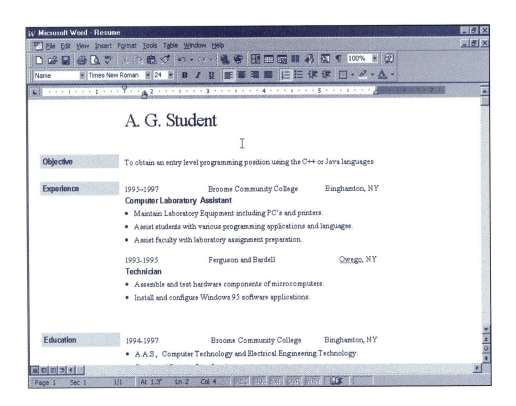

WYSIWYG means that what you see in Figure 14.1 is what you will get when you print the Word document.

All the controls for these WYSIWYG features are contained in toolbars or pull-down menus. Both new and experienced users can work efficiently with a few simple keystrokes or mouse clicks. This exercise will give you the Word experience necessary to create your own Word documents, such as technical reports, resumes, correspondence, and other written assignments.

The title bar located at the top of the Word window in Figure 14.1 shows the current document name on the left and the common Windows application options on the right: minimize window, maximize window, and close. The pull-down menus show the category names that provide access to all of Word's features. The status bar at the bottom of the Word window shows the status of the word processing session (page 1 of 1, cursor located at line 2, column 4).

By selecting a pull-down menu, all the options available on the menu are displayed. Figure 14.2 shows the View pull-down menu. The View menu provides a number of ways

FIGURE 14.2   Pull-down menu
for View

FIGURE 14.3   Additional menu items in the Toolbars submenu

to view the document while it is being prepared. For example, you can preview the actual page as it will look on the printer. Several small icons at the lower left of the Word window in Figure 14.1 provide shortcuts to change the view without having to open the View menu.

A pull-down menu can be selected with the mouse by left-clicking on the desired menu option, or by using keyboard commands such as the Alt-F to access the File menu. Arrows shown to the right of items on the pull-down menu indicate that more selections are available for those specific items. Figure 14.3 shows the View/Toolbars pull-down menu. Simply moving the mouse to the Toolbar option automatically displays the additional menu choices.

The toolbars allow the user to easily select and use the various features of Word by pressing one button. We will examine both pull-down menus and toolbars later in this exercise.

CREATING A NEW DOCUMENT

Microsoft Word is a member of the Microsoft Office product suite and can be accessed through the Microsoft Office options available using the Start menu or the Start Programs menu. As shown in Figure 14.4, Word can be accessed from the Start menu by selecting the New Office Document option. Figure 14.5 shows the New Office Document selection screen. This screen shows options for the complete Microsoft Office suite. For now, we will select the Blank Document icon available on the General tab to begin the Word editing session. To do this, select the Blank Document icon using the mouse and then press the OK button, or simply double-click on the Blank Document icon itself.

Microsoft Word can also be accessed directly by selecting Microsoft Word from the Programs menu or a Programs submenu. As you become more familiar with Microsoft Word and other Microsoft Office products, you will probably use a combination of these methods to open Word.

Once started, Word presents the user with a blank editing screen. Now it is just a matter of entering the text. We can begin to get comfortable with Word by entering a simple letter of correspondence. Enter the text shown in Figure 14.6.

You may notice several different things as you type on the Word screen. First, Word is constantly updating the current page information and examining each character as it is typed. If a word is misspelled, Word will underline it in red. If Word finds a grammatical or

FIGURE 14.4 Accessing
Word from the Start menu

FIGURE 14.5 New Office
Document screen

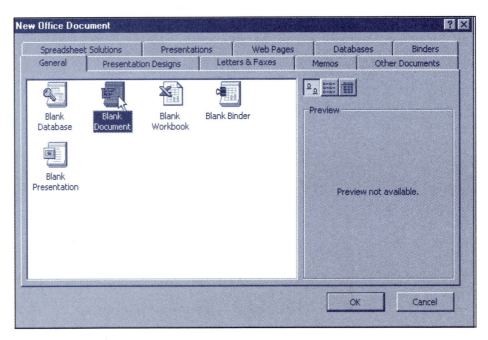

format error in a sentence, such as too many spaces between words in a sentence or problems with tense, Word will underline that in green. You are encouraged to become familiar with Word status indicators as you enter some text into the Word window.

Experiment with the bold, italic, centering, and font controls to format the text. This can be done to text as it is being entered, or after it has been entered. For example, to make a block of text bold, you could:

- Select the text and then left-click the bold button.
- Left-click the bold button (to turn bold on), enter the text, and then left-click the bold button again (to turn bold off).

Selecting text after it has been entered can be easily done with the mouse or keyboard. Position the mouse over the first character or line to be selected, left-click and hold, then drag the mouse over the last character or line to select. The selected text will be highlighted to make it stand out from the rest of the text. You can select text with the keyboard by positioning the cursor at the beginning of the text and holding the Shift key down while moving

FIGURE 14.6   Sample letter

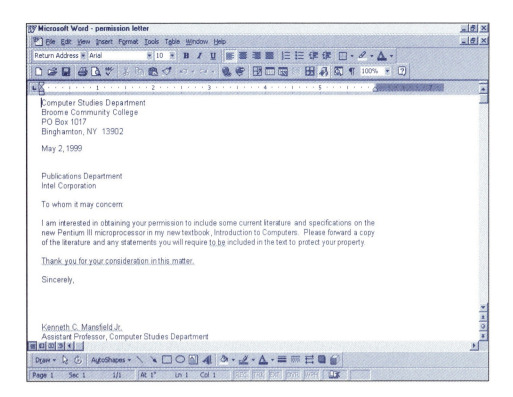

to the end of the selected text with the arrow keys. To deselect a block of text, left-click the mouse anywhere inside the editing window.

Many editing features require that a block of text be selected first, so it is worth the time invested to master this technique.

One last important point: do not forget about the Undo button (a small, blue counter-clockwise arrow). There will be times when you have made one or more errors while editing a document and the results are not what you intended. Left-clicking the Undo button will back up the editing process and undo your editing changes one by one.

EXITING FROM WORD

To exit from Word, two methods are available:

- Select Exit from the File pull-down menu, or
- Select the close option available on the Word window by clicking the box with an X in it located on the top right-hand corner of the screen.

If the current document in the editing window has not been saved yet, Word will ask you to choose whether to save the changes in your file. This dialog box is shown in Figure 14.7. If you want to save your document, select the Yes option and then verify the File name displayed in the File name field on the Save As dialog box, shown in Figure 14.8. Word will automatically fill in File name with the first words you entered into the document. If this name is suitable, select the Save option. If you want to change the name, simply enter a new name in the File name field and left-click the Save option.

FIGURE 14.7   Close dialog box

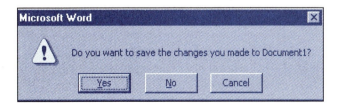

FIGURE 14.8 Saving a file

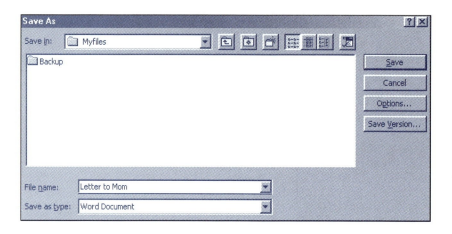

FIGURE 14.9 Standard Word toolbar

PULL-DOWN MENUS AND TOOLBARS

As we have already seen, pull-down menus allow access to the specific features of Word. These features are grouped into the following categories: File, Edit, View, Insert, Format, Tools, Table, Window, and Help. Examine each of the pull-down menus.

In addition to the pull-down menus, Word also uses icons to allow direct access to some of the most common features, such as file operations (save, open, and print); format operations such as text justification; and text attributes like bold, underline, and italics. The icons provide one-button access to features that were previously several mouse clicks or keystrokes away using the pull-down menus. The standard toolbar menu of icons is shown in Figure 14.9.

Many other toolbar menus are available and are probably already displayed on your Word screen. The toolbars are selected by choosing the Toolbars option from the View pull-down menu. This was previously illustrated in Figure 14.3.

The Formatting toolbar shows most of the common formatting options such as font size and selection; character attributes such as bold, italic, and underline; paragraph justification choices; and list and indentation options. As you view the toolbar, items that are currently selected show up as buttons that are depressed. For example, if left margin justification is selected, that button will also appear to be depressed. Similarly, as a word or sentence is bolded or underlined, these buttons will also appear to be depressed. These options are selected by left-clicking on them.

You can also customize the individual items that are displayed on each toolbar menu using the Customize option. You are encouraged to become familiar with all of the toolbar choices.

MOVING THE WINDOW

If you want to move the Word window on the Windows desktop, simply

1. Place the mouse cursor on the Microsoft Word window title bar located at the top of Word screen.
2. Press *and hold* the left mouse button.
3. Drag the window to its new place on the desktop. The window may be outlined as it is being dragged.
4. Release the left mouse button.

RESIZING THE WINDOW

You can change the size of the Word window, making it as large or small as Windows will allow. This is accomplished in the following way:

1. Place a mouse cursor on an edge or corner of the window. Notice that the mouse pointer changes when you are in this position.
2. Press *and hold* the left mouse button.
3. Drag the window edge or corner until the window is the desired size.
4. Release the mouse button.

To make the window as large as the entire screen, simply left-click on the maximize icon displayed at the top right-hand corner of the Word window.

SCROLLING THROUGH A DOCUMENT

Whenever a document window is resized, a scroll bar may be added or removed from the Word window. A scroll bar is present on a window whenever there is more information than can be displayed on the screen. On the Word screen, there may be two scroll bars: one to display the complete document from top to bottom and one to display the contents of the document from left to right. The mouse is used to control the scroll bars.

In addition to scrolling, the entire display can be updated by paging through the document. The Page Down and Page Up buttons on the keyboard do not actually move an entire page in either direction. Instead, they move the cursor a few inches forward or backward. Pressing and holding the Ctrl (Control) key while pressing Page Up or Page Down will move the cursor to the beginning of the next, or previous, page. Pressing Ctrl and End will take you to the end of the document. Pressing Ctrl and Home returns you to the beginning of the document.

SAVING THE DOCUMENT

As you are typing text into Word, it is always a good idea to save your work frequently to prevent accidental loss of data. To save your document, simply press the Save icon (the Floppy Disk) on the Standard Toolbar. The Save option is also located on the File pull-down menu. An AutoSave feature is built in that allows Word to save your document at predefined intervals such as every 10 minutes. Look in the Options submenu of the Tools pull-down menu. The Save tab contains all the parameters associated with the AutoSave feature.

If you have not named your document yet, you will be presented with the Save As dialog window previously shown in Figure 14.8. If your document has already been saved, Word will simply update the file already stored on the disk.

OPENING AN EXISTING DOCUMENT

There are several ways to open a Word Document. First, on the Start Menu, the "Open Office Document" selection will present a list of all Office documents. This is illustrated in Figure 14.10. Each of the files shown with a .doc extension is a Word document. To open one of these documents, simply double-click on the document, or select the document with a single click and use the Open option.

Another way is to open Word directly and use the Open option on the Toolbar or the File menu or by selecting one of the document names listed at the bottom of the File menu. This list contains the most recent files accessed by Word.

The opened document is displayed beginning with the first page. New material may be added to the document by left-clicking at the desired insertion point.

FIGURE 14.10 Opening a file
in Word

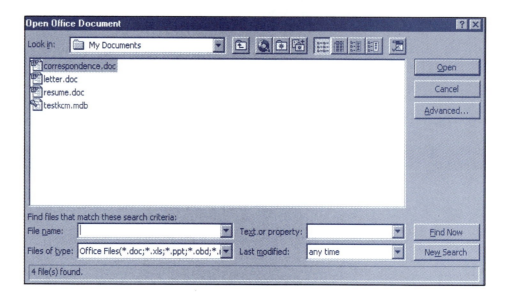

PRINTING DOCUMENTS

The Print option is available on the Standard Toolbar and the File pull-down menu. Using either one of these methods will cause Word to print the current document using the default Windows printer. The Print Preview option will display a graphic image of how the hard copy output should look after it is printed. This can help save paper and produce better looking output.

Word will print the entire document, a selected range of pages, or just the current page. Word can even print to a file instead of the printer.

OTHER WORD-PROCESSING APPLICATIONS

Another popular word-processing application is WordPerfect. WordPerfect began as a DOS application, and has grown into a powerful Windows application, comparable with Microsoft Word.

If you do not want to spend any money, two built-in word processors are available in Windows. They are Notepad and Wordpad. Notepad is text-based. Wordpad, like Microsoft Word, allows text formatting and other WYSIWYG features, but is not as powerful as Word.

**TROUBLESHOOTING
TECHNIQUES**

The Word help facility available from the pull-down menu or from the Office Assistant provides instant access to the features of Word. Figure 14.11 shows the help options available. You are even allowed to enter a question, such as "How do I insert page numbers," and Word help will look up the associated topic.

FIGURE 14.11 Word Help
pull-down menu

This self-test is designed to help you check your understanding of the background information presented in this exercise.

True/False

Answer the following questions *true* or *false*.

1. Word displays documents in WYSIWYG format. _True_
2. Word checks spelling and grammar as the text is entered into the document. _True_
3. Word menu options can be accessed only using keyboard commands. _False_
4. The Word toolbars can be customized by the user. _True_
5. You can forget to save your document and leave Word, losing all your changes. _false_

Multiple Choice

Answer the following questions by selecting the best answer.

6. The Word Office Assistant is used
   a. To access Word help features.
   b. To control how Word automatic features are configured.
   c. By Word to manage the user interface.
7. A submenu is indicated by
   a. The underscored character in the submenu category.
   b. The underscored character in the submenu category and the triangle located on the right margin in the pull-down menu.
   c. An icon shown next to the pull-down category.
8. To go to the beginning of a new page, press
   a. Page Down.
   b. Shift-Page Down.
   c. Ctrl-Page Down.
9. Word documents have the extension
   a. .doc.
   b. .txt.
   c. .wrd.
10. To make an existing block of text bold you must first
    a. Select it.
    b. Format it.
    c. Highlight it.

Completion

Fill in the blanks with the best answers.

11. The Toolbars are located on the _____ pull-down menu.
12. The spelling and grammar check can be activated using the _____ icon.
13. The buttons in the lower left corner of the Word document window are used to change the _____.
14 Print _____ shows a graphic display of what the printed page will look like.
15. Sentences underlined in green have _____ errors.

1. Open a new Word Document. Enter your name on the first line, and the title "Word Processing Experience" on the second line. Beginning on the third line, write a small paragraph explaining your experience using a word processor. Save the document.
   a. Select the name and title portion of the text and then try each of the formatting tools to see how the format of the text changes.
   b. Try the same operations on the paragraph portion of the text.
2. Experiment with tables. Try to duplicate the one shown in Table 14.1.

TABLE 14.1 Sample table

Input	Output	Memory	CPU
64-bit digital	64-bit digital	256 MB RAM	Pentium III
Four 16-bit	Two 16-bit, four	20 GB HD	500+ MHz
analog	8-bit analog		

3. Insert an image into your Word document. This is done by selecting Insert/Picture, Insert/File, or Insert/Object. You can even put the image on the clipboard and paste it into your Word document.
4. Save a Word document as an HTML file and then view the file with a browser. Does it look the same?

1. Locate information in Word about a mail merge. List the number of subtopics.
2. Enter a few words and deliberately spell a few of them wrong. Does Word underline them in red? What happens if you right-click the mouse over one of the incorrect words?
3. Run the Resume Wizard (File/New/Other Documents) to create a resume of your work experience and employment goals.

REVIEW QUIZ

Under the supervision of your instructor,

1. Enter and leave Microsoft Word.
2. Resize and move the Word window.
3. Create a new document.
4. Save a document.
5. Open and modify existing documents.
6. Spell check your document.
7. Use bold and underline format tools.
8. Print documents.
9. Use online help.

15 Spreadsheets

Joe Tekk was just putting the finishing touches on a spreadsheet he had been working on. Joe thought he was spending too much money and had decided to set up a budget. He didn't have any experience with spreadsheets before, but since RWA Software had standardized on the Microsoft Office product, he used the Excel application program (which had never been run except for a quick look when Microsoft Office was originally installed).

Joe was quite impressed with how easy it was to set everything up. Now it was just a matter of keeping track of receipts, entering the information, and looking at the results. As Joe thought about the spreadsheet, he wondered what he would be doing when he found out where all of his money was going.

PERFORMANCE OBJECTIVES

Upon completion of this exercise, you will be able to

1. Understand the terminology associated with spreadsheets.
2. Navigate around the spreadsheet window.
3. Create a spreadsheet.
4. Use several built-in spreadsheet functions.

BACKGROUND INFORMATION

An electronic spreadsheet is the equivalent of a traditional accounting worksheet. Although both of these documents (the spreadsheet and the worksheet) organize the data into a matrix of rows and columns, the electronic worksheet is much more flexible and easier to work with than its paper counterpart. The flexibility of a spreadsheet is demonstrated with the ability to apply the spreadsheet to every area of technology. For example, an electrical engineering student might use a spreadsheet to store data for a range of values, such as voltage or current, that are tracked over a period of time. A mechanical engineering student may be interested in storing data about the placement of components on a printed circuit board where the tolerance measurements are critical and must be reviewed frequently to ensure a high level of quality control. An engineering science student may use a spreadsheet to implement a formula in physics to determine the results of an equation, thereby eliminating the need to write a computer program.

FIGURE 15.1 Microsoft Excel spreadsheet window

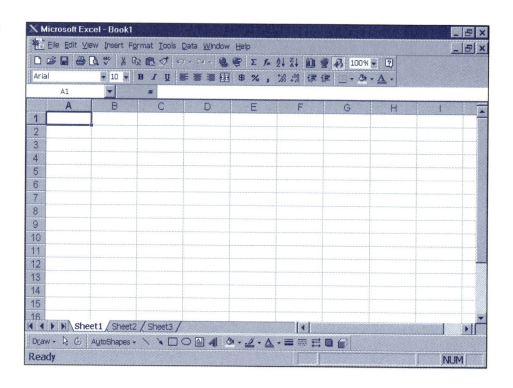

Before we get involved in the actual development of a spreadsheet of our own, it is necessary to learn about the terminology of a spreadsheet. Figure 15.1 shows a Microsoft Excel spreadsheet. The Microsoft Excel window contains several different components such as the title bar, pull-down menus, toolbars, status bar, and, lastly, the actual worksheets where all of the activities actually take place in the spreadsheet.

The smallest element in a spreadsheet is called a **cell.** A cell is defined by the intersection of one row and one column. As you can see from Figure 15.1, each column in an Excel spreadsheet is identified by a letter and each row is identified by a number. Therefore, specifying one column letter and one row number uniquely identifies each spreadsheet cell.

Each cell in a spreadsheet may contain text, a number, or a formula. Any text in a cell is commonly referred to as a label. Each label describes the contents of one or more cells. Examples of some common cell labels include months of the year, days of the week, or a list of any other set of items or objects being tracked in the spreadsheet. Numbers in a spreadsheet cell may be integers or floating-point numbers. Formulas in a spreadsheet may contain hard-coded values, references to other cells in the spreadsheet, built-in formulas, or custom calculations.

SPREADSHEET PLANNING

The process of building a spreadsheet consists of several different steps. These steps are similar to the steps used to develop a database application or the development of a computer program. In general, the basic steps are as follows:

1. State the purpose of the spreadsheet.
2. Design a model of the spreadsheet.
3. Build the electronic version of the model.
4. Test the spreadsheet.
5. Document the spreadsheet.

First, the purpose of the spreadsheet identifies what the spreadsheet is supposed to do, what inputs are required, and what type of output is required from the spreadsheet such as reports or graphs. With a clear purpose defined for how the spreadsheet should work, it becomes a simpler task to get started in the actual spreadsheet development.

Second, it is necessary to develop an outline or model indicating what the spreadsheet should look like and how the spreadsheet should function. The model should identify the basic form of the spreadsheet, indicating the number of cells necessary to hold the data. The model should also identify the different types of calculations that must be performed. With the simple model written on paper, the actual process of coding the spreadsheet can begin.

Third, the actual development of the spreadsheet is performed in steps. Each part of the spreadsheet must be tested to verify that the individual elements are correct and working properly. By testing each part of the spreadsheet individually as you go, problems are eliminated as they are encountered, without having to worry about fixing all the errors at the end of the project. In a small spreadsheet, it may not be necessary to perform the work in steps, but as the size of a project grows, testing in stages becomes more and more important.

Testing of the spreadsheet involves entering test data and determining whether the results are correct. The results are compared with manually calculated test data, usually developed when creating the spreadsheet model. When the results of the spreadsheet agree with the calculated test data, the spreadsheet can be documented as necessary.

The documentation process involves preparing a list of the instructions necessary to work with the spreadsheet or actually describing how the spreadsheet operates. Most spreadsheet software allows for comments to be added to individual cells to indicate what decisions were made and why. With these basic steps in mind, let us examine the details of the spreadsheet window to see how these development steps can be followed.

BUILDING A SPREADSHEET

As an example, let us build a spreadsheet to compute the distance that a body falls in feet per second for the first 10 seconds of free fall given by the equation:

$$S = \frac{1}{2} at^2$$

where

S = the distance in feet
a = acceleration due to gravity (32 ft/sec^2)
t = time in seconds

The purpose of the spreadsheet is to determine the distance in feet that the body falls during the first 10 seconds of free fall. The spreadsheet model therefore must contain all the elements necessary to calculate the free-fall distance. Although this may seem like a simple problem, it is necessary to calculate a few of the values for S that we expect to see when the data is displayed using the spreadsheet. Using a pocket calculator, the constant value for a and the individual values of t, the actual values can be computed very easily.

Let us begin the actual development of the spreadsheet using Sheet1 in the Excel workbook by entering the spreadsheet title and the column headings for both the time and the distance traveled. As this information is entered, it is displayed in the selected cell as well as the formula bar. We will discuss the formula bar in detail soon. The heading information is shown in Figure 15.2. The spreadsheet designer may choose any location on the spreadsheet to contain the heading information. In general, all the decision-making regarding the placement of information such as titles, headings, and data are made so that the user can enter spreadsheet data easily and at the same time make the spreadsheet visually appealing.

Next, it is necessary to supply the initial value for time, a zero, which is entered into the appropriate cell location under the time column heading as indicated in Figure 15.3. Notice that all that is necessary to enter the zero in cell C5 is to use the arrow keys or mouse to select the cell, and then to press the zero key and then the Enter key. Notice that when a particular cell is selected, the border around the cell is thicker and darker than the other cells in the spreadsheet.

To supply the remaining values for the time, the user may simply enter the remaining digits 1 through 10 in the appropriate cell locations C6 through C15, although a better choice would be to create a formula that performs the same job. To create a formula, it is

FIGURE 15.2 Adding a title
and heading to the spreadsheet

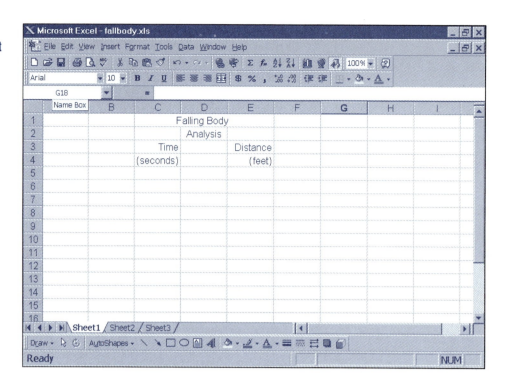

FIGURE 15.3 Entering the
initial value for time

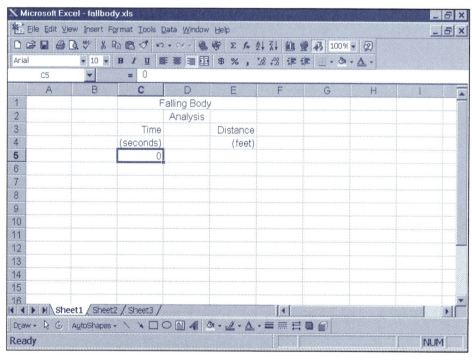

necessary to select the cell C6, which is where the formula is to be stored. Then, it is necessary to enter an = (equal sign) at the beginning of the formula followed by the specific formula to execute. A formula to increase the cell size by one is as follows:

$$=C5 + 1$$

entered in cell location C6. This formula causes the cell C6 to look at the value stored in location C5 and then add 1 to it. Figure 15.4 illustrates this process. Notice that the contents of the cell are displayed in the formula toolbar simultaneouly. Also notice that the size of the formula toolbar display is much longer than the size of a cell and allows for very long formulas to be examined. Note that it is also possible to create and edit formulas using the formula toolbar.

FIGURE 15.4 Entering a formula for the time values

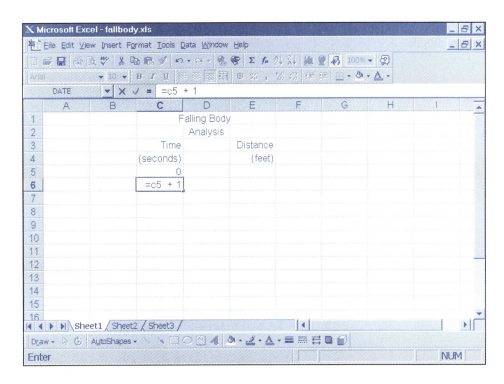

FIGURE 15.5 Value in cell C6 produced by the formula

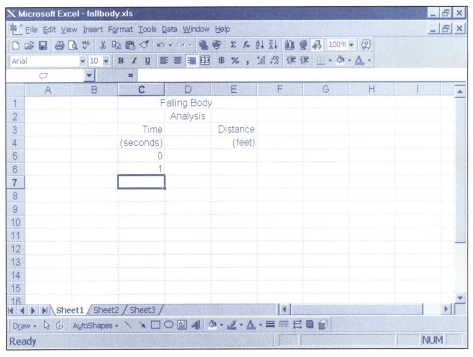

After the formula is entered, it is necessary to press the Enter key. This causes Excel to execute the formula and display the results inside of the specified cell, as is indicated in Figure 15.5. Notice that a 1 is now displayed in cell C6. After the formula in cell C6 is executed, cell C7 automatically becomes the next selected cell.

To continue the list of numbers in the time column, it is necessary to enter formulas in the remaining cells C7 through C15. Fortunately, it is not necessary to reenter the formula in each of the cells. Instead, the formula in cell C6 can be copied and then pasted into the remaining cells C7 through C15. To perform the copy process, it is necessary to select the formula in cell C6 and copy its contents to the Windows clipboard. For example, a right-click of the mouse on cell C6 displays the menu shown in Figure 15.6, where the copy option is selected. Following the copy process, the cell C6 is displayed in an outline box.

FIGURE 15.6 Preparing to copy a formula

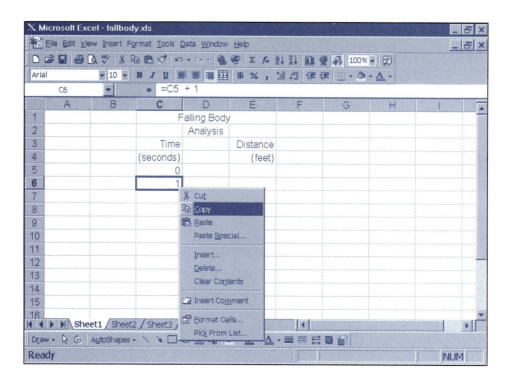

FIGURE 15.7 Copying the contents of a cell

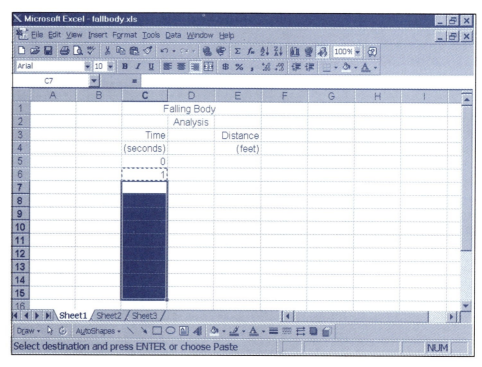

The last step in the copy process involves selecting the cells to which the formula should be copied to. These cells, C7 through C15, must be selected as shown in Figure 15.7. The first selected cell, C7, is displayed using the thicker lines around the cell and the remaining cells C8 through C15 are displayed in black. Notice that the status bar displays the simple instructions "Select destination and press ENTER or choose Paste." Following these instructions and pressing the Enter key causes the formula to be pasted into the selected cells as illustrated in Figure 15.8. Aside from the cell location C5, which contains a zero, the rest of the cells C6 through C15 use a formula to determine their values.

Each cell in the distance column where a number will be displayed must contain the distance S using the formula $S = {}^{1}/_{2} \, at^{2}$. We can begin by selecting the first cell location in the distance column where the new formula is to be entered, which is E5. The elements of the

FIGURE 15.8 Results of the pasted formula are displayed

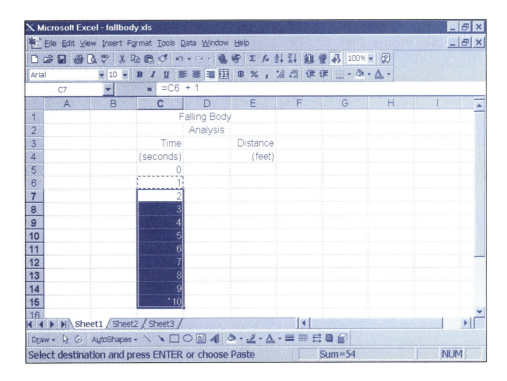

FIGURE 15.9 Entering the distance formula

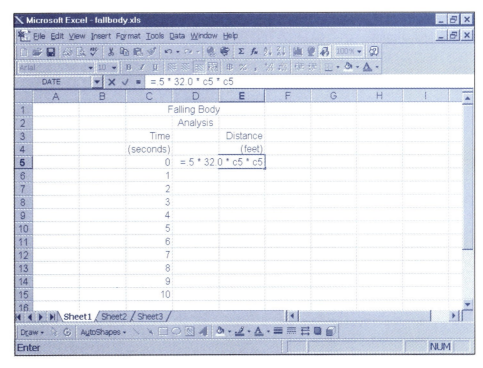

formula consist of constant values (0.5 and 32.0) and the variable *time,* which is located in the adjacent cell C5. Using this information, the formula

$$=0.5 * 32.0 * C5 * C5$$

is used to compute the first distance calculation, which is shown in Figure 15.9. After the Enter key is pressed, the results of the calculation are displayed and the next cell in the column becomes the currently selected cell, as indicated by the window displayed in Figure 15.10. Note that the value 0 displayed in cell E5 is correct since the value of time is zero. Now, all that remains is the process to copy the formula from cell E5 to the other cells in the distance column E6 through E15. The steps to perform this process are the same as we used to copy the formulas in the cells C6 through C15. First, the cell containing the formula, E5,

FIGURE 15.10   Distance formula calculation results

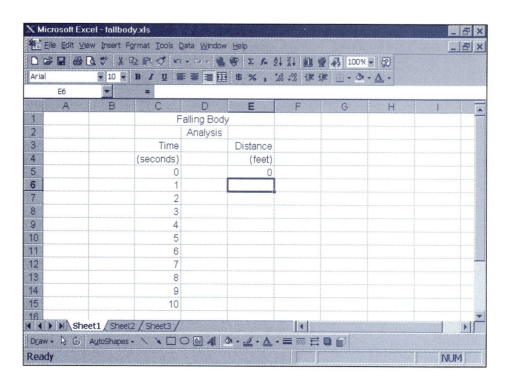

FIGURE 15.11   Completed spreadsheet display

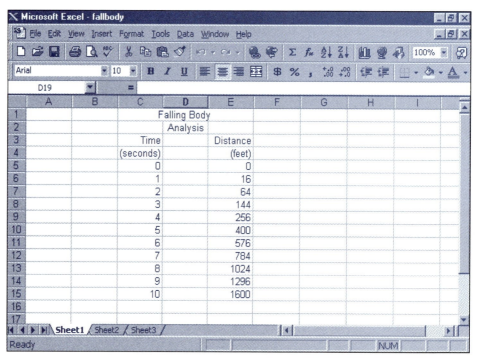

must be copied to the clipboard. Then copy the selected formula to the new cells E6 through E15. After the formulas have been pasted, notice that the new values for the distance traveled are displayed in the appropriate cell locations. Figure 15.11 shows the completed worksheet. Now it is necessary to verify that the values are correct by choosing a few sample elements in the list and manually performing the same calculations that were implemented in the formula. If the results are the same, the spreadsheet is working properly.

The last step in the development of a spreadsheet involves producing any required documentation and/or instructions that might be helpful to the user. For this example, no additional documentation or instructions would be necessary.

VIEWING THE OUTPUT FROM A SPREADSHEET

After a spreadsheet has been developed, it is common to print a hard copy of the results; this is accomplished by selecting the Print option from the File menu or by pressing the Print icon on a toolbar. Microsoft Excel also provides several chart and graph options to view the spreadsheet data. Continuing with the previous example, a graph can be created by selecting the Chart Wizard icon located on the toolbar. Figure 15.12 shows the first window of four displayed by the Chart Wizard. As you can see from the figure, many different chart options exist. First, it is necessary to choose the chart type from the list of items displayed along the left side of the display. The defaults selected by the Chart Wizard for "Column" are an appropriate selection for the Falling Body Analysis. You are encouraged to experiment with all the various charting options available. By selecting the Next button, the second step is displayed by the Chart Wizard.

In step 2, it is necessary to select the range of data to be displayed in the chart. The default selection

$$=sheet1!\$A\$1:\$E\$15$$

includes the numbers we want to examine. Microsoft Excel has also chosen the Column button for the series option. You may notice that the data displayed inside the Chart Wizard Window shown in Figure 15.13 is actually using the data from the worksheet. By selecting the next button, we can move to step 3.

The next step involves labeling the various parts of the graph. We have the option of entering a chart title, category for the X-axis, and the value for the Y-axis. This process is illustrated in Figure 15.14. Excel does not provide any default values for these options and they must be entered manually. After the necessary text has been entered, press the Next button to move to the last step.

The last step in the Chart Wizard process determines where the chart should be displayed. Figure 15.15 indicates that there are two choices: as a new sheet or as an object in an existing sheet. The default option selected by Word is as an object in Sheet1. Rather than choose the default value, it is desirable to select the option to create the chart in a new sheet. By selecting the Finish button, the Chart Wizard creates the chart as a new sheet in the spreadsheet, as shown in Figure 15.16.

FIGURE 15.12 Select the chart type using the Chart Wizard

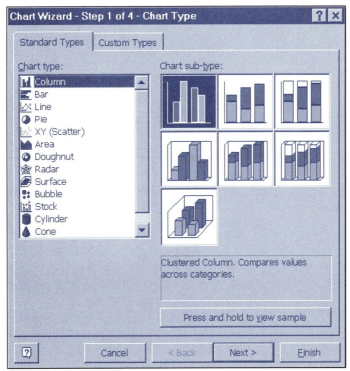

FIGURE 15.13   Choose the chart data source

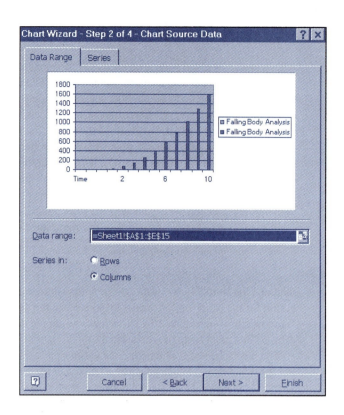

FIGURE 15.14   Specify the Chart options

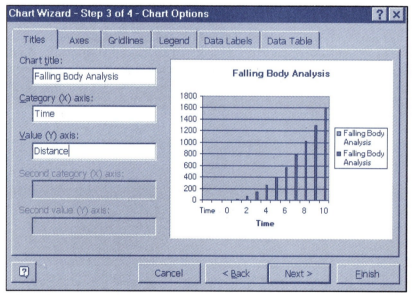

FIGURE 15.15   Select the location to store the chart

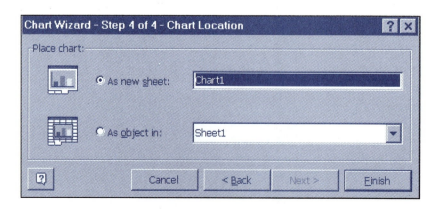

FIGURE 15.16   New sheet for the spreadsheet

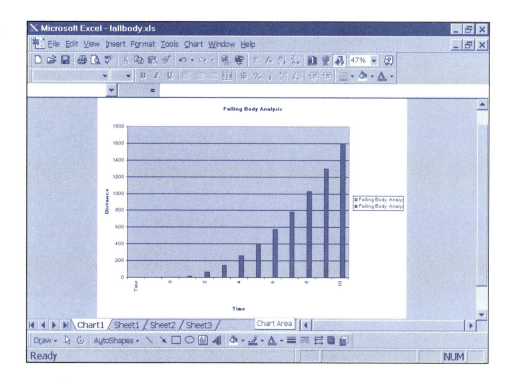

FIGURE 15.16   New sheet for the spreadsheet

Many of the options available when using the Chart Wizard can be used to create very useful output. The old saying that a picture is worth a thousand words applies to spreadsheets too. Countless options are provided to the spreadsheet user to allow various outputs to be created.

Many other output-formatting options are also available, including page orientation, headers, and footers.

FORMULAS

In the previous spreadsheet example, it was not necessary to develop sophisticated formulas to solve the problem. Many times, however, it is necessary to develop formulas that are somewhat sophisticated. Fortunately, Excel is well-equipped to handle these situations by using any of the built-in functions. Figure 15.17 shows the Paste Function window, which lists each function by category. Each category contains a list of related functions. A

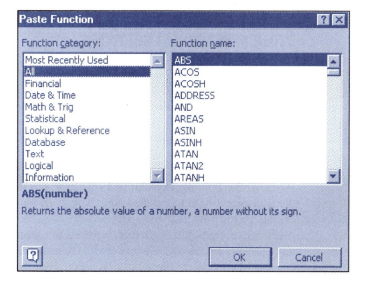

FIGURE 15.17   Partial list of Microsoft Excel built-in functions

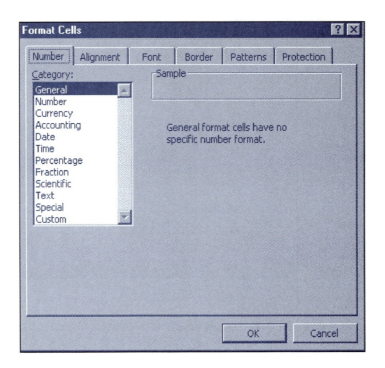

FIGURE 15.18 Cell format options

description of the selected function is provided as well, indicating what numbers or variables are needed to use the function. When developing a spreadsheet, it is useful to know about the different types of built-in functions. You are encouraged to examine each of the function categories and each of the individual functions within the category.

SPREADSHEET APPEARANCE

The appearance of a spreadsheet is an important part of the development process. Microsoft Excel provides the ability to format each cell or a range of cells in many different ways. For example, it might be necessary to center a heading over a range of cells, or center the data in the cells. The Format pull-down menu allows access to change the format of a cell, row, column, or the entire sheet. Figure 15.18 illustrates the variety of cell formatting options. Notice that if the cell contains a number, any of the items displayed in the list may be selected. As a particular item is selected, a sample display is also provided to show what the data will look like. Once again, you are encouraged to examine each of the various formatting options available.

OTHER SPREADSHEET SOFTWARE

Several other spreadsheet software applications are also available in addition to Microsoft Excel, such as Quattro Pro and Lotus 123, to name just a few. Each of these packages provides features similar those contained in Excel. You can perform the same operations using any of these packages.

TROUBLESHOOTING TECHNIQUES

When troubleshooting problems in a spreadsheet, it is necessary to compare the results of the spreadsheet with calculations that have been manually calculated. It is a mistake to assume that just because the spreadsheet contains some numbers, they are correct. It is necessary to check at least one type of each calculation and verify its correctness. As discrepancies are discovered between the spreadsheet and the manual calculations, it is necessary to review the formulas as well as to review the manual calculations to determine whether they are correct.

This self-test is designed to help you check your understanding of the background information presented in this exercise.

True/False

Answer *true* or *false*.

1. The smallest element in a spreadsheet is called a cell. *True*
2. Cells may only contain numbers. *False*
3. The value of a cell may be used in an equation. *True*
4. It is not necessary to verify the calculations performed by a spreadsheet. *False*
5. A1:B5 represents a range of cells. *False*

Multiple Choice

Select the best answer.

6. The best way to write a spreadsheet is
   a. By scratch, composing it as you go.
   b. To make a plan of the design.
   c. To revise an existing spreadsheet.
7. What formula is required to add 15 to the value of cells A1 and B2 and store the result in a new cell?
   a. A1 + B2 + 15 =
   b. 15 = A1 + B2
   c. = A1 + B2 + 15
8. Charts are displayed
   a. On new sheets.
   b. As objects on existing sheets.
   c. Both a and b.
9. ABS stands for
   a. Absolute value.
   b. Average-based spreadsheet.
   c. Arithmetic bias.
10. After a formula is entered, the value of its cell
   a. Remains the same.
   b. Is updated based on the formula.
   c. Is set to zero by default.

Completion

Fill in the blanks with the best answers.

11. A cell's border is thicker and darker when it is _____.
12. A spreadsheet may contain one or more _____.
13. Graphs can be easily created using the Graph _____.
14. ABS, AND, and ATAN are examples of _____.
15. The first cell on a spreadsheet is _____.

1. Duplicate the Falling Body Analysis spreadsheet. Extend the time to 25 seconds. Use the spreadsheet to determine how long it takes the object to fall 10,000 feet.
2. Design a spreadsheet to keep track of the score in a bowling game.
3. Design a spreadsheet that analyzes a three-resistor series circuit. Figure 15.19 provides the required equations.
4. Use a spreadsheet to graph the data shown in Table 15.1.

FIGURE 15.19   For
Familiarization Activity step 3

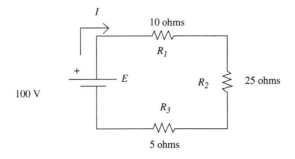

(a) Schematic of three-resistor series circuit

$$R_T = R_1 + R_2 + R_3$$

$$I = \frac{E}{R_T}$$

$$V_1 = I \cdot R_1 \qquad V_2 = I \cdot R_2 \qquad V_3 = I \cdot R_3$$

(b) Equations

TABLE 15.1   For
Familiarization Activity
step 4

Load (K ohms)	Power (mW)
1	23
2	45
3	89
4	121
5	167
6	126
7	98
8	61
9	39
10	27

QUESTIONS/ACTIVITIES

1. What limitations does the electronic spreadsheet have, if any?
2. Explain how a spreadsheet could be used to design a new part.
3. What is the advantage of using several sheets instead of one large sheet?

REVIEW QUIZ

Under the supervision of your instructor,

1. Understand the terminology associated with spreadsheets.
2. Navigate around the spreadsheet window.
3. Create a spreadsheet.
4. Use several built-in spreadsheet functions.

16 Presentation Software

Joe Tekk walked into the reception room of a large banquet hall. His friend and manager at RWA Software, Don Beers, was getting married. Joe quietly set up his laptop on a table near the door and started a PowerPoint presentation. Within 10 minutes, most of the guests were watching the presentation with delight, laughing and joking with each other. Joe had scanned pictures of Don, given to him by friends and relatives, and placed the images in a PowerPoint presentation. Then he added the many quotes and stories that went along with the pictures.

Don, watching with Joe, smiled happily, but mumbled to Joe under his breath, "I'm glad no one else is going to see this."

Joe glanced at Don and shrugged. "Sorry, Don, I saved the presentation in HTML format and posted it on your Web page."

PERFORMANCE OBJECTIVES

Upon completion of this exercise, you will be able to

1. View an existing PowerPoint presentation.
2. Create your own PowerPoint presentations (from scratch and from a template).
3. Print out a PowerPoint presentation (outline and slide formats).

BACKGROUND INFORMATION

Presentation software is used to create professional presentations of text, graphics, and even multimedia audio and video files. Any kind of report can be improved by the powerful features of presentation software.

MICROSOFT POWERPOINT

Microsoft PowerPoint is a powerful application that allows you to make dynamic presentations using a variety of resources. The presentation format can be tailored so that the information is presented in the style envisioned by the presenter.

PowerPoint presentations consist of *slides,* graphical objects that may contain text, images, or other types of objects (almost anything that can be cut and pasted). Presentations can be manually stepped through by the presenter, or automatically sequenced in a slide-show format. Individuals who do not have PowerPoint can still view presentations that are specially saved as stand-alone slide shows.

Let us examine many of the basic features of PowerPoint.

OPENING AN EXISTING PRESENTATION

When PowerPoint is started, the dialog box shown in Figure 16.1 is displayed. If you are creating a new presentation, the AutoContent Wizard is a good place to start. We will use the AutoContent Wizard later in this exercise. For now, we will examine an existing presentation. Left-clicking the radio button to Open an existing presentation and then left-clicking OK will produce the Open dialog box in Figure 16.2. Like any other file open dialog, you can navigate to the directory of your choice to locate your presentation.

Note the series of PowerPoint files in the file display. The one selected (*convert*) has its first slide displayed in the viewing window at the right. This is a nice way to find the presentation you are looking for.

Double-clicking on the file name, or left-clicking Open after selecting a file, opens the presentation. Depending on how the presentation was last saved, it may come up in outline form or in slide form. Figure 16.3 shows the Outline view of the *convert* presentation. A nice feature of the Outline view is that it can be printed. Then everyone watching the show has a detailed outline they can make notes on during the presentation.

FIGURE 16.1 Initial PowerPoint dialog box

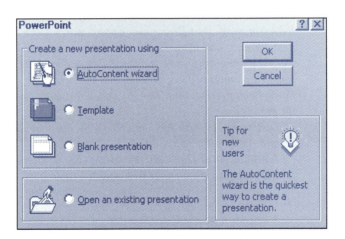

FIGURE 16.2 Opening a PowerPoint presentation

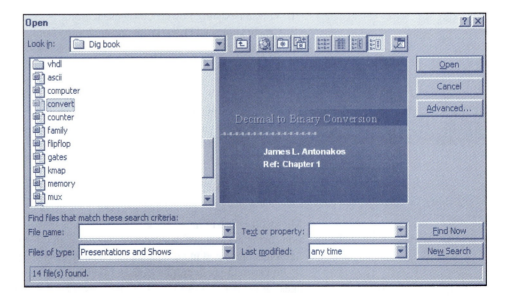

FIGURE 16.3   Outline view of
a presentation

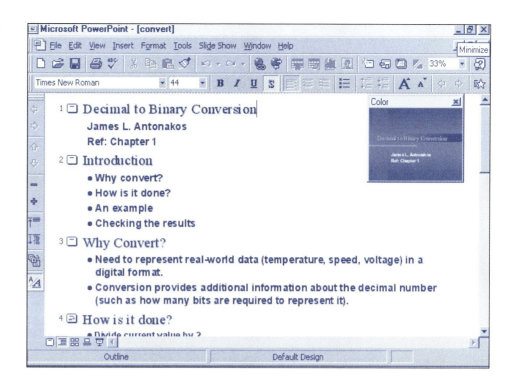

VIEWING A PRESENTATION

The Outline view shown in Figure 16.3 contains a number of features. Note the small color window to the right. It displays a reduced version of the current slide. The current slide is the slide where the cursor is positioned.

The small icons next to each slide number are also significant, as they indicate the type of slide presented. Notice that slide 4's icon is different from the other three slides. This is because slide 4 contains a graphic object as well as text. The other three slides only contain text.

The font of the outline is the same font that is used in the slides. So, if text appears bold or italic in the outline, it will appear bold or italic in the slide as well. The same goes for the font size and style.

To change the presentation view, use the View pull-down menu illustrated in Figure 16.4, or left-click one of the small view buttons at the lower left corner of the PowerPoint window. The Slide format replaces the text-based Outline view with the actual graphic slide. Figure 16.5 shows the Slide view of slide 1 in the *convert* presentation. The current slide is indicated in the status box at the lower left of the PowerPoint window. The total number of slides (6) in the presentation is also indicated.

The slide can be edited by left-clicking inside it and then making the necessary changes. Individual slides are accessed via the vertical scrollbar on the right of the display.

A window showing a small black-and-white version of the slide may also be displayed. It is enabled through the View pull-down menu. Its purpose is to show what the slide would look like when printed on a black-and-white printer.

To view the entire presentation, select the Slide Show option from the View pull-down menu or the View Show option in the Slide Show pull-down menu. Slides can be advanced during the slide show by pressing Space or Enter, or by left-clicking anywhere on the screen. Built-in tools for generating the timing of the slide show are included for presenters who require a specific pace for their presentations.

Right-clicking during a slide show will bring up the context-sensitive menu shown in Figure 16.6. From this menu you can move forward or backward in the slide show, adjust options, and even end the show.

FIGURE 16.4   View pull-down menu

FIGURE 16.5   Slide view of a presentation

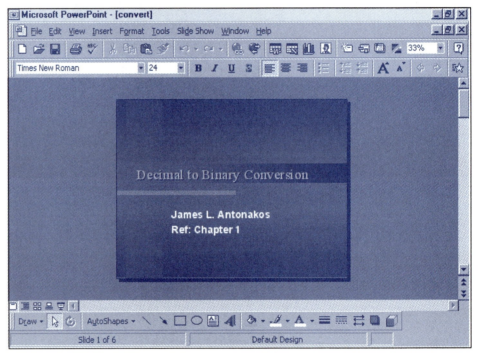

FIGURE 16.6   Slide show control menu

ADDING SLIDES TO A PRESENTATION

Slides are easily added to a presentation by selecting New Slide from the Insert pull-down menu. The slide is inserted after the current slide.

A large variety of slide types is available. One must be selected when you insert a new slide. Figure 16.7 shows the New Slide window. Left-clicking a slide icon causes a brief description to be displayed so that you know what you are getting. Note the wide variety of slide formats: text-only, object-only, text and object, multiple objects, and so on.

After choosing a slide, the new slide is displayed and set up for editing. For slides that contain objects, it may be better to edit the slide in Slide view than Outline view.

Figure 16.8 shows the new slide, which clearly indicates what area of the slide can be edited. The number of the slide is displayed in the status box (5 of 7). Other parameters, such as the font style and size (Times New Roman, 24 pt.), justification (left), and text formatting (bold), are also indicated.

To add text for the title, left-click inside the rectangular title box and begin entering text. Figure 16.9 shows the slide with its new title. Note that the title box is outlined differently once it has been selected. Any edge or corner of the box can be dragged to change the size of the box.

FIGURE 16.7 Choosing a new slide format

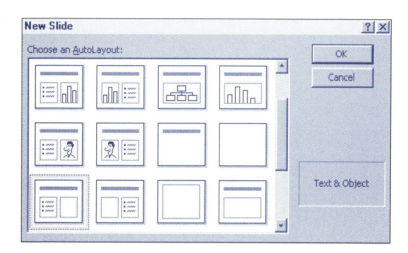

FIGURE 16.8 New text-only slide

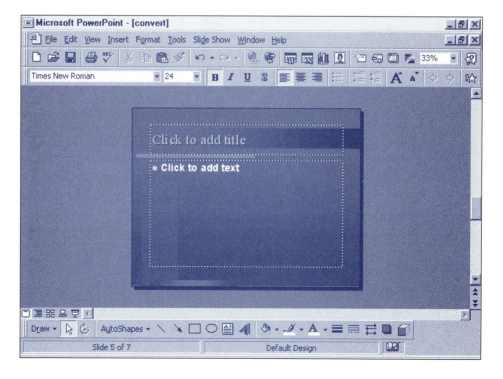

FIGURE 16.9   Adding the title to the new slide

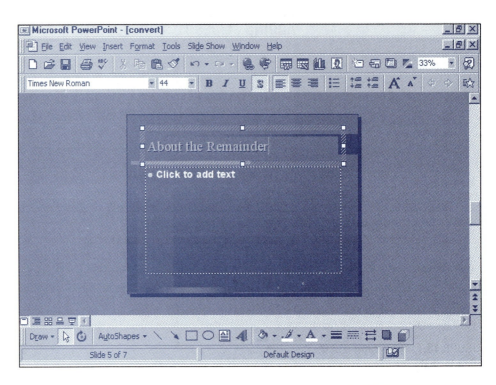

FIGURE 16.10   Adding text to the new slide

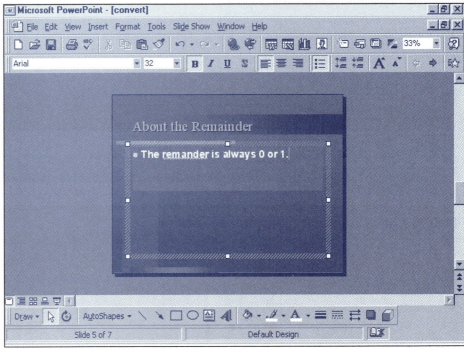

Text is added by left-clicking inside the text box, which may also be resized as necessary. Figure 16.10 shows the first line of new text. PowerPoint has underlined the misspelled word *remander* (should be *remainder*). This spell-checking ability is a nice feature and one more boost in the creation of a professional presentation. When multiple lines of text are added to a text box, they can be formatted as an ordinary paragraph of text, or as ordered or unordered lists by selecting the appropriate icons on the toolbar.

INSERTING OBJECTS INTO A SLIDE

A slide that contains one or more objects will indicate where to double-click to add an object. This is illustrated in Figure 16.11. The new slide contains a text box and an object

FIGURE 16.11   Slide
containing text and object
elements

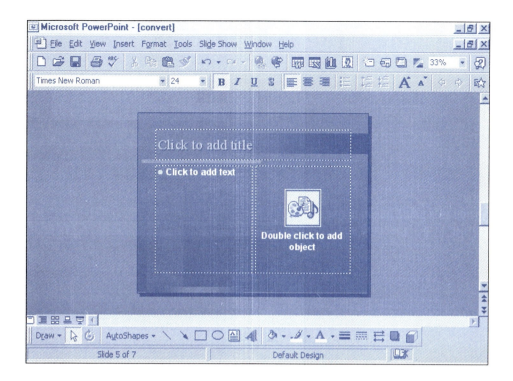

FIGURE 16.12   Insert Object
dialog box

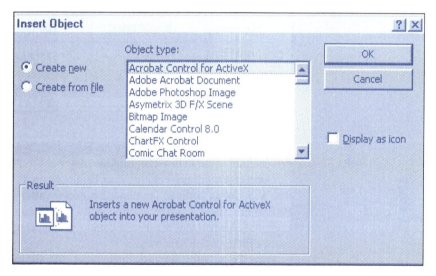

box. As usual, you can resize the object just as you can the text. Double-clicking the object icon brings up the window shown in Figure 16.12. Note that you can insert an existing object that has been saved in a file, or you can create a new object to insert. After the object has been inserted into the slide, double-clicking it will bring up the application associated with the object. This is useful if the object needs to be edited at a later time.

Figure 16.13 shows the finished slide containing an object created with an equation editor. Virtually any type of object, from clip art to animation files, can be inserted into a slide.

SAVING A PRESENTATION

PowerPoint lets you save your presentation in many different ways. Typical formats are Presentation, PowerPoint Show, and HTML (Web-based). PowerPoint Shows are presentations that always open up as slide shows.

Another useful save option is Pack-and-Go. In this option a viewer program (PPVIEW32.EXE) is packaged with the presentation (which is compressed). Individuals who do not have PowerPoint can view the presentation with PPVIEW32. The file extension

FIGURE 16.13 Slide
containing text and a graphic
object

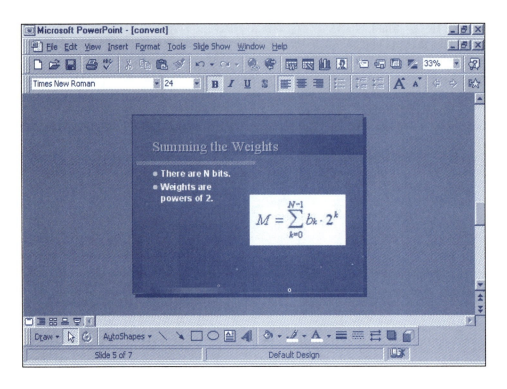

TABLE 16.1 File sizes for
various Save options

File name	Size (K bytes)	File Type
CONVERT.PPT	63.5	Presentation
CONVERT.PPS	62.5	PowerPoint Show
PRES0.PPZ	890	Pack-and-Go

for this save option is .PPZ. Examine Table 16.1, which shows all versions of the *convert* presentation saved in different formats.

Once saved, the presentation may be placed on a floppy disk to hand out, or posted on the Web (as an ordinary presentation file or after conversion to HTML format). The PRES0.PPZ file is saved to a floppy with a small setup program that extracts the viewer and presentation when they are installed on a different computer. Its large increase in size is due to the PPVIEW32.EXE code.

PowerPoint also supports three other file formats: WMF (Windows Metafile) for graphic slides, RTF (Rich Text Format) for outlines, and POT (PowerPoint Template).

PRINTING A PRESENTATION

As previously mentioned, PowerPoint presentations can be printed in a variety of formats. These formats include:

- Slides (which can be scaled)
- Handouts (two or more slides per page)
- Outline

The outline format is nice for handing out at the beginning of a presentation. This allows the audience members to take notes without being rushed.

CREATING A PRESENTATION

Recall from Figure 16.1 that we can use the AutoContent Wizard to create a new presentation. This is done when PowerPoint is started, but may also be initiated at any time by selecting the

FIGURE 16.14 Choices for new presentations

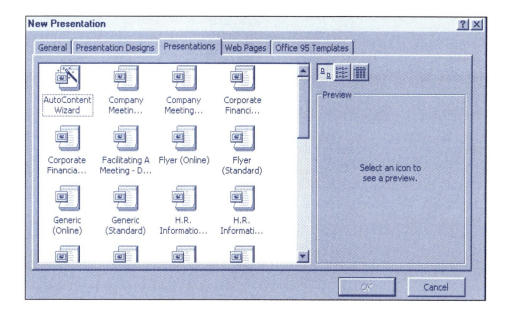

FIGURE 16.15 Initial AutoContent Wizard window

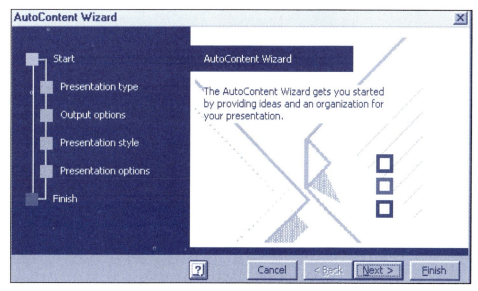

New option from the File pull-down menu, and then choosing AutoContent Wizard in the Presentations section of the New Presentation window shown in Figure 16.14. Note the additional predesigned presentations. These are called *templates,* and contain fully developed presentations, with specific wording (such as "Company Meeting Title" and "Presenter") included to indicate where to customize the presentation by entering your own information.

The AutoContent Wizard starts with the screen shown in Figure 16.15. The first thing to do is choose a presentation type, such as General, Corporate, Personal, or Sales/Marketing. This selects a template to use for the presentation.

Then the following output options must be specified:

• On-screen presentation
• Black-and-white slides
• Color slides
• 35-mm slides
• Handouts

And these presentation options:

• Title
• Name
• Additional information

Once all the information has been entered, the AutoContent Wizard builds the presentation.

A new presentation can also be started by left-clicking the New icon in the Standard toolbar. Here there are no choices except for what type of slide will be used for slide 1.

OTHER SOFTWARE

Many companies offer presentation software products. It is worth the time spent browsing the Web for presentation software and multimedia authoring tools. Here is a short list of several popular products:

- Multimedia Toolbook (www.asymetrix.com): Multimedia authoring tool. Create "books" that contain active objects (such as buttons that can be clicked). Includes audio, video, and animation.
- Corel Presentations (www.corel.com): An easy-to-use presentation software program with many of the same features that are found in PowerPoint.
- Harvard Graphics (www.harvardgraphics.com): PowerPoint will convert Harvard Graphics presentations into its own format.
- Lotus Freelance (www.lotus.com): Powerful presentation package containing support for Web publishing, networking, database access, and group presentations. PowerPoint will convert Freelance presentations into its own format.

TROUBLESHOOTING TECHNIQUES

There may come a time when you are editing a slide in outline format and cannot perform some simple editing step such as left-clicking to select an insertion point. Whatever the difficulty, it may help to switch display modes and try the same step when viewing the graphical slide. This is especially true for slides that contain objects and text. The outline does not make it clear what needs to be done to access the object, but it is very obvious on the graphical version.

Also, when saving your presentation for others to view, bear in mind that they may not have the same fonts installed on their machine as you do. You may want to have the fonts embedded in your presentation to avoid this situation.

SELF-TEST

This self-test is designed to help you check your understanding of the background information presented in this exercise.

True/False

Answer *true* or *false*.

1. PowerPoint presentations are strictly text-based. ~~True~~ *False*
2. The Slide Show format is used to display slides larger than normal for editing purposes. *True*
3. Pack-and-Go presentations contain their own viewers. *True*
4. PowerPoint presentations can be saved in HTML format. *True*
5. Title and text areas of a slide may be resized as necessary. *False*

Multiple Choice

Select the best answer.

6. The outline view of a presentation
   a. Can be printed.
   b. Contains all text and graphics in the presentation.
   c. Neither a nor b.
7. Which presentation file has the extension .PPT?
   a. Presentation.
   b. PowerPoint Show.
   c. Pack-and-Go.

216

8. Which presentation file is the largest if the same presentation is saved many different ways?
   a. Presentation.
   b. PowerPoint Show.
   c. Pack-and-Go.
9. New slides are inserted
   a. Before the current slide.
   b. After the current slide.
   c. At the end of the presentation.
10. Which file extension is not supported by PowerPoint?
    a. .RTF.
    b. .PPT.
    c. .PCX.

Completion

Fill in the blanks with the best answers.

11 PowerPoint presentations consist of one or more _____.
12. A printed presentation containing two or more slides per page is called a(n) _____.
13. PowerPoint performs _____ checking on text entered into a slide.
14 New presentations are easily started using the _____ Wizard.
15. Fully developed presentations that require customization are called _____.

FAMILIARIZATION ACTIVITY

1. Design a PowerPoint presentation that explains how to add two three-digit decimal numbers. Show an example, such as 143 plus 893 equals 1036.
2. Prepare a PowerPoint presentation on a subject of your own choice. Include several different slide styles.
3. Search the Web for PowerPoint presentations posted by others. Select three that you find interesting and merge them into a single presentation.

QUESTIONS/ACTIVITIES

1. What information is not shown in the Outline form of a presentation?
2. Compare the file sizes of a presentation when saved as a Presentation and as a PowerPoint Show.

REVIEW QUIZ

Under the supervision of your instructor,

1. View an existing PowerPoint presentation.
2. Create your own PowerPoint presentations (from scratch and from a template).
3. Print out a PowerPoint presentation (outline and slide formats).

17 Planning Tools

INTRODUCTION

Joe Tekk was sitting as his desk planning the network upgrade to Windows 2000. He had spent a lot of time trying to keep all of the application software on the servers up to date so that a major system upgrade would be as easy as possible.

Don, Joe's manager, stopped by Joe's desk to ask how things were going regarding the upgrade. "Hey, Joe. Do you still think the systems will be up on Monday morning at 5:00am as we discussed?"

"Yes," Joe replied, "According to my plan, it looks like everything should be done by 4:00pm on Sunday afternoon. I've broken it down into five major steps. When the last one is done, the upgrade will be complete."

Don smiled before he walked away, saying "I'll be home this weekend. Don't forget to call me if things turn out differently than you planned. I've got to know if we can't transfer those files on Monday morning."

PERFORMANCE OBJECTIVES

Upon completion of this exercise, you will be able to:

1. Explain the features of a planning tool.
2. List the steps necessary to complete a project.
3. Create a Gantt chart.
4. Update the project status.

BACKGROUND INFORMATION

Planning Tools have become a popular way to control or manage virtually any type of project. A project can be a simple step-by-step progression from start to finish over the course of several weeks or months, or it may consist of a series of tasks, some of which can be performed in parallel, with various milestones that may be completed over the course of several days, months, or possibly even years. In this exercise, we will examine the some of the tools that are useful to planning any type of project.

WHAT ARE PLANNING TOOLS?

From a traditional standpoint, planning involves the allocation of resources to perform a certain task. Examples of resources are: money, time, personnel, equipment, and space.

Personnel may include an engineer who would plan the various phases of building a house, or a bridge, or any other type of structure. A computer scientist may need to plan how the various elements of a complex software system may be designed. Virtually every type of profession can benefit by creating a plan. The challenge of planning involves dealing with the problems of managing and tracking many different tasks at the same time.

These are precisely the reasons that computerized planning tools have been developed. A computer program can be used to minimize the required effort to maintain, update, and monitor the progress of any project. The remainder of this exercise is devoted to outlining the capabilities of Microsoft Project, a planning tool for the personal computer running the Windows 95/98/NT/2000 operating system. The initial Microsoft Project window is displayed in Figure 17.1. Notice that there are the pull-down menu items and toolbars located at the top of the window. Context sensitive help is displayed for each of the toolbar icons by simply moving the mouse pointer over any of the icons. The display selection area, shown along the left hand side of the window allows the user to choose how the data is to be displayed on the screen. The Gantt chart display mode is selected by default, and causes the timeline to be displayed on the right side of the screen. Let us proceed by examining the first step in the process: developing a plan.

DEVELOPING A PLAN

To develop a plan that can help manage resources and control costs it is necessary to identify the critical information necessary to accomplish the task. In order to use Microsoft project, we need to perform these steps

- Define the project.
- Design and build a plan.
- Track and monitor progress on each task as well as the whole project.
- Adjust the schedule to reflect the actual progress.
- Bring the project to closure.
- Study the differences between the planned and actual completion dates.

FIGURE 17.1 Initial Microsoft Project Window

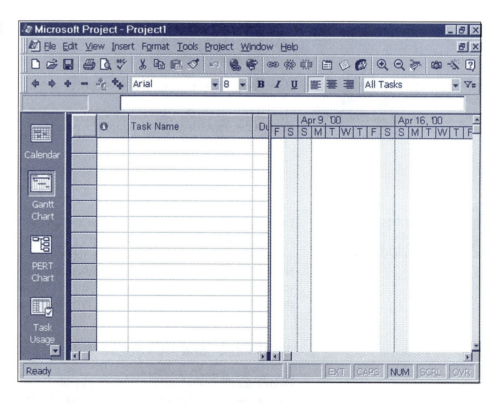

Using Microsoft Project, each one of these items is part of the planning process and requires additional review. Each one of these steps requires various amounts of time and effort to complete depending on the type of project being developed. We will discuss each of these important steps with an emphasis on the individual steps within each of these areas of concern. First, let us take a look at what is required to actually define the project.

DEFINE THE PROJECT

The process to define the project is one of the most important steps. It is here that all of the various parts of the project must be identified. It is a good idea to be as thorough as possible when trying to identify all of the different phases that make up the complete project. In general, it is necessary to perform the following tasks when defining a project:

• Identify all of the tasks to be completed.
• Develop a baseline schedule from which to determine progress.
• Communicate the schedule information.
• Delegate responsibility for each of the tasks.
• Get periodic status updates and compare this information against the baseline schedule.
• Produce status reports to keep everyone informed.

Although these tasks would appear to be relatively simple in nature, keep in mind that most projects consist of many tasks. Some of these tasks may be performed in parallel with other tasks, while others may need to be completed before other tasks may begin. It is extremely important to identify the relationships between the tasks so that the project proceeds on schedule. With a clear understanding of the tasks, it is possible to develop the baseline schedule that will be used to measure progress.

In order to complete any type of project (large or small), it is necessary to communicate the schedule information to all of the individuals who are responsible for completing one or more of the tasks. In practice, this means delegating responsibility for each of the tasks. In this way, it is possible to have one point of contact for each of the tasks. Note that within a task, it is possible to identify subtasks that explain individual processes that must be finished to complete the task.

Let us consider a simple example project to fabricate an electrical circuit. The individual tasks and task durations that are necessary to perform the circuit fabrication are shown in Table 17.1.

TABLE 17.1 Electronic Circuit Fabrication Schedule

Task Number	Task Name	Task Duration (days)
1	Planning and design	5
2	Get the parts that are required	1
3	Breadboard the circuit	3
4	Test and Troubleshoot the circuit	2
5	Design the printed circuit board	5
6	Etch and drill the printed circuit board	2
7	Populate the printed circuit board	1
8	Final troubleshooting and test	2
9	Assemble the circuit in a fabricated chassis	1

During the actual work being performed on a task it is very important to stay on top of the project progress. Due to many different factors, it is possible for the tasks in a project to take more time or less time than originally allocated in the baseline. The reasons for the changes are varied, but can be attributable to poor planning, problems with equipment and resources, weather, and a host of many other items.

By maintaining a measure of control on the project as a whole, it is possible to identify problems up front, possibly before they may be a problem that puts a project in jeopardy of missing critical deadlines. The success or failure of a project depends on every one of the individual tasks to be completed. This is exactly the role of project management software; to keep track of every important detail.

Let us continue by defining the steps necessary to use the Microsoft Project program to manage a project. The steps that must be completed are:

1. Design and build a plan.
2. Track and monitor the project progress.
3. Adjust the schedule as required.
4. Bring the project to closure.
5. Evaluate the actual project dates as compared to the baseline.

We will examine each of these topics as it relates to our sample project.

DESIGN AND BUILD A PLAN

There are several steps involved when using the Microsoft Project planning tool. In general, following items must be performed.

- Create new planning file.
- Specify either the start or finish date.
- Enter task information.
- Identify task duration.
- Assign resources.

To create a new planning file, New Project can be selected from the toolbar or from the File pull down menu. This causes the Project Information screen to be displayed as shown in Figure 17.2. Notice that Project Information is for 'Project2' since Project1 was opened by default. It contains entries for the project start date or finish date depending on the 'Schedule from' selection box. The Schedule from box can be set to the start date or end date as required. In addition, the current date, a status date, and a calendar selection are displayed. The Calendar can be set to 'Standard' (the default), 24 hour, or night shift.

We can begin to create the plan by entering the actual tasks that must be performed to complete the project. As shown in Figure 17.3, to enter a task, all that is necessary is to enter the name of the task in the task name field. When the ENTER key is pressed, three noticeable events occur on the project screen. First, the current cell is moved to the next line

FIGURE 17.2   New Project Dialog Box

Project Information for 'Project2'

Start date:	Thu 4/6/00		OK
Finish date:	Thu 4/6/00		Cancel
Schedule from:	Project Start Date		
	All tasks begin as soon as possible.		Statistics...
Current date:	Tue 4/4/00		
Status date:	NA		
Calendar:	Standard		

the in the Task Name column. Second, a task number is assigned to that task and displayed in the shaded area to the left of the task description. Since this is the first task being entered, the task number is 1. Third, the task is displayed on the Gantt chart. Notice that it has been allocated a duration of one day, the default duration.

Since the planning stage of the Circuit Fabrication project is allocated five days, the task must be modified. To do this, it is necessary change the task information. By right clicking on the task name and selecting 'Task Information...' the Task Information menu is displayed. This is shown in Figure 17.4. Several tabs are displayed at the top of the window that show the categories that contain properties that may be modified. On the General tab, the task name, duration, percent complete, priority, and start and finish dates may be updated. For our Planning and Design task, it is necessary to change the duration to 5 days. By pressing the small up arrow next to the duration field and pressing ENTER, the display is updated as illustrated in Figure 17.5.

Notice that the Gantt chart now indicates that a five day duration is required and the days over the weekend are not counted.

FIGURE 17.3   Planning and Design default information

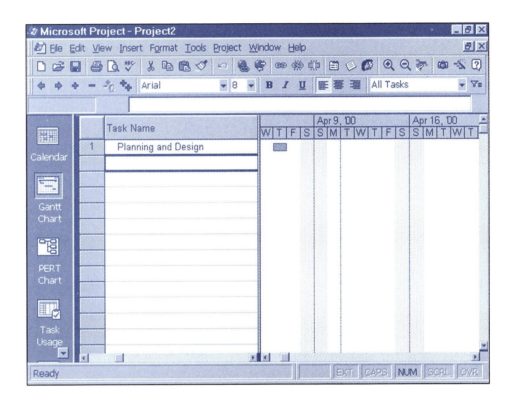

FIGURE 17.4   Planning and Design Task Information

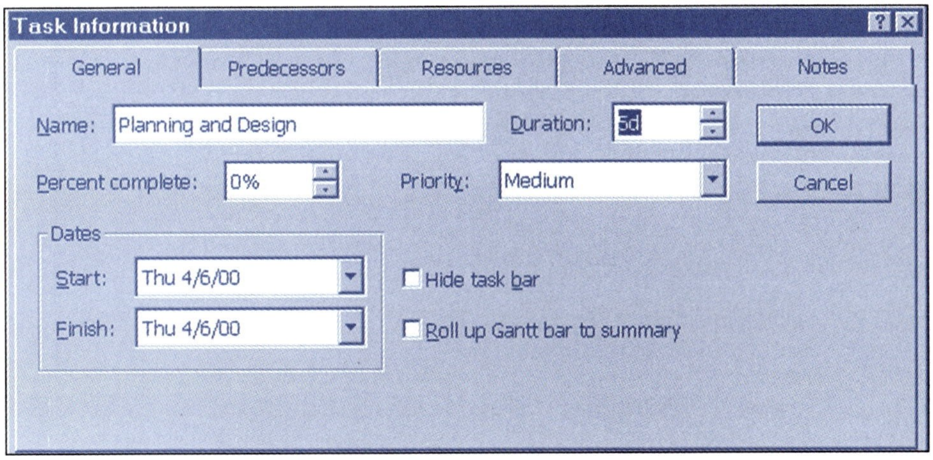

Now we are ready to add the next task. It has been determined that it will take one day to obtain all of the necessary parts to build the circuit. Therefore, we can simply add a task called Get Parts. In the Circuit Fabrication example, when a task is added it must then be linked to the previous task since the planning and design phase must be completed before we can get the parts. In general it is not always necessary to link tasks in this manner. To link the tasks together, two steps are required. First, it is necessary to select the tasks to be linked. This is accomplished by holding the control button down and left clicking on the two tasks to be linked. Next, it is necessary to press the ' Add Link' button on the taskbar. This is the icon that looks like a chain link. Note that is also possible to choose 'Add Link' from the pull down menu. The result of linking these two tasks together is shown in Figure 17.6.

FIGURE 17.5   Modified Planning and Design task display

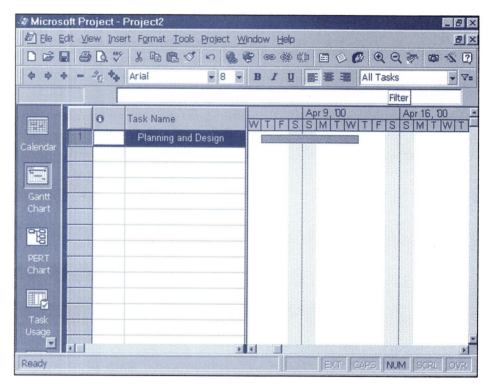

FIGURE 17.6   Planning and Design task being linked to Get Parts

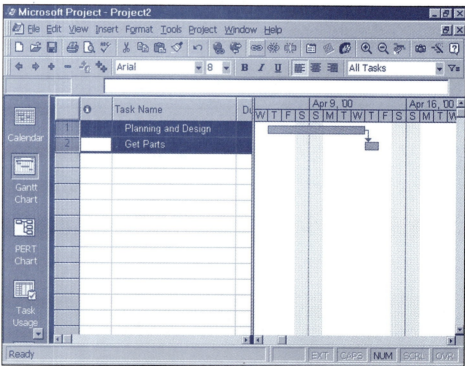

To complete the Circuit Fabrication project plan, the add new task procedure must be performed for the remaining tasks in Table 17.1. After each new task is created, it must then be linked to the previous task.

Figure 17.7 shows the completed Circuit Fabrication project. Each of the required tasks is linked to the previous task. Notice that the information column (the circle with an 'i' in it to the left of the Task Name column) shows that information is available for the task. Also note that a small black bar is displayed in the first task on the Gantt chart next to 'Planning and Design'. This indicates the percentage of the task has been completed. Also notice that the project has been saved on the disk using a more descriptive name 'Circuit Fabrication'.

In a more complex project, it may be necessary to assign resources in order to complete each task. This may involve items such as equipment and/or people. Microsoft Project is able to handle the most demanding type of planning assignments.

TRACK AND MONITOR THE PROJECT PROGRESS

After a project is built, it is necessary to monitor the progress of the project. This is accomplished by keeping information up to date, comparing the baseline schedule to the actual project progress, and by updating the schedule as necessary. To update the tasks within the project, all that is necessary is to edit the 'Percent complete' field on the Task Information window. Refer to Figure 17.4.

While tracking and monitoring a project, a small amount of effort is required to keep the project files up to date. By doing this, a significant amount of information can be read from the data. This may be the difference between a successful on-time project completion and an over-budget dismal failure. For example, it may become necessary to lengthen the project schedule due to a problem that has just been identified. Similarly, if the project is running ahead of schedule, it may be possible to shorten the schedule. In either case, the information that can help make the decisions is available within the project data.

On large projects, tracking and monitoring the project status can be useful when trying to keep expenses and resource expenditures to a minimum. If a problem is identified early enough, the overall impact to the project schedule may be minimized.

FIGURE 17.7   Electronic Circuit Fabrication Schedule

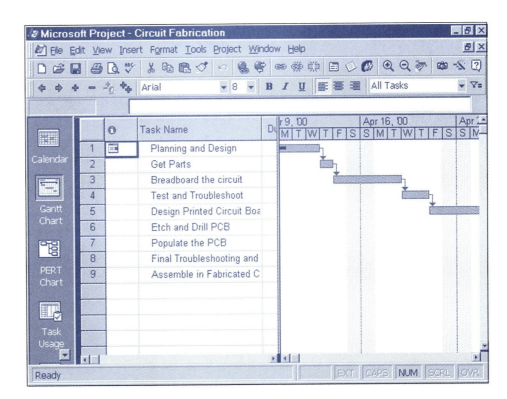

BRING THE PROJECT TO CLOSURE

The last step in project management involves finishing the project. This involves learning as much as possible from the problems and mistakes that were made. Emphasis may be placed on identifying the source of each problem if it can be identified. By spending some time trying to identify problems now, a final problem resolution may be identified and used in the future to prevent similar occurrences.

A RECURRING TASK PROJECT

Unlike the Circuit Fabrication project that consisted of a series of serial processes, many different projects require tasks to be performed in parallel. For example, let us take a look at a small assembly line project. The assembly line consists of manufacturing parts that must be machined, sanded, and painted. Each of these tasks takes one day. After the first three days, the first part comes off the assembly line. Let us examine how a recurring task is used to make three parts on the assembly line.

Using Microsoft Project, it is possible to add a recurring task. This is accomplished by selecting 'Insert Recurring Task...' from the Insert pull down menu. The Recurring Task Information window shown in Figure 17.8 is displayed. By default, a weekly recurring task in selected. For the assembly line project, our recurring tasks occur every day. To make a change from weekly to daily, all that is required is to change the 'This occurs' box from weekly to daily. When the daily option is selected, the menu presented in Figure 17.9 is displayed. Notice that the first step in the process, machining the part, has a duration of 1 day on workdays only. Also notice that three occurrences have been selected, once for each of the three parts to be produced.

The next task, sanding the parts, can only be performed after the parts have been machined. The third task, painting the parts, can only be performed after the parts have been sanded. The one day delay between each of these tasks is accomplished by adjusting the start date for tasks two and three. Figure 17.10 shows the tasks and Gantt chart after each of the recurring tasks have been entered.

Notice that a plus sign precedes each of the recurring tasks in Figure 17.10. This indicates that the individual details for each element of the recurring tasks are not displayed separately although the Gantt chart shows each of the three parts during each phase of the project. Figure 17.11 shows how the tasks can be expanded to show each part. You are encouraged to explore the power of recurring tasks on your own. You may find many different uses for them.

FIGURE 17.8   Creating a Recurring Task

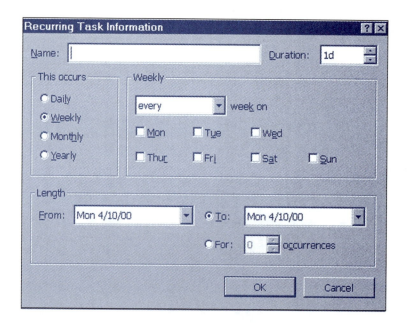

FIGURE 17.9   Daily Recurring
Task Information Display

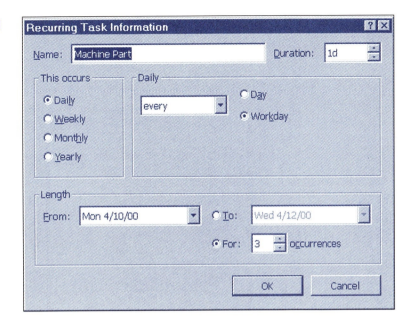

FIGURE 17.10   Assembly Line
(condensed)

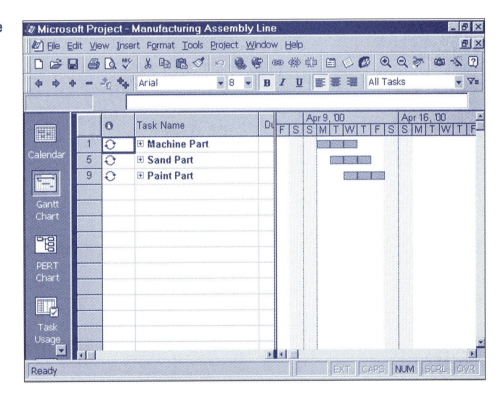

**TROUBLESHOOTING
TECHNIQUES**

It is important to remember that, when time is of the essence, or when money or resources can be saved, it is smart to take advantage of the many help topics that are available for review. Whether setting up a plan to create a project or trying to produce a custom report, there are many typical problems and questions that may be asked by both novice and experienced planners.

Figure 17.12 shows the Getting Started screen that is available in the help system. It provides step-by-step instructions on how to use many of the features available as well as definition of terms and other useful information.

In addition, each of the typical steps used to create a project in Microsoft Project can be explored from the 'Project Map', also available from the help system. The project map is displayed in Figure 17.13.

FIGURE 17.11   Assembly Line (showing all subtasks)

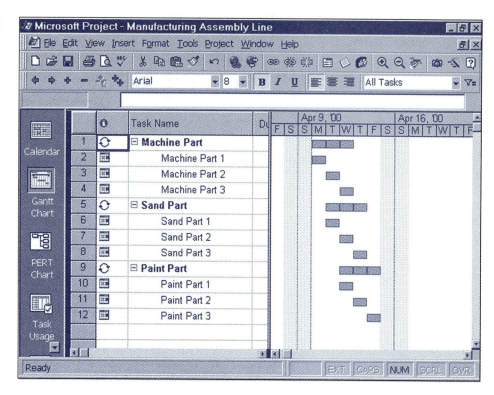

FIGURE 17.12   Learn while you work Help Mode

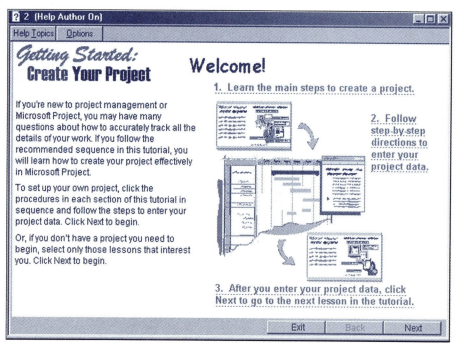

You are encouraged to explore all of the features that are available in Microsoft Project. You will discover that any project, no matter how large or small, can be planned with a minimum of effort.

SELF-TEST

This self-test is designed to help you check your understanding of the background information presented in this exercise.

FIGURE 17.13   Microsoft
Project Navigation Map

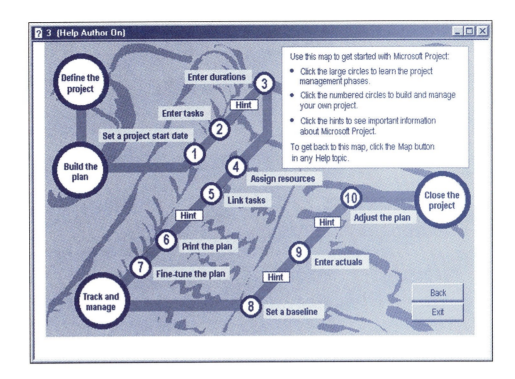

True/False

Answer *true* or *false*.

1. The Gantt chart display mode is selected by default. *True*
2. Defining the project is an optional step. *True*
3. When a new task is added it must be linked to the previous task. *True*
4. A task may contain subtasks. *False*
5. The default duration for a task is one week. *False*

Multiple Choice

Select the best answer.

6. Planning involves allocating resources and
   a. time.
   b. effort.
   c. personnel.
7. The first step in a new project is to
   a. define the project.
   b. enter all chart information.
   c. gather statistics.
8. Progress is tracked on a
   a. notepad.
   b. baseline schedule.
   c. Gantt chart.
9. You must specify the
   a. start date.
   b. task duration.
   c. both a and b.
10. A recurring task
    a. is a single task entered more than one time.
    b. describes a task that occurs more than one time.
    c. cannot occur more than 100 times.

Completion

Fill in the blank or blanks with the best answers.

11. Microsoft Project is a _____ tool.
12. The Calendar can be set to Standard, 24 hour, or _____ _____.
13. All tasks are assigned a task _____.
14. Tasks may be _____ to previous tasks.
15. The Project _____ is used to explain the major steps in completing a project.

1. Sketch a Gantt chart on a sheet of paper. The chart should describe all your household chores for one week.
2. Turn your sketch into a real Gantt chart using Microsoft Project.
3. Design a plan that indicates the necessary steps required to complete your educational curriculum.
4. Design the Circuit Fabrication example presented in Table 17.1.
5. Design the Assembly Line recurring task example.

1. Why not just sketch a plan for a project on a sheet of paper? Why bother with planning software?
2. Search the web for additional planning applications.

Under the supervision of your instructor,

1. Explain the features of a planning tool.
2. List the steps necessary to complete a project.
3. Create a Gantt chart.
4. Update the project status.

18 Communication Tools

INTRODUCTION

Joe Tekk checked his email for the fifth time in ten minutes. He was getting impatient, having gotten used to the quick turn around at his favorite software download site.

He checked it for a sixth time and was excited to see the email he had been waiting for, an account verification response. The email came with a small attachment containing a custom client application used with the software download site. Joe installed the client software and connected to the site to verify its operation.

Then Joe scanned several recent photographs and emailed their images to his sister, sent copies of a new network application he wrote to several friends, and finally handled the nine email messages waiting in RWA Software's help desk mailbox.

Satisfied, Joe went to lunch. When he returned 45 minutes later, he had eighteen additional emails to go through.

PERFORMANCE OBJECTIVES

Upon completion of this exercise, you will be able to:

1. Describe the features of e-mail communication software.
2. Configure an electronic mail client.
3. Send and receive electronic mail.
4. Organize electronic mail messages into folders.

BACKGROUND INFORMATION

Communication tools are at the heart of the personal computer revolution. In this exercise, we will explore Electronic Mail (commonly referred to as E-mail), one of the most common communication tools available. This exercise will cover the basic features of electronic mail, how to configure client software, how to send and receive electronic mail, and how to organize E-mail messages on a computer that is connected to the Internet.

WHAT IS E-MAIL?

In the early days of computer networking, a simple electronic mail program was used to exchange text messages between two people. Since then, electronic mail has evolved into personal communication tool that can be used to

- Send a message to several recipients.
- Send a message that contains text, graphics, and even multimedia audio and video files.
- Send a message that a computer program will respond to such as a mailing list program or mail exploder.

Electronic mail combines the speed of electronic communication with features similar to the postal mail service. The major difference between the postal mail service and e-mail is that a computer can transmit a message across a computer network almost instantly.

When using electronic mail, several common features are available to the computer user. For example, it is possible for every user to:

- Compose an e-mail message.
- Send an e-mail message.
- Receive notification that an e-mail message has arrived.
- Read an e-mail message.
- Forward a copy of an e-mail message.
- Reply to an e-mail message.

Let us begin our examination of E-mail by looking at how E-mail actually works on the Internet.

HOW E-MAIL WORKS

In order for a computer to use e-mail, it is necessary to install an E-mail client. Electronic mail uses the client-server method to allow for mail to be exchanged. Client computers exchange messages with a server that is ultimately responsible for delivering the e-mail messages to the destination.

On the server computer each user is assigned a specific mailbox. Each electronic mailbox or E-mail address has a unique address. It is divided into two parts: a mailbox name and a computer host name which are separated using an "at" sign, @ such as:

```
mailbox_name@computer_name
```

Together, both of these components provide for a unique E-mail address.

The mailbox portion of the address is often made from a user's name. The host name part of the address is chosen by a network administrator. For example, Joe Tekk has the e-mail address joetekk@stny.rr.com. From the example, this indicates that joetekk is the mailbox name and stny.rr.com is the computer name. Notice that Joe Tekk's e-mail address ends in .com. The .com indicates that stny.rr is a commercial organization. You will observe that the last three characters of an e-mail address will normally end with a limited number of Domain name categories. These categories are shown in Table 18.1.

E-mail messages are actually exchanged using the client-server environment illustrated in Figure 18.1. Note that both of computers in Figure 18.1 are called E-mail servers. When the mail message is exchanged, the mail transfer program on the sending computer tem-

TABLE 18.1 Common Domain Names

Domain Name	Assigned Group
com	A company or commercial organization
edu	An educational institution
gov	A government organization
mil	A military organization
net	Network service provider
org	Other organizations
Country code	A country code

FIGURE 18.1   How E-mail is
exchanged between servers

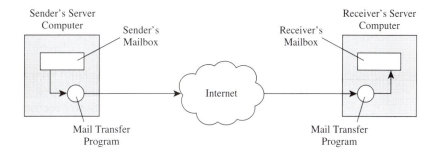

porarily becomes a client and connects to the mail transfer program running as a server on
the receiving computer. In this way, depending on whether mail is being sent or received,
the mail transfer program acts as a client or a server.

FORMAT OF E-MAIL MESSAGES

The format of an e-mail message exchanged between the servers is quite simple. Each message consists of ASCII text that is separated into two parts. A blank line is used as the separator between the parts. The first part of the message is called a *header*. A header consists of a keyword, followed by a colon and additional information. Some of the most common header keywords are shown in Table 18.2.

The second part of the message is called the *body* and contains the actual text of the message. Note that because it is possible to attach many different file types to an e-mail message, a special scheme called Multimedia Internet Mail Extensions (MIME) was developed. This provides a way for binary programs, graphical images, or other types of files to be attached to an e-mail message. The sending computer encodes the message into text for transport and the receiving computer decodes the message back into the original form.

E-MAIL CLIENT SOFTWARE

One of the most popular client software E-mail client programs is Microsoft Outlook Express. It is installed as a part of the Windows operating system. There are usually several different ways to access the Outlook Express program. For example, there may be an icon on the desktop that may be double clicked, a small Outlook Express icon may be placed on the taskbar, or it may be a program that can be selected from the Windows start menu. In any case, after the Outlook Express program is started, the computer user is presented with a screen display similar to Figure 18.2.

In order to use the Outlook Express program, it must be configured properly. This configuration consists of providing user information such as the user name, organization, e-mail address, and reply address. These items are located on the General E-mail properties tab as

TABLE 18.2   Typical E-mail
header keywords

Header Keyword	Description
To	The mail recipients e-mail address
From	The senders e-mail address
Cc	List of carbon copy addresses
Bcc	List of blind carbon copy addresses
Date	The date when the message was sent
Subject	The subject of the message
Reply-to	The address to which a reply should be sent

shown in Figure 18.3. It is also necessary to identify the server computer to which the client will connect to send and receive mail. This information is found on the Servers tab of the Mail Properties window shown in Figure 18.4. There is a server associated with both incoming and outgoing mail.

Incoming mail uses POP3, the Post Office Protocol, whereas the outgoing mail server uses SMTP, the Simple Mail Transport Protocol. Notice that this is the screen where the user enters their Incoming Mail Server account name and password. As an added convenience, it is possible for Outlook Express to save or remember the password for future use.

Sometimes it is necessary to change some of the mail parameters. For example, it may be necessary to change the server timeout value of the mail program, or change the setting

FIGURE 18.2   Microsoft Outlook Express displaying a message

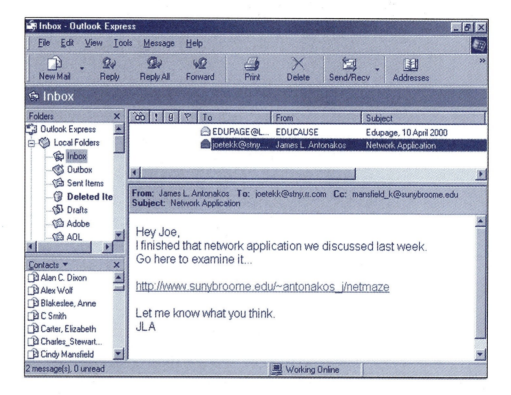

FIGURE 18.3   Outlook Express General E-Mail Properties

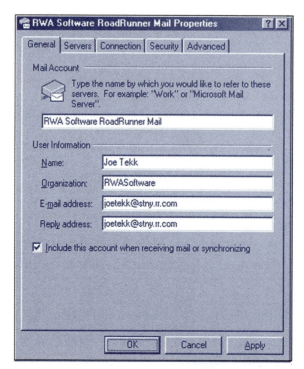

FIGURE 18.4   Outlook Express
Mail Servers

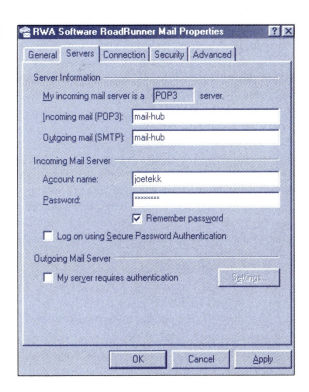

FIGURE 18.4   Outlook Express
Mail Servers

FIGURE 18.5   Outlook Express
Advanced Mail Properties

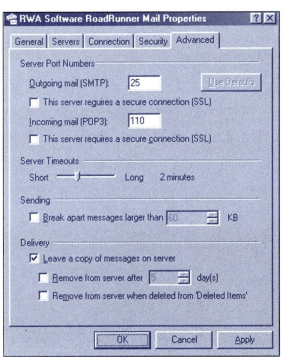

which determines if a copy of a mail message is to be left on the server computer after it has been transferred to the client. As you can see from Figure 18.5 there are several different settings which can be modified. It is always a good idea to leave the settings alone unless there is a good reason to change them.

SENDING AN E-MAIL MESSAGE

Let us consider an example where Joe Tekk creates the e-mail message shown in Figure 18.6. Joe uses the e-mail client to send a message to windy@alpha.com.

The message is sent by Joe to the e-mail server at stny.rr.com. The mail server at stny.rr.com forwards the message to the e-mail server at alpha.com where the user Windy can read that message. Figure 18.7 illustrates how the E-mail message is sent using the Microsoft Outlook Express client program. Notice that SMTP is used to transfer the message everywhere except for the client connection at the destination, which uses POP3.

RECEIVING AN E-MAIL MESSAGE

E-mail messages are received by the server and stored in the Inbox inside of a users mailbox until it is read. For example, Figure 18.8 shows a message from the Java Developer Connection mailing list. After the message has been read, it can be deleted or saved. If a message is saved, it is normally moved to a folder other than the Inbox. This allows for mail to be stored in user-defined categories. To create a new folder for the Java Developer Connection message, simply right-click on the Local Folders in the folder list and select New Folder. To move the message into the folder, drag it from the Inbox message list to the appropriate folder. This provides for an easy way to keep track of all related messages.

Note that Outlook Express provides the capability to store as many messages as necessary (as long as there is enough disk space available) although it is a good idea to keep the mailbox clean.

E-MAIL ERROR MESSAGES

There are several reasons why an error message may be generated when trying to send E-mail. Two of the most common errors stem from the user incorrectly specifying either the mailbox name, or the computer name. In either case, a message will be sent back to the

FIGURE 18.6   Creating a New E-Mail Message

FIGURE 18.7   Sending and Receiving E-Mail

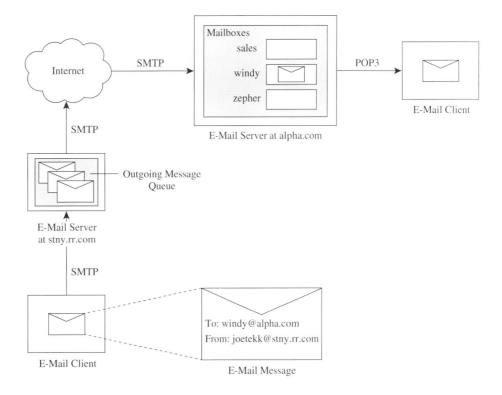

FIGURE 18.8   Reading a message in the Outlook Express Inbox

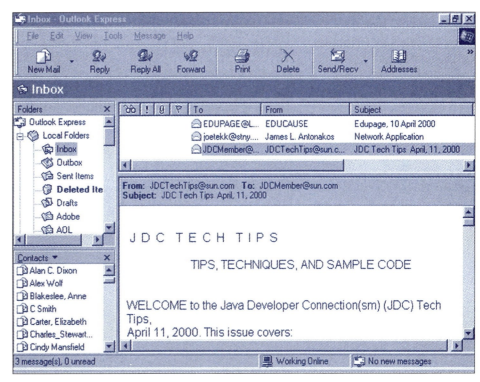

sender indicating what type of error has occurred. Figure 18.9 illustrates an error with the mailbox portion of the address while Figure 18.10 indicates a problem with the computer portion. Other problems with the mail will have their own specific message which may help to resolve the problem.

ACCESS TO E-MAIL USING THE WWW

Some E-mail servers allow access to the mail system using a World Wide Web browser. The browser acts the same as an E-mail client that allows a user to send and receive E-mail

FIGURE 18.9   E-mail message indicating an invalid recipient

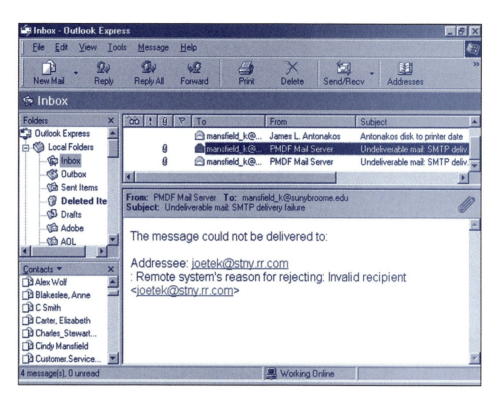

FIGURE 18.10   E-mail message indicating an invalid host/domain name

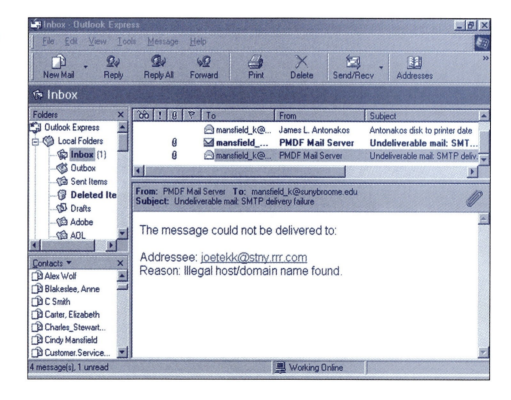

messages. Figure 18.11 shows a screen shot of the Microsoft Outlook Web Access screen which uses the Microsoft Exchange Server. Note that a username and password are required in order to access any mail files.

FIGURE 18.11   Accessing the E-mail server using the web

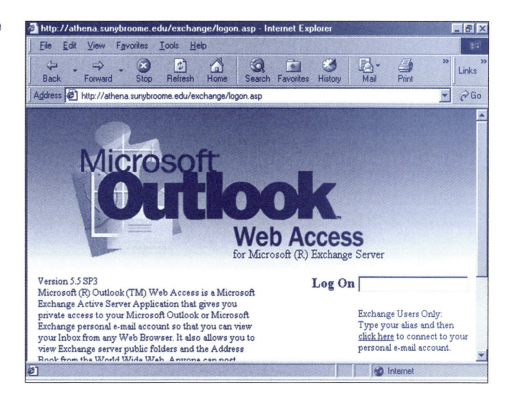

FIGURE 18.11   Accessing the E-mail server using the web

TROUBLESHOOTING TECHNIQUES

One reason it might be helpful to know a few basic POP3 commands has to do with a real-world situation where several email messages were queued up behind an email with a very large (over 4 MB) attachment. Unfortunately, a network router problem creating frequent packet losses prevented the email with the attachment from being properly transferred to the recipients email client. To get at the queued up email messages, the user used TELNET to connect to the POP3 server and deleted the email message containing the large attachment. This allowed the remaining messages to transfer to the email client. Since the messages were small, they transferred quickly, with only a slight delay introduced by the router problem.

SELF-TEST

This self-test is designed to help you check your understanding of the background information presented in this exercise.

True/False

Answer *true* or *false*.

1. An E-mail address consists of three parts.
2. The header and body of an E-mail message are separated by a blank line.
3. MIME provides a way to send binary attachments to an E-mail message.
4. Microsoft Outlook Express is an E-mail server program.
5. E-mail messages can be read using a World Wide Web browser.

Multiple Choice

Select the best answer.

6. In order to keep all related E-mail messages together
   a. Create a message bucket.
   b. Create a message folder.
   c. Keep all messages in the Outbox.

7. The mailbox portion of an E-mail address is typically
   a. A Server.
   b. A Client.
   c. A user name.
8. Outlook Express is a _____ E-mail program.
   a. Client.
   b. Server.
   c. Both a and b.
9. E-mail messages are delivered on the Internet using the
   a. POP3 protocol.
   b. SMTP protocol.
   c. Both a and b.
10. To create a folder, it is necessary to right-click on the
   a. Inbox folder.
   b. Outbox folder.
   c. Local Folders folder.

Completion

Fill in the blank or blanks with the best answers.

11. MIME stands for _____ _____ _____ _____.
12. The TELNET application can be used to connect to a _____ server.
13. The _____ header keyword is used to send a carbon copy of an E-mail message.
14. After reading an E-mail message, it may be saved or _____.
15. The E-mail _____ is ultimately responsible for delivering E-mail messages to their destination.

FAMILIARIZATION ACTIVITY

1. Create and send an E-mail message to an invalid mailbox. What type of error message is generated?
2. Create and send an E-mail message to an invalid computer. What type of error message is generated?
3. Identify the name of the server computer for your E-mail address.

QUESTIONS/ACTIVITIES

1. Why is it necessary to organize E-mail messages into folders?
2. Search the web to locate information on POP3 commands. Make a detailed list.
3. Search the web for free E-mail client programs. Are there any limitations or restrictions on using the software?

REVIEW QUIZ

Under the supervision of your instructor,

1. Describe the features of e-mail communication software.
2. Configure an electronic mail client.
3. Send and receive electronic mail.
4. Organize electronic mail messages into folders.

19 Databases

Joe Tekk was frustrated. His manager, Don, had put him in charge of the electronics lab after one of his co-workers had retired. Unfortunately, Joe was having difficulty trying to keep all the necessary electrical components and other various parts in stock. Every time Joe turned around, someone was asking him for some part that turned out to be out of stock. This usually meant that the lab project needed to be put on hold because it typically took between seven and 14 days to complete the order process and get the stocks replenished.

Joe knew that he needed to get a better handle on the situation. As he explored his options, he soon determined that a database could be used to keep track of the various components and also automate the reorder process at the same time. Joe began to read some of the documentation about the Microsoft Access program, which was included with the Microsoft Office Professional package that everyone at RWA Software used. After just a few minutes, Joe realized Microsoft Access included a Database Wizard, which could help him develop the database with a minimum of effort.

As Joe began the database development, he identified all the electronic components and other parts that he had to keep track of. He then proceeded to use a Database Wizard to develop the tables, create the forms necessary to enter all the data, and produce the reports that he planned to use to keep up with the lab inventory. Before he knew it, the job was done.

Upon completion of this exercise, you will be able to

1. Understand the terminology associated with databases.
2. Navigate around the different components of a database.
3. Create a database using the Database Wizard.
4. Use a Report Wizard to create a new custom database report.

The database is a revolutionary way to store and access data. A database may contain information applicable to any field of technology. A more appropriate and descriptive term for a database is a "relational database." The database designer determines what relationships exist between each element of data. Traditionally, a database is used in the field of business technology to keep track of customers, maintain the inventory in the company warehouse, maintain credit accounts for a company's customers, and keep track of the employees of a company. The applications for databases in technology include inventory control, task scheduling, and service call management. The applications for a database are limited only by the imagination of the database designer.

Before we are able to explore the wide variety of database applications, it is necessary to develop a basic knowledge of database terminology. The terminology of databases begins with a general definition of what a database actually is. A database is a collection of related information. The data contained in the database is organized into a hierarchy consisting of tables, records, and fields. Within the database, the data is grouped into categories called tables. Within each table are the records, which contain the individual data fields. The smallest element that can be referenced in a database is a field.

To determine what relationships exist in the data contained in the database, it is necessary to indicate what fields in the database are designated as *key fields*. A relational database contains both primary and secondary keys. The keys in a database allow for specific records of information to be selected. Consider a part number as a key to a database. By entering a part number in the key field, all data associated with the particular part may be displayed.

DATABASE DESIGN

Whenever a new database must be developed, it is necessary to follow a set of steps that are designed to specify all the necessary requirements. In general, the steps that can be followed when building a database are:

1. State the purpose of the database.
2. Design a model of the database on paper.
3. Build the electronic version of the paper model.
4. Test the database.
5. Document your work.

Before a database can be built, it is important to understand what is expected to be accomplished. It is necessary to identify each of the elements or fields that will be included in the database and what relationships, if any, exist between the database elements. It is also helpful to produce a model of what the database will look like.

For example, a database needed to keep track of the individual components in an electronics laboratory inventory begins with the identification of the components. A review of the lab inventory might reveal that the components fall into several standard categories: resistors, capacitors, integrated circuits, and one other category, miscellaneous items. In addition to the actual components, it is also necessary to keep track of the component supplier information, purchasing information, and shipping information. Once a plan is in place, the actual entry of the database elements can begin.

USING AN EXISTING DATABASE

To use a database, a user simply runs a database program. One of the most popular database applications for the personal computer is Microsoft Access, which is purchased separately or as a part of the Microsoft Office Professional suite of products. When the Access database is started, a window is displayed similar to Figure 19.1. The user can select a database that already exists (the default selection), or the user can create a new database. The default selection as shown in Figure 19.1 lists the names of the databases that have been accessed. An existing database that has never been opened before can be located using the More Files selection option.

To open the Inventory Control database, simply select it and press the OK button. This causes Access to open the database and display the Main Switchboard as shown in

FIGURE 19.1   Opening an existing database

FIGURE 19.2   Inventory Control Main Switchboard

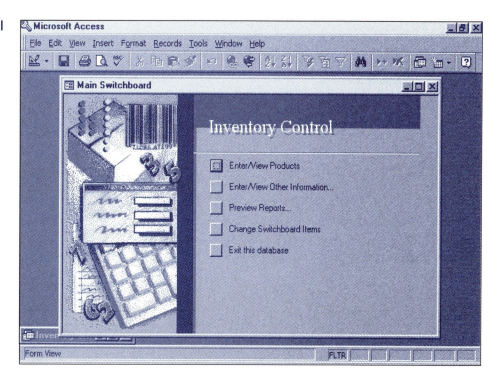

Figure 19.2. The Main Switchboard is called a *form.* A form contains either menu options or data entry screens. The Main Switchboard contains options to Enter/View Products, Enter/View Other Information, Preview Reports, Change Switchboard Items, and Exit the database. When selected, each of these buttons will cause another form to be displayed, either containing another menu or a data entry screen.

For the Inventory Control database, the items being tracked are called *products.* When a user selects the Enter/View Products option from the Main Switchboard, the Products window is displayed as illustrated in Figure 19.3. The first product, R100, a 100-ohm resistor, is automatically displayed. The ProductID is a unique number automatically assigned by Access. It is one of the keys to the database. The program user can enter data into the other fields on the form. Each product is given a name, description, category, lead time, and a reorder level. Purchase order information is also maintained for each product. Notice

FIGURE 19.3   The Enter/View
Products form

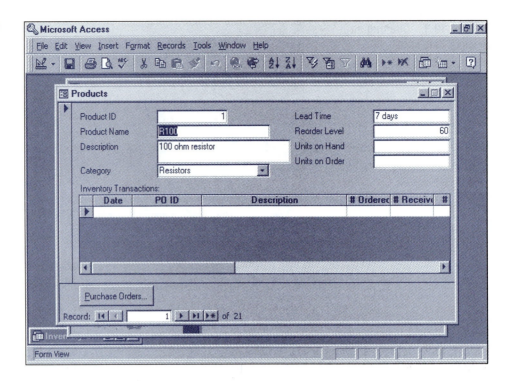

FIGURE 19.4   Adding a new
product

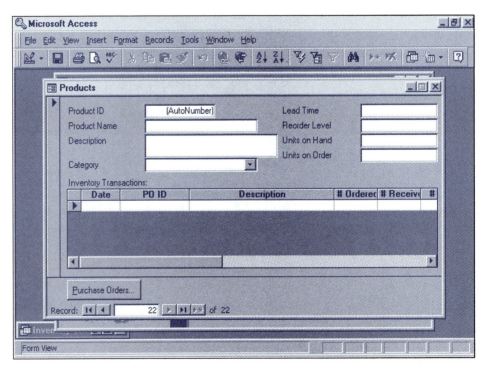

that R100 is record number 1 in the Products database. This is shown in the bottom left corner of Figure 19.3, along with a few control buttons. Using these buttons, the user can move back and forth between all of the products very easily. Two of the buttons (the arrowheads with bars next to them) provide access to the beginning of the list and the end of the list, respectively, with the click of one button. Two other buttons (the arrowheads without the bars) move one record at a time. All new records are entered at the end of the list.

To get to the end of the list and enter a new record, all that is necessary is to press the button that contains the right arrowhead with the asterisk next to it; this causes the screen shown in Figure 19.4 to be displayed. Notice that the Product ID field contains the text

FIGURE 19.5   Other information contained in the database

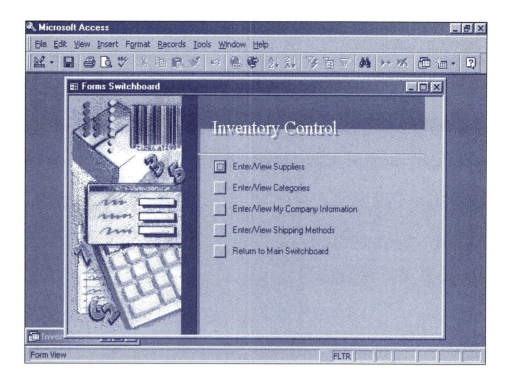

message (AutoNumber), which is how the sequential record numbers are assigned, and the record number is set to a value of 22, the next number that was available. The database user can then enter the necessary new product information into the database. After the products have been updated, or added, the window is closed and the Main Switchboard is redisplayed.

As previously stated, in an electronics laboratory inventory database, in addition to the components and miscellaneous parts, it is also necessary to keep track of the component suppliers, component categories, shipping information, and other company data. Selecting the Enter/View Other Information option, which is located on the Main Switchboard, provides access to enter or view each of these items. Figure 19.5 shows the Forms Switchboard menu. Data about suppliers, categories, company information, and shipping methods can be added to or modified as required using steps similar to those used when working with the Products form.

RELATIONSHIPS BETWEEN DATA ELEMENTS

With all the data elements contained in the Inventory Control database, it is possible for many different relationships to exist between them. For example, products are associated with a particular category of component, and purchase orders are associated with different suppliers and shipping methods. Figure 19.6 shows the relationship between the various elements in the database. Each of the boxes shown in the Relationships window is a data table. The data in a database is stored in tables. The lines between the data elements indicate what relationships exist.

In order to keep the database as simple as possible, relationships are created between two tables using data elements of the same name in both tables. For example, a supplier identification number stored with the purchase order is the same supplier identification number used in the supplier table. Relationships between the data elements are classified as one-to-many, many-to-many, and one-to-one. Notice that the links in Figure 19.6 indicate what type of relationship exists by specifying next to each table either a 1 or an infinity sign ∞.

The one-to-many relationship is the most common type of relationship. In a one-to-many relationship, a record in Table 1 can have many matching records in Table 2, but a

FIGURE 19.6 Different relationships between data elements

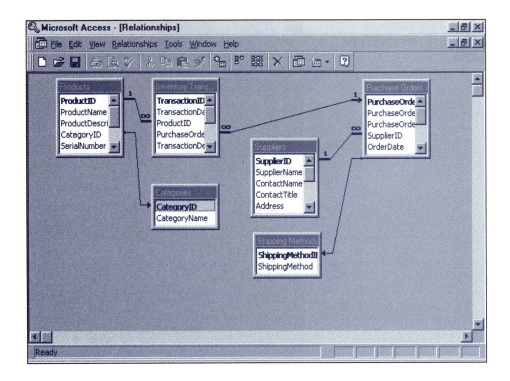

record in Table 2 has only one matching record in Table 1. In a many-to-many relationship, a record in Table 1 can have many matching records in Table 2, and a record in Table 2 can have many matching records in Table 1. In a one-to-one relationship, each record in Table 1 can have only one matching record in Table 2, and each record in Table 2 can have only one matching record in Table 1. This type of relationship is not common, because information related in this way would usually be defined using only one table. The kind of relationships that can be created depends on how the fields are defined. For example,

- A one-to-many relationship is created if only one of the related fields is a primary key or has a unique index.
- A many-to-many relationship is actually two one-to-many relationships that require use of a third table whose keys consist of the keys from the original two tables.
- A one-to-one relationship is created if both of the related fields are primary keys or have unique indexes.

When the definition for each field is correct, relationships are created by simply selecting an element in one table and dragging the link to the corresponding element in the other table.

DATABASE INTERNAL STRUCTURE

Each of the tables, which are used to store the database information, is maintained using the Inventory Control database. Notice that the Inventory Control database is the minimized window located in the bottom left corner of Figure 19.2 through Figure 19.5. When this window is restored, all the internal structure of the database can be examined or modified. The Inventory Control database window is shown in Figure 19.7. Each of the tabs displayed in the window are used to access the specific database elements such as tables, queries, forms, and reports.

The Tables tab lists all eight of the tables associated with the Inventory Control database. The program user can choose from three different options: open a table, design a table, or create a new table. Figure 19.8 shows the result of selecting the Open option. The Datasheet View lists each record of data that is stored in the table in a format

FIGURE 19.7   Inventory Control database window

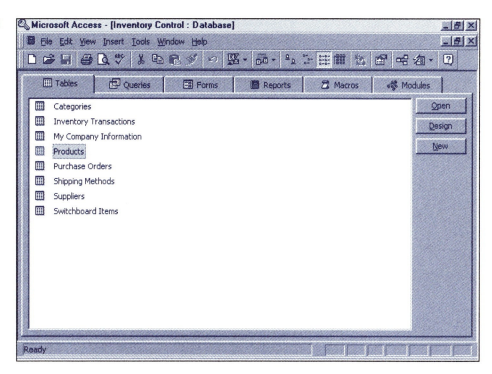

FIGURE 19.8   Datasheet View of the Products Table

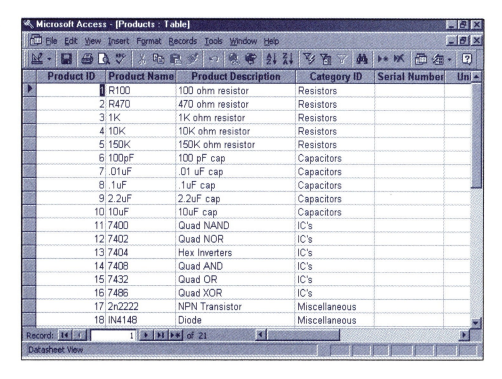

similar to a spreadsheet. Using the Datasheet View, data may be entered, modified, or deleted.

Choosing the Design option, the user is presented with the Products Table design window as shown in Figure 19.9. Notice that each of the field names is associated with a data type and a description. The common data types are listed in Table 19.1. In addition, each field is associated with different types of properties.

For example, the field properties for the ProductID specified by the AutoNumber data type include field size, new values, format, caption, and indexed indicator. You are encouraged to explore the field properties for each of the various data types available.

FIGURE 19.9   Products Table
element details

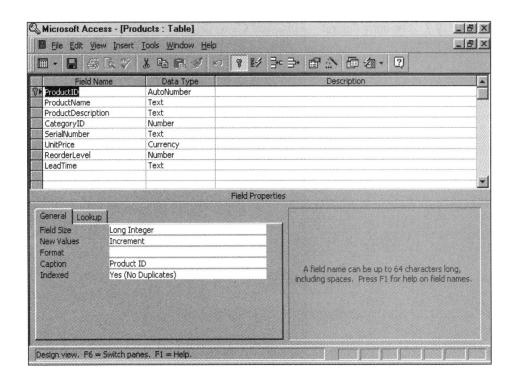

TABLE 19.1   Common
data types

Data Type	Field Contents
Text	Text or numbers
Memo	Text or numbers
Number	Numbers only
Date/Time	Date/Time values
Currency	Currency values
AutoNumber	System assigned values
Yes/No	Yes or No

DATABASE FORMS

Database forms are a convenient way to enter and review information in a database. Figure 19.10 shows all the different forms associated with the Inventory Control database. The Switchboard form contains the menus, and the others provide access to the data. The Forms window provides the same options as the Tables window. By selecting the Open option, the form is opened and the data or a menu is displayed similar to Figures 19.2 through 19.5. The Design option allows for the form to be modified. Figure 19.11 shows the Products Form being edited. Notice how each of the data fields on the form is identified by a descriptive label. This helps the user to begin working with the different forms in the database very quickly. The form details are a very important part of the development process.

CREATING A DATABASE

As you can see from the existing Inventory Control example database, much thought, planning, and development time are spent creating a database. Recall from Figure 19.1 that there are two options available when creating a new database. The first option is to

**FIGURE 19.10   Inventory
Control database forms**

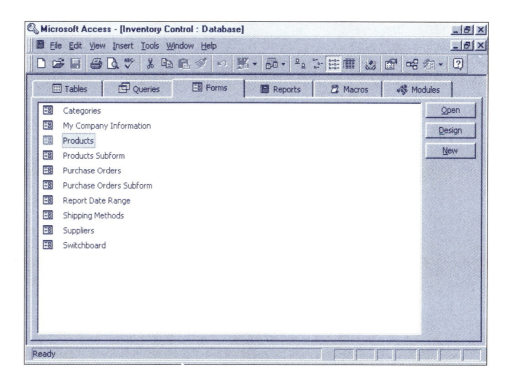

**FIGURE 19.11   Editing the
Products Form**

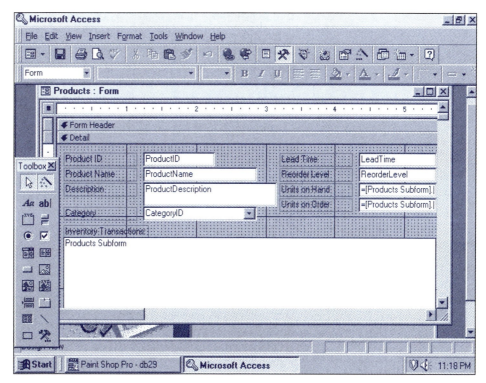

create a blank database. In a blank database, the development of each of the individual database components (tables, queries, forms, and reports) are handled by the database designer. The development of a blank database takes place in stages. It begins with the definition of all of the table elements and properties. Next, all the relationships between the various tables are defined. What remains is the creation of queries, forms, and reports that are required. These can be written in any order.

The second option available to the database designer is to use the Database Wizard to develop all the required components for standard types of database applications. Figure

FIGURE 19.12 Database Wizard custom application categories

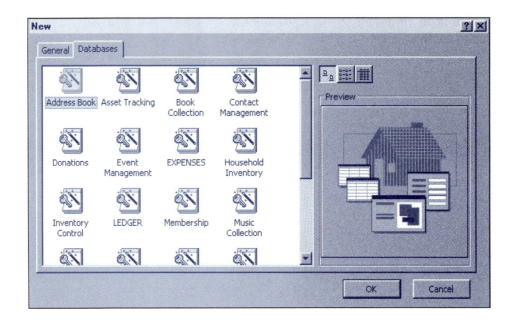

19.12 shows the Databases tab on the New database window. This list contains many of the typical personal uses for a database. If one of these databases does not fit your particular application, you might be forced to select the blank database and build it from scratch. Rather than resort to the blank database option immediately, review each of the database applications provided in the Database Wizard to see if one of them comes close. If so, it may be possible to modify the database produced by the wizard to meet many other applications not included on the list. For example, many of the items in the list refer to some type of collection, whether it is books, music, or a video collection. It is possible that one of these existing applications might be adapted to other types of collections as well.

When choosing one of the Database Wizard applications, Access will automatically create all the tables, forms, and reports that may be required. All that is necessary by the user is to begin entering the data. You are encouraged to review all the Database Wizard applications and sample databases. Note that Access can also provide a sample set of test data after the database is created. This provides an example of what the database looks like and what the reports look like.

DATABASE REPORTING AND QUERIES

Output from a database may take on many different forms. Traditionally, a database report was printed out on paper. The paper reports could then be distributed to anyone who needed a copy. In an effort to save resources, it is now more common to view a report right on the screen. This saves the cost of the paper but might also make it more convenient and up-to-date. Note that it is also possible to produce a report for distribution using the Internet.

Figure 19.13 shows the reports that are included in the Inventory Control database. As their titles suggest, they provide information about cost comparisons, product purchases by supplier, a product summary report, and a product transaction detail report. Other reports can be added as required using the one of the wizards. Figure 19.14 illustrates choosing the Chart Wizard to help create a new report that will be displayed as a graph. As you can see from the list, many different reporting capabilities are available.

Queries in a database provide access to a subset of the information contained in the database based on parameters selected by the user. When a query is executed, only the records that meet the query specifications are selected. For example, in the Inventory

FIGURE 19.13 Inventory
Control report selection screen

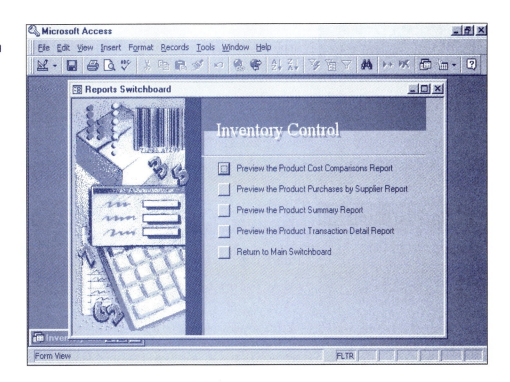

FIGURE 19.14 Select the Chart
Wizard to graph supplier data

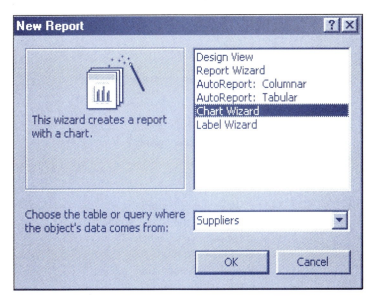

Control database, a query could be used to select only products in the database that need to be ordered. A query may be created manually or using one of the wizards.

OTHER DATABASE SOFTWARE APPLICATIONS

Many different database application programs are available for use on the PC. Reasons to choose one product over another are based on many different types of decisions. Some of the items that might be considered include cost, license fees, additional resource requirements such as hard disks and memory, transaction speed, and data security. Here is a short list of several popular products:

- Foxpro
- DBASE
- Oracle

The best type of troubleshooting advice for a database is to plan the database well. For example, it can sometimes be difficult to add new elements to a database after the database file already contains data. Occasionally, a process to migrate the data from one file format to another is required.

When each data element is created, review the field properties to ensure that each of the items contain appropriate values and that the key field settings contain the appropriate values. Remember that the relationships between the data elements and tables are determined by the key field values.

SELF-TEST

This self-test is designed to help you check your understanding of the background information presented in this exercise.

True/False

Answer *true* or *false*.

1. Databases organize information by tables.
2. The smallest element in a database is a key.
3. The keys for two different items can be identical.
4. Information in a database is organized into a hierarchy.
5. Database reports must be created using a wizard.

Multiple Choice

Select the best answer.

6. Relationships are created between
   a. Keys.
   b. Tables.
   c. Fields.
7. Which is not a relationship?
   a. One-to-many.
   b. Many-to-many.
   c. Many-to-none.
8. The AutoNumber feature
   a. Is applied to the whole database.
   b. Is applied to each table.
   c. Is applied to a field.
9. A blank database contains
   a. Empty tables connected in a standard way.
   b. No tables at all.
   c. Tables containing test data with no relationships.
10. The smallest element in a database that can be referenced is a
    a. Table.
    b. Record.
    c. Field.

Completion

Fill in the blank or blanks with the best answers.

11. The Main _____ is displayed when a database is opened.
12. Two different kinds of keys in a database are _____ and _____.
13. A database is also called a _____ database.
14. Each record of data stored in a table is displayed in the _____ view.
15. Information is entered into a database using a _____.

1. Design a database that will keep track of IP address assignments for a small network.
2. Create a database using the Database Wizard. Select one of the built-in database templates.

1. Ask a professional from a local business to visit your class and explain how their company uses a database.
2. Search the Web for database products and/or applications. What are the common features of each?

Under the supervision of your instructor,

1. Understand the terminology associated with databases.
2. Navigate around the different components of a database.
3. Create a database using the Database Wizard.
4. Use a Report Wizard to create a new custom database report.

UNIT III Technical Applications

20 Electronics Applications

Joe Tekk balanced a large electronics book on his lap while using his hands to operate the mouse and keyboard of his office computer. Joe was busy entering a schematic he found in the book, dragging and dropping components and connecting them with wires, all on the screen of his display.

Don, Joe's manager, came into Joe's office and asked what he was doing.

"I'm getting ready to simulate a circuit I found in this electronics book. It's for that class I'm teaching at night."

Don watched while Joe connected virtual instruments to make measurements. Then, after simulating the circuit several times, Joe saved the circuit to his hard drive, and put a second copy on a floppy disk.

"What is the second copy of the circuit file for?" Don asked.

Joe smiled. "I'm putting the circuit file on my web page so my students can download it."

PERFORMANCE OBJECTIVES

Upon completion of this exercise, you will be able to

1. Discuss the advantages of electronic simulation.
2. Explain how a circuit can be setup and simulated with Electronics Workbench (multiSIM).

BACKGROUND INFORMATION

The purpose of this exercise is to examine the features of an electronic circuit simulation application called Electronics Workbench (or simply multiSIM). With this application, you can setup a virtual electronic circuit, complete with a power source, resistors, capacitors, switches, indicator lights, or even transistors and integrated circuits. After completing the schematic of the circuit, you can also connect different virtual instruments as well, such as voltmeters, ammeters, or even an oscilloscope for viewing waveforms. Then, with the click of a button, actually simulate the electronic circuit, in real time. The results of the simulation are very accurate, practically eliminating the need to setup a real circuit in a laboratory, or wire it up on a circuit board. Plus, the entire circuit can be saved as a file and loaded and worked with again at later time. It is not so easy to leave a circuit unattended on a laboratory workbench. Furthermore, the price of a single copy of Electronics Workbench is much less than the cost of the instruments and supplies required for an actual laboratory workbench, and the virtual instruments will not go out of calibration like their real-world counterparts.

PART I: GETTING STARTED

Figure 20.1 shows a screen shot of the multiSIM simulation window. A simple series circuit is being simulated. The ammeter indicates 5.000 mA of current is flowing.

The title bar of the window contains the name of the circuit file being simulated (series1). The rocker switch near the upper right corner of the circuit window is used to start and stop the simulation. While the simulation is running, the simulation time is displayed in the status area at the bottom of the simulation window.

Down the left side of the simulation window are 14 buttons that provide access to the built-in parts bins containing all of the components that can be used to construct a circuit. Holding the mouse pointer still over a button will cause a small pop-up window to appear showing the name of the parts bin (such as Sources, Diodes, or Indicators).

Normally, multiSIM colors components of a circuit based on their properties. For example, active components such as transistors are colored green, whereas passive components (resistors, capacitors) are colored blue. Wires are usually colored black. You can change the assigned colors, or select a different color scheme for the background (black on white, white on black, etc.). This is accomplished by left-clicking Edit and then User Preferences to get to the color selection window. For the purposes of discussion in this exercise, all components are colored black.

PLACING COMPONENTS

Components are placed in the circuit window by selecting them from the desired parts pin and then left-clicking on the desired screen position. When the mouse pointer moves over a parts bin button, the parts bin will automatically pop up and allow you to choose a component. For example, to select and place a DC voltage source, do the following:

1. Move the mouse pointer over the Source button (the top button on the left side of the simulation window).

FIGURE 20.1   Simulating a simple circuit using multiSIM

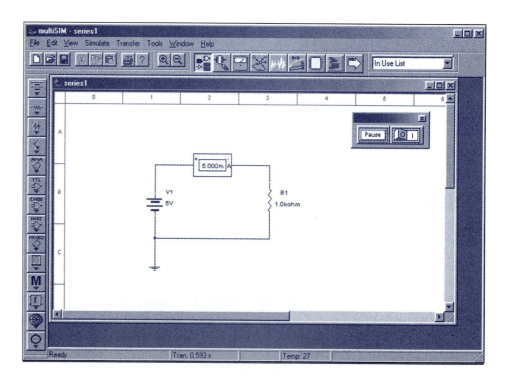

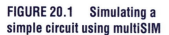

2. After the parts bin pops up, move the mouse pointer over the icon for the DC voltage source (the second icon in the first column) and left-click once.
3. The mouse pointer will change to indicate you are in the middle of a placement operation. Move the mouse pointer to the location where you would like the DC source and left-click again to place the source on the screen.

Figure 20.2 shows the DC source after it has been placed. Note that its voltage is automatically set to 12 V. It is often necessary to change the voltage value of the source. To do this, left double-click on the DC source. The Battery properties window will open up, as shown in Figure 20.3. Enter '5' in the Voltage box and click OK.

FIGURE 20.2   Placing a DC voltage source

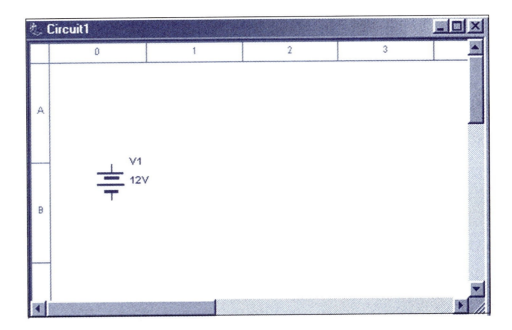

FIGURE 20.3   Battery properties

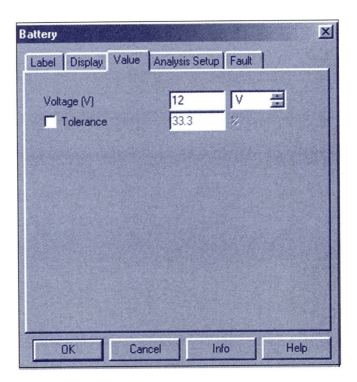

ADDING A RESISTOR

Choose a resistor from the Basic parts bin (the second button down on the left side of the simulation window). The resistor icon is the first icon in the first column.

The properties window for the resistor opens up automatically when the resistor is selected. This window is shown in Figure 20.4. It is necessary to change the value of the resistor before it is placed. This is done by selecting the desired ohm amount in the Component List. Set the resistor value to 1.0 K ohms and then place it in the circuit window. To change the value of a resistor after it has been placed, left double-click on it to bring up the properties window and then choose Replace.

FIGURE 20.4 Setting the resistor properties

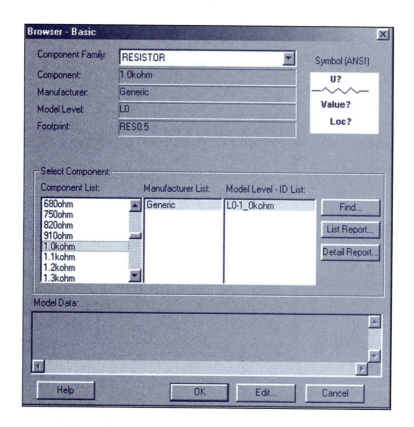

FIGURE 20.5 Placing a resistor

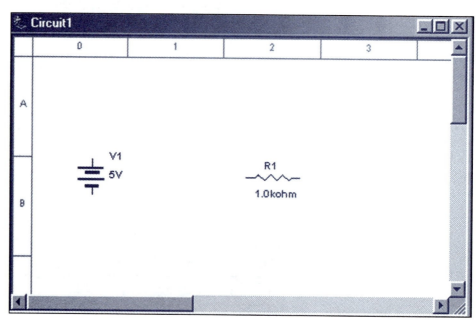

The resistor is oriented horizontally after placement, as indicated in Figure 20.5. To change its orientation to vertical, select the resistor with a single mouse click and then press Control-R on the keyboard to rotate it. Note that the component identifier (R1) and its value (1.0kohm) will also move around when the resistor is rotated. If necessary, grab the identifier or value (left click and hold) and drag it to the desired location and release the left mouse button.

ADDING A VOLTMETER

Voltmeters are used to measure voltage. They must be connected in parallel to work properly. The voltmeter is found in the Indicators parts bin (the button with the red figure 8 on it, the fifth button from the bottom). It is the first icon in the first column (it has a V in it). When the voltmeter is selected, it properties window will automatically open up. Choose the VOLTMETER_V setting to orient the voltmeter vertically. Figure 20.6 shows the resulting circuit window after the voltmeter is placed.

Note that a component may not be in the exact position where you want it after placing it. To move the component, grab it and drag it to the new location.

ADDING AN AMMETER

Ammeters are used to measure current and must be connected in series. Select an ammeter from the Indicators parts bin (it has an A in it) and place it on the screen as indicated in Figure 20.7.

Now that we have placed all the components for our circuit we must connect them with wires.

ADDING WIRES

To add a wire to the circuit, move the mouse pointer over the component terminal and left click once. This sets the starting point of the wire. Then move the mouse pointer to the location for the ending point of the wire. As you do this, a dashed line representing the wire will appear and follow the mouse pointer around. Left click again to make the second wire connection. This process is illustrated in Figure 20.8.

If you make a mistake adding a wire, or simply want to delete a wire that has already been added, move the mouse pointer over the wire, right click on it and choose Delete from

FIGURE 20.6 Adding a voltmeter

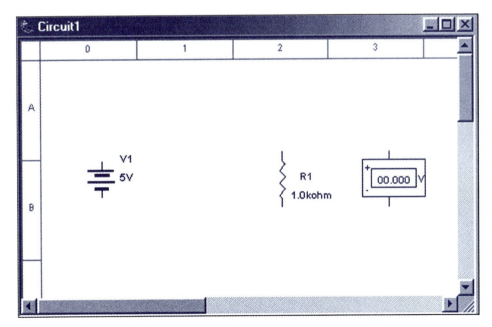

FIGURE 20.7   Adding an ammeter

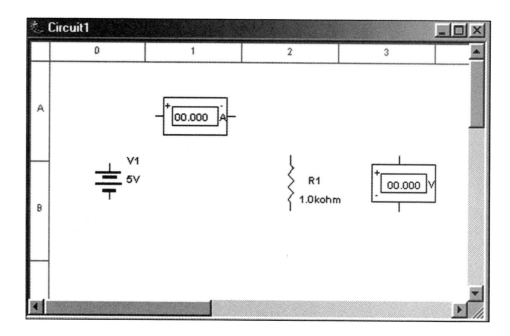

FIGURE 20.8   Adding a wire

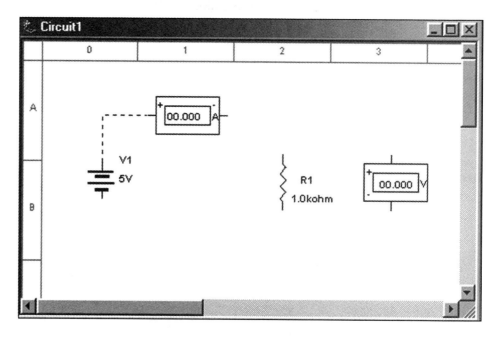

the pop-up menu. If you are in the middle of adding a wire, right clicking will erase it. Also, there will be times when it is difficult to connect a wire to a component that already has a wire connected to it. In this case, drag the connected wire to make it longer to give more room for an additional connection. Left click once on the wire to select it, then drag portions of the wire to adjust it.

Add the remaining wires to the circuit so that it looks like Figure 20.9.

ADDING A GROUND TERMINAL

Most circuits require a *ground* connection to properly simulate. The ground symbol is found in the Sources parts bin (the first icon in the first column) and is usually connected to the negative side of the voltage source. Figure 20.10 shows the final circuit with the ground terminal included.

FIGURE 20.9   All wires have
been added to the circuit

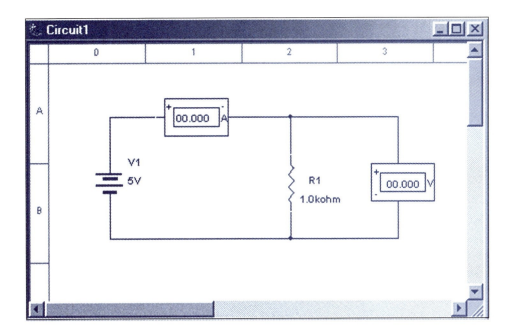

FIGURE 20.10   Adding the
ground terminal

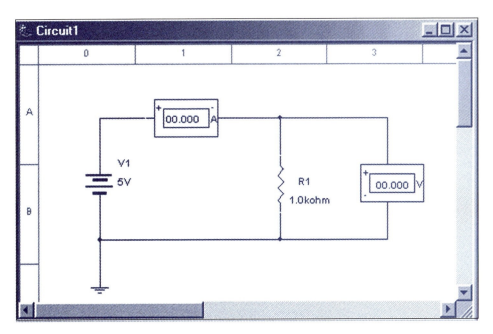

SIMULATING THE CIRCUIT

The circuit can be simulated by left clicking the power switch. The results are indicated by the numbers displayed in the voltmeter and ammeter, as indicated in Figure 20.11. The circuit can be saved by choosing Save from the File menu, or by clicking on the icon of a floppy disk near the upper left corner of the simulation window. In a similar fashion, you can also load a circuit that has been previously saved.

THE TWO-RESISTOR SERIES CIRCUIT

Refer to
*Thought
Project #14*

Figure 20.12 shows a two-resistor series circuit. You will setup and simulate this circuit, as well as the remaining ones in this exercise, when you perform the Familiarization Activity.

One characteristic of a series circuit is that the voltages measured across each resistor add up to the supply voltage. The voltmeter values in Figure 20.12 (4.002 V and 5.998 V)

FIGURE 20.11   The results of the simulation

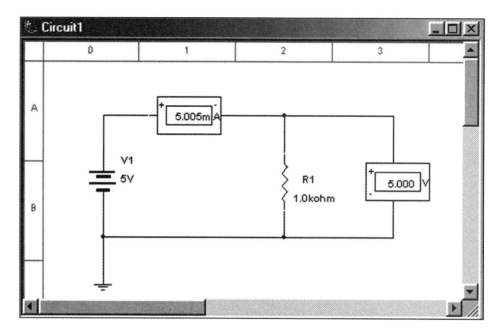

FIGURE 20.12   Two-resistor series circuit

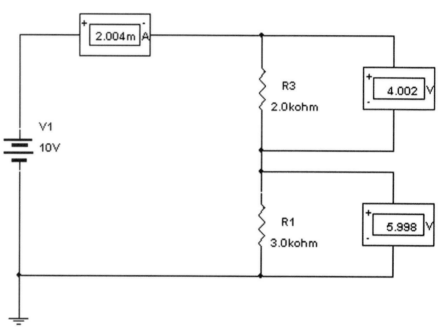

indicate that this is true. Another characteristic of a series circuit is that each resistor has the same current flowing through it (there is only one current in a series circuit). If you multiply the circuit current (2.004 mA, which equals 0.002004 A) by the individual resistor values (2.0 K ohms and 3.0 K ohms, or 2000 ohms and 3000 ohms) you will get the indicated voltage values.

THE TWO-RESISTOR PARALLEL CIRCUIT

Refer to
*Thought
Project #14*

The same two resistors from the series circuit in Figure 20.12 can be connected differently to form a two-resistor parallel circuit. This type of circuit is shown in Figure 20.13.

Note that three ammeters are used to measure the circuit currents, and that each ammeter has a different current measurement displayed. Furthermore, the sums of the currents through R1 and R2 add up to the current in the ammeter at the top of the circuit. This is an

FIGURE 20.13   Two-resistor
parallel circuit

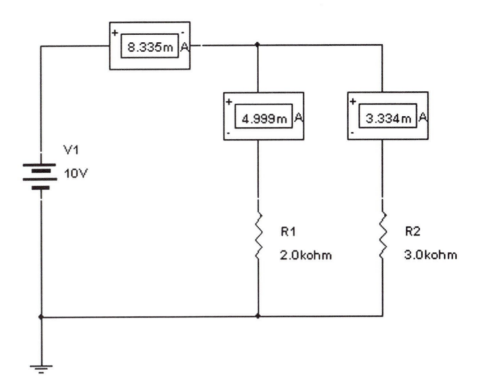

important characteristic of parallel circuits. The sum of the individual resistor currents (also called *branch* currents) add up to the current supplied by the voltage source.

Another characteristic of a parallel circuit is that there is only one voltage present (10 V in this case). If you divide 10 V by the value of R1 (2000 ohms), you get 0.005 A (5 mA). Note that the ammeter for R1's branch measures 4.999 mA. Check the current in R2 using this same method. You should get 3.333 mA.

PART II: ADDING A NEW TWIST: TIME

The DC voltage source, together with a resistive circuit, produces a steady current that does not change. There are, however, other electrical and electronic components that have characteristics that produce a time varying change in voltage or current. One of these components is the capacitor.

THE CAPACITOR

Refer to
*Thought
Projects
#26 and #27*

The capacitor is essentially used to store electrical energy. When a capacitor is connected in series with a resistor and a DC voltage source, as indicated in Figure 20.14, the capacitor 'charges' from 0 V to the DC supply voltage (10 V in this case) over a certain period of time. This time is based on the time constant of the R-C circuit. The time constant equals the product of R and C. For the circuit in Figure 20.14, the time constant is 4.7 K ohms times 5.1 micro-Farads, or just under 24 milli-seconds (0.024 seconds). The capacitor will become fully charged after five time constants, which equals 120 milli-seconds. Because the capacitor voltage is constantly changing (causing a change in current as well), it is not appropriate to measure the capacitor voltage with a voltmeter. Instead, an instrument called an oscilloscope is used.

THE OSCILLOSCOPE

The oscilloscope is an instrument that is used to display a voltage waveform. Refer back to Figure 20.1. There is a group of nine large buttons across the top of the simulation window.

FIGURE 20.14   Resistor-capacitor charging circuit

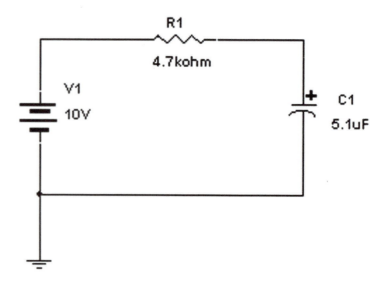

FIGURE 20.15
Instruments available in
multiSIM

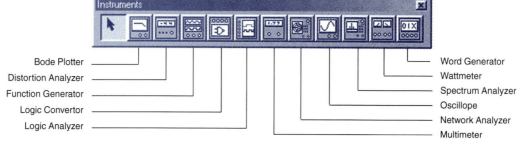

Bode Plotter
Distortion Analyzer
Function Generator
Logic Convertor
Logic Analyzer

Word Generator
Wattmeter
Spectrum Analyzer
Oscillope
Network Analyzer
Multimeter

Left clicking on the third button from the left opens up the Instruments window shown in Figure 20.15. Any technician would be jealous of the instruments provided by multiSIM, since many of them are typically very expensive.

The oscilloscope is very useful for displaying time-varying waveforms. The oscilloscope provided by multiSIM has two channels, each capable of displaying its own waveform. Figure 20.16 shows how the oscilloscope is connected to the R-C circuit. Notice that the DC voltage source has been replaced by an AC square wave source set to 10 V and a frequency of 5 Hz. This means that the source will output 10 V for 100 milli-seconds, and 0 V for 100 milli-seconds. This will allow the capacitor to charge for 100 milli-seconds and discharge for 100 milli-seconds, over and over again. This effect will be captured with the oscilloscope.

To view the oscilloscope controls, left double-click the oscilloscope and it will open up into the instrument panel shown in Figure 20.17. The Timebase Scale (currently 10 ms/Div) controls how much time passes between horizontal divisions on the display screen. The Channel A and Channel B Scale values (both set to 5 V/Div) control the vertical size of the waveform as it is displayed.

Simulate the circuit for a few seconds and watch what happens on the oscilloscope display. Then stop the display and use the horizontal scroll bar in the oscilloscope display window to view the waveform over time. Figure 20.18 shows a sample display. Experiment with the vertical settings (V/Div and Y position) to get the desired display.

FIGURE 20.16   Using the oscilloscope

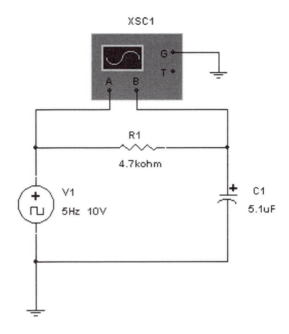

FIGURE 20.17   The oscilloscope controls

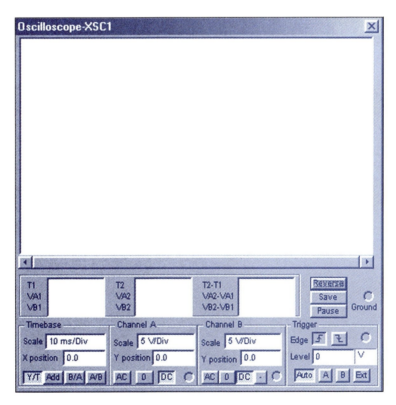

CAPACITIVE REACTANCE

When an AC signal is applied to a capacitor, it exhibits opposition to current similar to resistance, but the amount of opposition depends on the frequency of the AC signal. This opposition is called capacitive reactance. To determine the reactance of a capacitor, divide the voltage across the capacitor by the current through it. Figure 20.19 shows a 1 micro-Farad capacitor being operated at 5000 Hz. Dividing the capacitor voltage (0.225 V) by the circuit current (7.071 mA) gives 31.8 ohms of capacitive reactance. Note that the voltmeter and ammeter must be changed to read AC values. This is accomplished by left double-clicking on the meter, and then choosing the AC mode from the Value tab.

Refer to
*Thought
Project #40*

267

FIGURE 20.18   R-C charge/
discharge waveforms

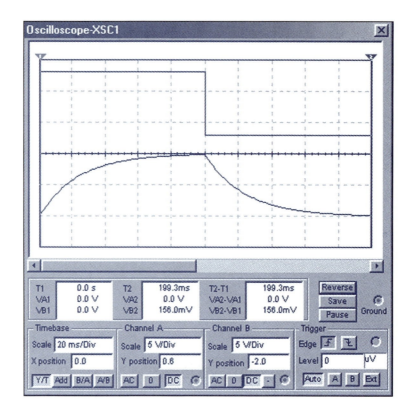

FIGURE 20.19   R-C circuit with
AC voltage source

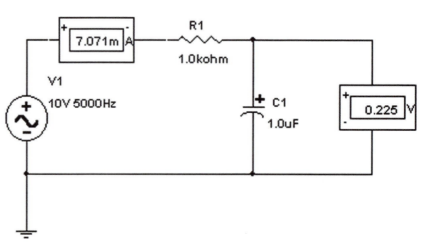

PART III: SPECIAL COMPONENTS

In this last part we examine the operation of four additional components. Three of these components are called semiconductors, due to the special way they must be operated. These are the diode, transistor, and light emitting diode. Before looking at these components, let us take a look at one more AC component, the transformer.

THE TRANSFORMER

The transformer is an AC device that relies on magnetic fields to couple a signal from the input side (the primary) to the output side (the secondary). The number of turns of wire in the primary and secondary determine if the output voltage is higher (a step-up transformer) or lower (a step-down transformer) than the input voltage.

Refer to
*Thought
Project #38*

Figure 20.20 shows a simple step-down transformer circuit. The input voltage of 10 V (7.07 VAC) is stepped down to 0.616 VAC. The number of turns in the primary of the transformer is 141. The secondary contains 1622 turns (this information is available in the transformer menu that pops up when a transformer is selected). Dividing 1622 by 141 gives 11.5. Thus, the *turns ratio* of the transformer is 11.5 to 1. So, dividing the input voltage of 7.07 VAC by 11.5 gives 0.615 VAC, practically the same result found during simulation.

Transformers are used in power supplies to help convert the high voltage 120 VAC from the wall outlet to a lower AC voltage that can be rectified (by a diode) and regulated.

THE DIODE

<table>
<tr><td>Refer to
Thought
Project #30</td><td>The diode is the most basic type of semiconductor, containing a single P-N junction. An important property of the diode is that it only allows current to pass through it in one direction. This happens when the diode is forward biased. This means that the anode of the diode is more positive than the cathode. Figure 20.21 shows two diode circuits. In Figure 20.21(a) the diode is forward biased and is conducting current. The anode (the triangular part) is 0.7 V more positive than the cathode.</td></tr>
</table>

In Figure 20.21(b) the diode is *reversed biased*. Notice that no current is flowing now (the 1.77 micro-ampere reading is considered to be zero current) and that the resistor

FIGURE 20.20   Step-down transformer

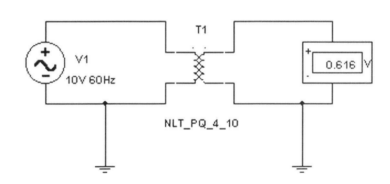

FIGURE 20.21(a)   Forward-biased diode

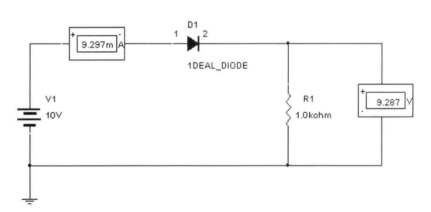

FIGURE 20.21(b)   Reverse-biased diode

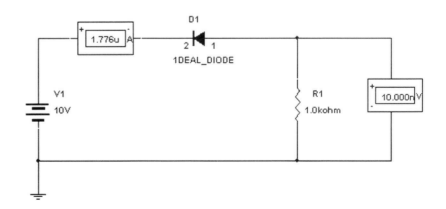

FIGURE 20.22   Transistor
biasing circuit

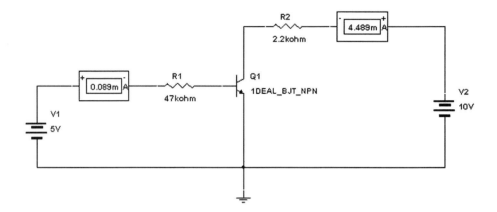

FIGURE 20.23   Light emitting
diode circuit

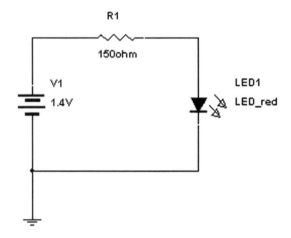

voltage is practically zero (10 nano-volts). This property of the diode makes it useful in circuits called rectifiers, where an AC voltage is converted into a pulsating DC voltage through the action of a diode.

THE TRANSISTOR

Refer to
*Thought
Project #30*

The transistor is essentially a current amplifier. A small amount of input current is able to control a large amount of output current. The simple transistor biasing circuit shown in Figure 20.22 indicates that an input current of 89 micro-amperes produces an output current of 4.489 mA. This is a current gain of 50.4. An individual transistor may have a current gain anywhere from 20 to 300.

THE LIGHT-EMITTING DIODE

Refer to
*Thought
Project #30*

The light emitting diode (LED) is a special diode that gives off light when it is turned on. Figure 20.23 shows a simple LED circuit. During simulation, the two arrows coming out of the LED glow if the LED is on. LEDs are used in all sorts of electronic equipment, from toys to alphanumeric displays.

TROUBLESHOOTING TECHNIQUES

The educational version of Electronics Workbench multiSIM allows *faults* to be introduced into a circuit. For example, a 10 K ohm resistor may have a fault added to it that makes it look open, shorted, or leaky (in parallel with a leakage resistance). The fault can be hidden so that the user analyzing the circuit does not know there is a problem. This forces the user to apply proper troubleshooting skills to determine what is wrong. Even a fault as simple as an open or shorted resistor can provide a useful troubleshooting challenge.

SELF-TEST

This self-test is designed to help you check your understanding of the background information presented in this exercise.

True/False

Answer *true* or *false*.

1. The multiSIM program is not very accurate.
2. Only a handful of electrical components are simulated in multiSIM.
3. In a two resistor series circuit, the resistor voltages always add up to the supply voltage.
4. In a two-resistor parallel circuit, the branch currents are always the same.
5. A diode only conducts current in one direction.

Multiple Choice

Select the best answer.

6. To rotate a component, select it and
    a. press 'R'.
    b. press Alt-R.
    c. press Control-R.
7. Voltmeters must be connected in
    a. series.
    b. parallel.
    c. both ways are correct.
8. Ammeters must be connected in
    a. series.
    b. parallel.
    c. both ways are correct.
9. The ground terminal is found in the
    a. Sources parts bin.
    b. Basic parts bin.
    c. Indicators parts bin.
10. Which instrument is used to view an AC waveform that varies with time?
    a. multimeter.
    b. logic analyzer.
    c. oscilloscope.

Completion

Fill in the blank or blanks with the best answers.

11. A component that stores energy is the _____.
12. The AC 'resistance' of a capacitor is called capacitive _____.
13. Diodes and transistors are examples of _____.
14. VOLTMETER_V indicates that the voltmeter will be oriented _____.
15. Circuits in multiSIM typically require a _____ terminal to properly simulate.

FAMILIARIZATION ACTIVITY

Part I

1. Setup the series circuit in Figure 20.11 and simulate it. What is the current?
2. Change the resistor value to 10 K ohms. What is the new current?
3. Change the source voltage to 50 volts. What is the new current?
4. Setup the series circuit in Figure 20.12 and simulate it. Do the two resistor voltages add up to the source voltage?

5. Change the first resistor to 4.7 K ohms. Do the two resistor voltages add up to the source voltage?
6. Change the source voltage to 25 volts. Do the two resistor voltages add up to the source voltage?
7. Setup the parallel circuit in Figure 20.13 and simulate it. Does the source current equal the sum of the branch currents?
8. Change the first resistor to 4.7 K ohms. Does the source current equal the sum of the branch currents?
9. Change the source voltage to 25 volts. Does the source current equal the sum of the branch currents?

Part II

1. Setup the RC charging circuit in Figure 20.16 and simulate it. How long does it take the capacitor voltage to reach 10 volts? Use the oscilloscope to view the voltage source and capacitor waveforms.
2. Change the capacitor value to 10 uF. How long does it take the capacitor voltage to reach 10 volts?
3. Change the resistor value to 2.2 K ohms. How long does it take the capacitor voltage to reach 10 volts?
4. Setup the circuit in Figure 20.19 and simulate it. Divide the capacitor voltage by the measured current. This is the capacitive reactance. Divide the voltage source value by the measured current. This is the impedance of the circuit.
5. Change the frequency to 1 KHz. What is the new capacitive reactance? What is the new impedance?

Part III

1. Setup the transformer circuit in Figure 20.20. Verify the simulation results shown in the figure.
2. Setup the diode circuit of Figure 20.21(a). Place a voltmeter across the diode to measure the voltage across it. What is the diode voltage?
3. Change the power supply voltage to 25 V. What is the diode voltage now?
4. Repeat steps 2 and 3 for the reverse-biased diode.
5. Setup the transistor circuit in Figure 20.22. Verify that it works properly.
6. Change R1 to each of the following values, recording the currents each time:

   - 33 K
   - 22 K
   - 10 K
   - 2.2 K
   - 1 K

   Does the output current always equal the input current multiplied by the transistors gain?

7. What is the minimum voltage that will cause the LED to turn on in the circuit of Figure 20.23?

QUESTIONS/ACTIVITIES

1. Search the web for other electronic simulation applications.
2. What disadvantages are there to electronic simulation?

REVIEW QUIZ

Under the supervision of your instructor

1. Discuss the advantages of electronic simulation.
2. Explain how a circuit can be setup and simulated with Electronics Workbench (multiSIM).

21 CAD Applications

The furniture in Joe Tekk's new office glowed brightly. His desktop computer moved from his desk to the floor, then back to the desk. Networking wires hovered, changing their connections occasionally. His printer, telephone, and FAX machine changed places several times, eventually ending up on a new work desk that suddenly appeared. The walls pulled away from each other, adding several square feet of floor space.

Don, Joe's manager, watched these activities for a while and then asked "What are you doing Joe?"

Joe looked up from his computer screen, where the image of his new office still glowed brightly. "Oh, hi Don. I'm designing my new office with that CAD application you gave me."

PERFORMANCE OBJECTIVES

Upon completion of this exercise, you will be able to
1. Describe the typical operations performed when using CAD techniques.
2. Use Actrix Technical to create a drawing.

BACKGROUND INFORMATION

The purpose of this exercise is to examine the features of a technical drawing application called Actrix Technical. The drawings you are able to create using Actrix Technical are constructed using CAD (Computer Aided Design) techniques. Actrix Technical provides you with *templates*, built-in electronic stencils that contain all of the drawing shapes you might use to draw a computer network, the layout of an office, a flowchart, or an air-conditioning system, to name just a few. You may also draw simple shapes, such as lines, arcs, circles, and rectangles, to create your own custom drawing.

PART I: GETTING STARTED

Figure 21.1 shows a screen shot of the Actrix Technical window. A simple drawing of a system of pipes, including pressure gauges, is shown. This drawing was created using shapes from a template called Plumbing and Piping. Every shape found in the template can be accessed through the Content Explorer (the window on the right side of the screen). Shapes chosen from the Content Explorer are dragged-and-dropped into the drawing

FIGURE 21.1 Drawing pipes using Actrix Technical

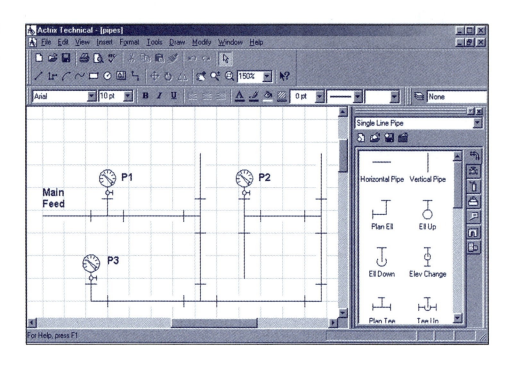

window. Once there, they can be resized, rotated, grouped with other shapes, or simply moved to a new location in the drawing. Notice the seven tabs along the right side of the Content Explorer. The first one has a small icon of a pipe on it. Each tab represents a *catalog* in the associated template. Catalogs can be selected by left-clicking on their tab, or by choosing their name from the pull-down list box at the top of the Content Explorer.

If the Content Explorer window ever gets closed by accident, it can be reopened by choosing Content Explorer from the View pull-down menu.

STARTING A NEW DRAWING

To start a new drawing, choose New from the File pull-down menu, or left-click on the New button in the standard toolbar. This will open up the New Drawing window shown in Figure 21.2. There are eight categories to choose from when beginning a new drawing. Spend some time exploring what types of shapes are found in each.

Left double-click on the Blank Template selection to start a new drawing. You can also select the template by left-clicking the template and then the OK button. Figure 21.3 shows the initial Blank Template drawing. Note the lack of shapes in the Content Explorer. How can anything be drawn without shapes to drag-and-drop? If you look at the buttons right above the "Arial 10pt" font box, you will see eight basic drawing buttons. In order, they have the following functions:

1. Line
2. Polyline
3. Arc
4. Spline
5. Rectangle
6. Circle
7. Text
8. Connector

Let us examine how to work with some of these basic shapes.

FIGURE 21.2   New Drawing window

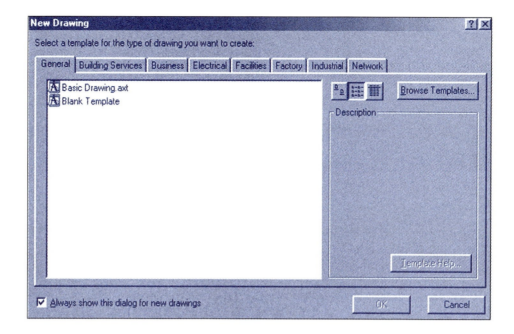

FIGURE 21.3   New Blank Template drawing

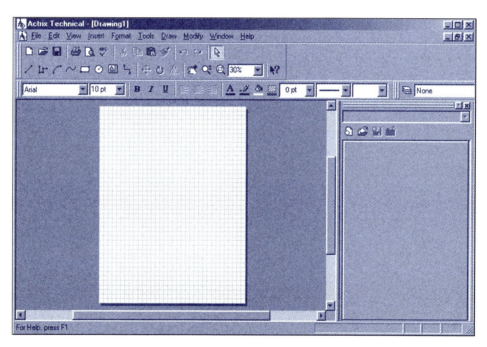

USING THE BASIC SHAPES

To select a basic shape, left-click once on the desired button. For example, left-click the Line button. To draw a line in the drawing window, move the mouse to the desired starting point of the line, then left-click and hold to anchor the starting point. Move the mouse to where the end of the line should be and release the left mouse button. The line will be visible the entire time you are moving the mouse to the end point.

The properties of the line can be adjusted after you have drawn it by right-clicking on the line and choosing Properties from the popup menu. Figure 21.4 shows the properties window for the line. Left-clicking in any of the entries in the right column allows you to adjust the associated property. For example, left-clicking in the Right column for Pattern allows the type of pattern to be selected. The pattern is used to draw the line, and may be solid, dashed, dotted, or even a combination.

Practice with the basic shapes. See if you can determine how to draw all the shapes shown in Figure 21.5. The dotted line at the top of the figure has had its Pattern and Weight properties increased. Remember that left-clicking in the right column of any property allows you to adjust it.

The dashed line with the arrow has had its Pattern, Weight, and End Arrow properties adjusted.

The circle was placed onto the drawing surface *before* the rectangle. Since a portion of the circle is covered by the rectangle, we see that there is an ordering to the way shapes have been placed into the drawing. The small squares around the border of the rectangle indicate that it is the currently selected shape. A shape is selected by left-clicking on it once. To deselect a shape, left-click on a blank portion of the drawing window (or on a different shape), or press the Escape key. The shape can be made larger or smaller by moving the mouse close to one of the small squares (midpoints, corners), grabbing it, and dragging the mouse to resize the shape. The mouse pointer will change into a different set of arrows during this operation.

FIGURE 21.4   Property window for a line

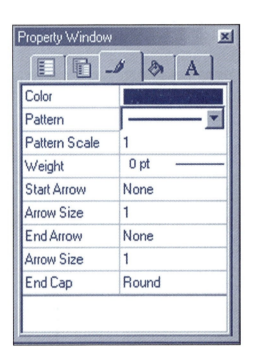

FIGURE 21.5   Simple shapes

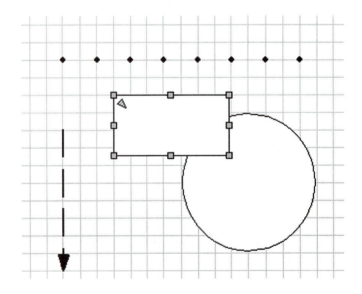

276

Drawing the circle requires three steps:

1. Left-click the Circle button.
2. Move the mouse to the desired center location of the circle. Press and hold the left mouse button.
3. Drag the mouse to increase/decrease the radius of the circle.

As with the line, the circle properties may be adjusted at a later time to change its appearance. Drawing the rectangle is performed in a similar manner, with the mouse specifying the upper-left and lower-right corners. It is even possible to make the rectangle transparent, allowing the circle to show through. In the Fill Properties window, select No Fill to make the rectangle (or the associated shape) transparent.

To move any of the shapes in a drawing, first select the shape with a left click. The mouse pointer should change to a small crosshair. Left click and hold to drag the shape to a different location.

ADDING TEXT

It is easy to add text to a drawing. This is a nice feature, providing us with the ability to label portions of our drawings with meaningful information.

To add text to a drawing, left-click on the Text button. Then, move the mouse to the position where the text will be placed and left-click again. A small, outlined rectangle will appear. You can begin typing in your text right away. If necessary, change the font and size and set any other text properties, such as bold or italic, before you left-click inside the drawing. You can always change the text properties after the text has been entered by selecting it and then making the required changes.

Figure 21.6 shows the text '143' added to the drawing of basic shapes. Note that the default font and size of Arial, 10pt has been changed to Century Schoolbook, 36pt and has bold emphasis.

ROTATING A SHAPE

There will be many times when it is necessary to rotate a shape found in a catalog. Here are the steps to take:

1. Right-click on the shape to open the popup menu.
2. Select Rotate from the menu. The mouse pointer will change into a small, clockwise arrow.

FIGURE 21.6   Adding text to the drawing

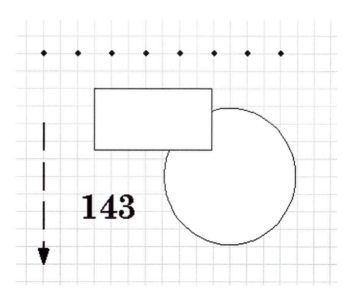

3. Left-click-and-hold in the drawing area near the shape.
4. Move the mouse to increase/decrease the angle of rotation. Hold the Shift key down to step the angle of rotation in fixed amounts of 15 degrees.
5. Release the left mouse button when the shape has been rotated the desired amount.

Figure 21.7 shows a simple shape being rotated counterclockwise. Using the shift key is a great help when rotating shapes 90 degrees.

DELETING A SHAPE

To delete a shape, select it and then press the Delete key, or select Cut from the Edit pull-down menu.

CHANGING THE SIZE OF THE VIEW

There are some templates that contain shapes that are too small to see the details of when they are placed into the drawing. To zoom in and see the details, select Zoom from the View pull-down menu. Figure 21.8 shows the Zoom window. You can select one of the prede-

FIGURE 21.7 Rotating an object

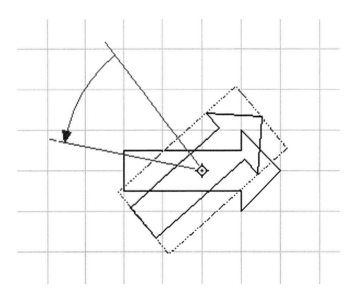

FIGURE 21.8 Zoom window

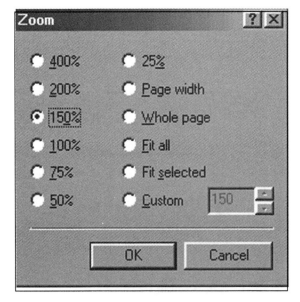

fined zoom sizes, or enter you own custom size. Choosing Zoom In/Out allows you to use the mouse to control zooming. Left-click and hold, then drag the mouse to change the zoom size.

WORKING WITH SNAPS

When one shape is connected to another shape, some assistance is provided in making the connection through the use of an automatic *snap* feature. For example, when the end of a piece of pipe gets close to a Tee connector, or to another piece of pipe, the two shapes will snap together. This is an aid to an individual who may not yet have the necessary mouse skills to maneuver shapes.

The snap properties can be changed (larger or smaller snap distance) or even turned off by choosing AutoSnap from the Tools pull-down menu. Figure 21.9 shows the AutoSnap properties window.

WORKING WITH DIMENSIONS

It is easy to add dimension arrows to a shape and have Actrix automatically indicate how wide, or long the shape is. Figure 21.10 shows a Desk from the Facilities Management– Small Layout catalog in the Business template. After the desk is placed, the Horizontal and

FIGURE 21.9   AutoSnap properties window

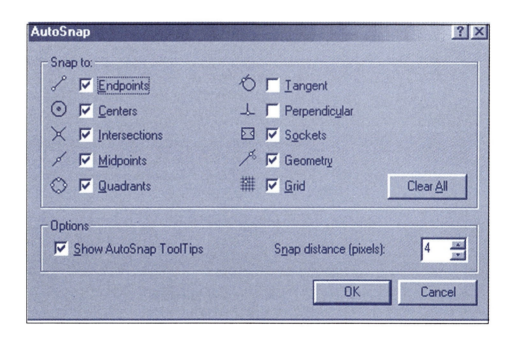

FIGURE 21.10   Annotating a shape with dimensions

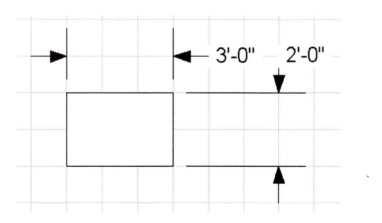

FIGURE 21.11   Checking the scale of the drawing

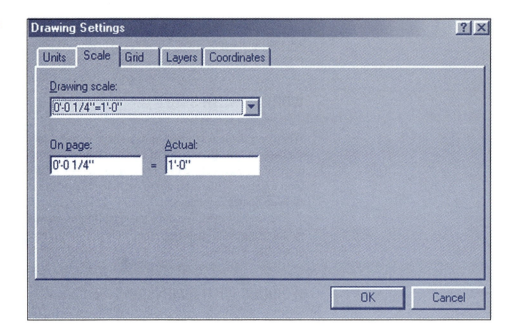

Vertical Dimension lines can be added by selecting the Annotations entry in the Content Explorer. For example, to add the Horizontal Dimension shape, drag it into the drawing window. Its size may be larger or smaller than the shape being dimensioned. There should be yellow endpoints on the dimension lines. Grab the left endpoint and drag it to the upper left corner of the desk. It should snap into position. Then grab the endpoint on the right dimension line and drag it to the upper right corner of the desk. You will notice that the size displayed by the Horizontal Dimension lines changes to reflect the actual distance measured. This distance is based on the number of grid squares in the dimension as well as the *scale* of the drawing. To check or set the scale, select Scale from the Format pull-down menu. Figure 21.11 shows the Scale settings for the drawing in Figure 21.10. Do you see why the horizontal dimension is 3 feet? The last step is to grab the yellow dot between the arrow heads and drag it to adjust the overall size and position of the lines and arrows.

The Vertical Dimension lines are added in a similar manner.

SPECIAL OPERATIONS

Actrix Technical provides many more features than those presented here. A short list of them is as follows:

1. Filling shapes with colors.
2. Adding layers to a drawing.
3. Importing/exporting AutoCAD files.
4. Creating new shapes and catalogs.
5. Checking the spelling of a text block.

It is worth the time invested learning about these features, and the additional capabilities of Actrix.

PART II: ARCHITECTURAL DRAWINGS

In this part we examine two drawings that are concerned with architecture. The first drawing, shown in Figure 21.12, shows the same system of pipes we saw in Figure 21.1. The individual shapes used in the drawing are:

- Horizontal and vertical pipes
- Three pressure gauges

FIGURE 21.12   System of
pipes

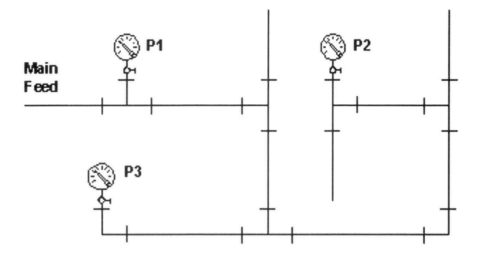

FIGURE 21.13   Layout of a
small office

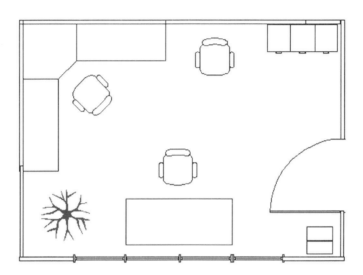

- Two Plan Ell (corner) connectors
- Five Plan Tee connectors

Some text has also been added to indicate the main feed to the system. Note that one of the Plan Ell corner connectors must be rotated to be used in the drawing. Also, several of the Plan Tee connectors must be rotated.

This drawing is started by selecting the Plumbing and Piping - Small Layout template from the Building Services tab of the New Drawing window.

Next, Figure 21.13 shows a drawing of a typical office. The shapes used in this drawing are:

- Single Workstation (upper left corner of the office)
- Two additional chairs (Common chairs)
- Three filing cabinets (top right corner)
- A tree (lower left corner)
- A rectangular table
- A Coat Rack (behind the door)

In addition, there are several other shapes used to complete the outer walls of the office. These are the Horizontal and Vertical walls, Double-hung Windows, and the door (Single door). Many of the wall shapes must be lengthened or shortened to connect them.

This drawing is started by selecting the Facilities Management—Small Layout template from the Facilities tab of the New Drawing window.

PART III: TECHNICAL DRAWINGS

In this last part we examine two additional drawings, both of a technical nature. The first drawing is of a flowchart, a diagram that indicates how a particular task is to be performed. The flowchart is shown in Figure 21.14. There are four different types of shapes used in the flowchart. They are the:

- Terminator
- Process
- Decision
- Connector circle

Text has been added to all of the shapes except one (the Connector circle). This is done by left double-clicking inside the shape and then entering the text. In addition, the 'Yes' and 'No' text has been added outside the Decision shape. The lines connecting the shapes have arrowheads on them to clearly indicate the flow of the process.

This drawing is started by selecting the Basic Flowchart template from the Business tab of the New Drawing window.

The last drawing we will examine is shown in Figure 21.15 and represents the design of a simple Ethernet computer network. The components of the network are as follows:

- Six Desktop Computers
- One PC Server

FIGURE 21.14   Sample flowchart

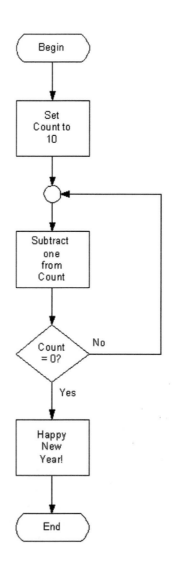

FIGURE 21.15 **Computer network**

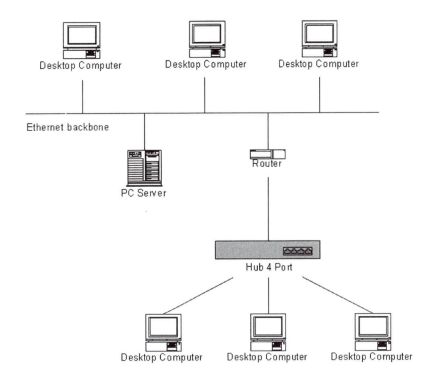

- One Router
- One 4-port Hub
- One Ethernet backbone cable

All of the text for these shapes are built in. The only additional text that must be added to the drawing is the "Ethernet backbone" label.

Some additional points should be made about Figure 21.15. First, the size of the Ethernet backbone cable has been reduced (its original size when first placed is larger). Second, simple lines were added to make the network connections between the lower three Desktop Computers and the Hub.

This drawing is started by selecting the Networks - Small Diagram template from the Network tab of the New Drawing window.

TROUBLESHOOTING TECHNIQUES

It is very easy to make a simple mistake during a drawing. For example, you may rotate a shape too many degrees, select the wrong font, or place a shape improperly. If any of these unfortunate situations occur, simply left-click on the Undo (the counter-clockwise arrow button located under the Draw and Modify pull-down menu items). The Undo button is capable of backing up several steps in the drawing, allowing you to return to a good drawing state.

SELF-TEST

This self-test is designed to help you check your understanding of the background information presented in this exercise.

True/False

Answer *true* or *false*.

1. Actrix Technical is only used for drawing industrial parts.
2. Each tab on the Content Explorer represents a catalog of shapes.
3. There are ten basic drawing buttons.

283

4. Actrix can import and export AutoCAD files.
5. Once a shape is placed into the drawing its properties can not be changed.

Multiple Choice

Select the best answer.

6. In a flowchart, the decision shape looks like a
   a. square.
   b. rectangle.
   c. diamond.
7. When two shapes come close to each other, they may
   a. bounce away from each other.
   b. snap together at predefined locations.
   c. deform so that they never touch.
8. When adding text, you must
   a. left-click to select a text position.
   b. Right-click to select a text position.
   c. Only use one of four predefined text locations.
9. The shapes used to create the computer network in Figure 21.15 are found in the
   a. Industrial template.
   b. Electrical template.
   c. Network template.
10. CAD stands for
   a. Circuit Aided Design.
   b. Common Architectural Drawing.
   c. Computer Aided Design.

Completion

Fill in the blank or blanks with the best answers.

11. The _____ Explorer shows the shapes contained in a catalog.
12. To select a basic shape, _____-click once on the desired button.
13. To delete a shape, _____ it and then press the delete key.
14. To rotate a shape by 15 degree amounts, hold the _____ key down during rotation.
15. One shape may _____ another shape.

FAMILIARIZATION ACTIVITY

Refer to *Thought Project #26*

Part I

1. Draw the basic shapes shown in Figure 21.5.
2. Draw the dimensioned desk shown in Figure 21.10.
3. Use the line shape to draw a battery and a resistor. Connect them to make a series circuit.
4. Use the arc shape and the line shape to draw a capacitor.

Part II

1. Draw the system of pipes shown in Figure 21.12.
2. Draw the office layout shown in Figure 21.13.

Part III

1. Draw the flowchart shown in Figure 21.14.
2. Draw the computer network shown in Figure 21.15.

1. Why not just sketch a new drawing on a sheet of paper? What advantages does Actrix Technical give you?
2. Search the web for other CAD-related design software.

Under the supervision of your instructor

1. Describe the typical operations performed when using CAD techniques.
2. Use Actrix Technical to create a drawing.

Appendix: Microcomputer Hardware

This appendix provides an overview of microcomputer hardware devices.

One of the most common ways of getting information into a computer is through a keyboard. A computer keyboard consists of separate keys that, when tapped, send specific codes to the computer. Essentially, such a code tells the computer that a key is depressed, what key is being depressed, and when the key is no longer depressed.

Another device used for getting information into a computer is a computer **mouse.** A mouse is simply a device that moves a cursor to any desired area of the screen. The computer always knows at what position on the screen the cursor is located. On the mouse itself, there are buttons (usually three). When a button is depressed, this—along with the position of the cursor on the screen—gives the computer specific information. Usually the screen contains information as to what that particular area of the screen means to the mouse user. For instance, it could mean to begin or terminate a process. Figure 1 shows actions of a keyboard and a mouse.

As you can see from Figure 1, both these devices are input devices. The major disadvantages of the keyboard are the typing skill required to use it and the need to know specific key sequences to initiate computer actions (such as DIR in order to get the directory listing of a disk).

The disadvantages of the keyboard are overcome through the use of a mouse. Using the mouse does not require any typing skills or knowledge of special key sequences (such as the DOS commands). Windows essentially requires only the use of the mouse for system interaction. Windows allows you to execute multiple programs simultaneously, and quickly change from one application to another, with a single mouse click.

THE KEYBOARD

A new keyboard, used with most PCs, contains 101 keys. The function keys are located horizontally along the top of the keyboard, where there are now 12 of them. The ESC key is at the upper left. The keyboard contains duplicate cursor-movement and other similar keys. This is sometimes referred to as the **enhanced keyboard** (Figure 2).

Identifying Keys

There are four ways of identifying a key on an IBM keyboard: by the **character** on the cap of the key, by the **character code** associated with each key-cap character, by

FIGURE 1 Actions of a
keyboard and a mouse

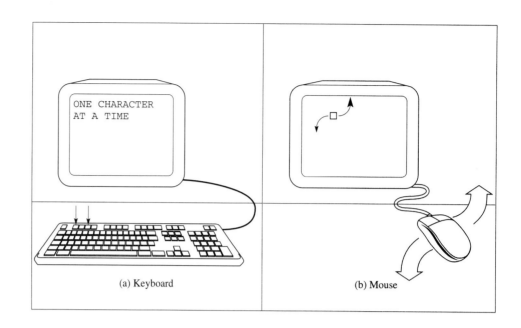

(a) Keyboard

(b) Mouse

FIGURE 2 The 101-key
enhanced keyboard

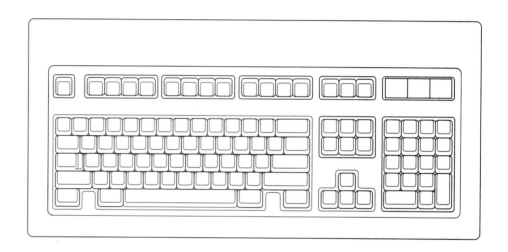

FIGURE 3 Four ways of
identifying a key on an IBM
keyboard

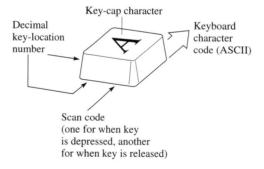

the scan code of the key, and by the decimal key–location number. These are illustrated
in Figure 3.

During the power-on self-test, the first part of the keyboard scan code is displayed if
there is a problem with that particular key. Figure 4 shows the scan codes for the IBM 101-
key keyboard, and Figure 5 shows the key-location numbers for the same keyboard.

As shown in Figure 5, each key is assigned a decimal number that is used as a key-
location reference on most IBM drawings. These numbers are used only as convenient

guides for the physical location of the various keys and bear no relationship to the actual characters generated by the corresponding keys.

Keyboard Servicing

Outside of routine cleaning of the keyboard, there is little you can do to service it. In many cases, the keyboard assembly is a sealed unit. The major hazards to a keyboard are spilled liquids. Periodically you can use a chip puller to pull the keytops off the keyboard. (Be sure to have a similar keyboard to use as a reference when replacing these key caps.) Then hold the keyboard upside down and blow it out with compressed air.

The keyboard is connected to the computer through a cable to the **keyboard interface connector.** This connector is shown in Figure 6.

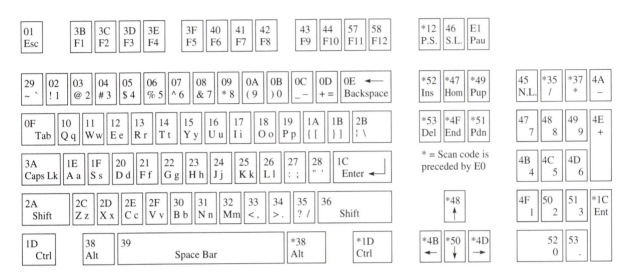

Enhanced Keyboard Scan Codes

FIGURE 4   Scan codes for 101-key keyboard

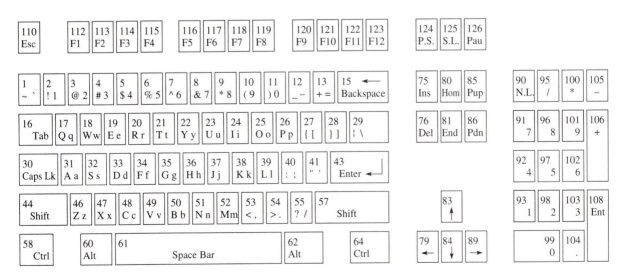

Enhanced Keyboard Location Numbers

FIGURE 5   Key-location numbers for IBM 101-key keyboard

289

FIGURE 6   Keyboard interface
connector (socket)

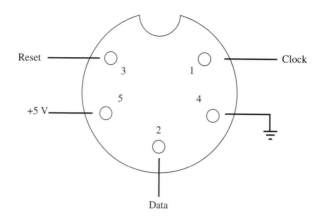

You can use a voltmeter to test the operation of the keyboard interface connector. Voltages between all pins and pin 4 of the connector should be in the range of 2 to 5.5 V DC. If any of these voltages are wrong, the problem is usually in the computer's system board. If these voltages are correct, the problem may be in the keyboard or its connector cable. Some keyboards have one or more switches (on the bottom side) to make them compatible with the computer to which they are connected. Check these switch settings as well as the documentation that comes with the keyboard.

If you find that only one key is malfunctioning, you can check the small spring on the key. Simply remove the key cap, under which you will see a small spring. Try pulling the spring slightly and then replace the key cap. You can check the cable continuity by carefully removing the bottom plate of the keyboard and observing how the cable interfaces with the computer.

Because new keyboards are so inexpensive, it is usually cheaper to replace a bad keyboard.

Today, you can purchase custom keyboards that have themes (a *Star Trek* keyboard), keyboards with infrared transmitters and no cables, or a keyboard with the keys arranged into two groups, one for each hand. You can even buy a keyboard with a built in scanner.

THE MOUSE

There are basically two types of mouse drivers used with PCs: one is a **serial mouse** and the other is a **parallel mouse.** The serial mouse interfaces with the computer through the serial port; the parallel mouse interfaces through a parallel port. In the PC, the mouse is typically connected to the 9-pin male plug on the COM1 serial port. A 9- to 25-pin adapter is available if the serial port has a 25-pin connector.

When a serial port is not available, a **bus mouse** can be interfaced with the computer using its own **bus interface** board. Figure 7 shows a typical computer mouse.

FIGURE 7   Typical
computer mouse

290

TRACKBALLS

A trackball is similar to a mouse except the device does not move. Instead, the user pushes a round trackball around inside its case, allowing the same movement as a mouse but not requiring a mousepad or large surface for movement. Many laptop and notebook computers have trackball mouse devices built in.

VIDEO MONITORS AND VIDEO ADAPTERS

The computer display system used by your computer consists of two separate but essential parts: the monitor and the video adapter card as shown in Figure 8. Note from the figure that the monitor does not get its power from the computer; it has a separate power cord and its own internal power supply.

The video adapter card [Figure 8(b)] interfaces between the motherboard and the monitor. This card processes and converts data from the computer and allows you to see all the things you are used to seeing displayed on the screen.

FIGURE 8   The two essential parts of a computer display system

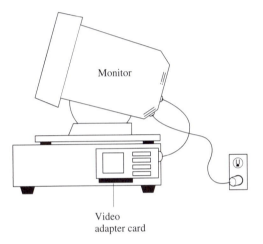

(a) Video adapter card with companion monitor

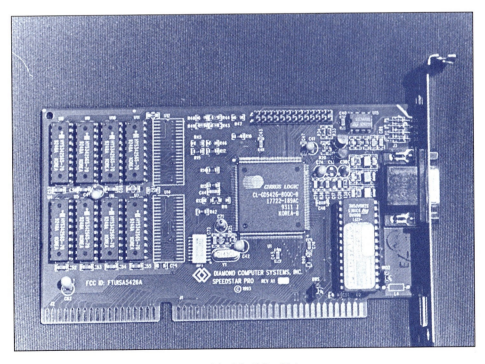

(b) SVGA graphics accelerator card (*photograph by John T. Butchko*)

291

It is very important to realize that there are many different types of monitors and that each type of monitor essentially requires its own special video adapter card, as shown in Figure 9. Connecting a monitor to an adapter card not made for it can severely damage the monitor or adapter card, or both.

Monitor Servicing

Very seldom is the computer user expected to repair a computer monitor. Computer monitors are very complex devices that require specialized training to repair. These instruments contain very high and dangerous voltages that are present even when no power is being applied. The servicing of the monitor itself is, therefore, better left to those who are trained in this specialty.

What you need to know is what kinds of monitors are available, their differences, and how they interface with the computer. Then you need to know enough about hardware and software in order to tell if a problem that appears on the monitor is in the monitor itself, its adapter card, the computer, or the monitor cabling—or is simply a lack of understanding about how to operate the computer.

Monitor Fundamentals

All monitors have the basic sections shown in Figure 10. Table 1 lists the purpose of each of the major sections of a computer monitor.

FIGURE 9   Necessity of each computer monitor having its own matching adapter

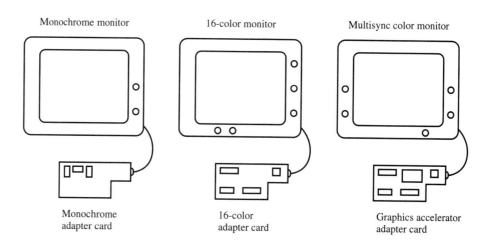

FIGURE 10   Major sections of a computer monitor

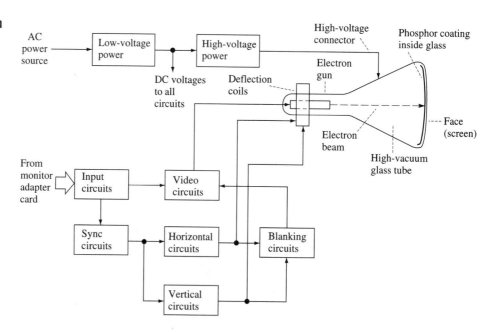

TABLE 1   Major sections of a
computer monitor

Section	Purpose
Glass CRT	The cathode-ray tube (CRT) creates the image on the screen. It is so named because the source of electrons is called the cathode and the resulting stream of electrons is called its rays (cathode rays).
Electron gun	Generates a fine stream of electrons that are attracted toward the glass face of the CRT (the screen) by the large positive voltage applied there.
Phosphor coating	A special kind of material that emits light when struck by an electron beam.
High-voltage power source	Supplies the large positive voltage required by the CRT to attract the electrons from the electron gun.
Deflection coils	Generate strong magnetic fields that move the electron beam across the face of the CRT.
Horizontal circuits	Generate waveforms applied to the deflection coils, causing the electron beam to sweep horizontally across the face of the CRT from left to right.
Vertical circuits	Generate waveforms applied to the deflection coils, causing the horizontal sweep of the electron beam to move vertically across the face of the CRT from top to bottom and creating a series of horizontal lines.
Blanking circuits	Cause the electron beam to be cut off from going to the face of the CRT so that it isn't seen when the electron beam is retracing from right to left or from bottom to top. (This is similar to what you do when writing. You lift your pen from the surface of the paper after you finish a line and return to the left side of the paper to begin a new line just below it.)
Video circuits	Control the intensity of the electron beam that results in the development of images on the screen.
Sync circuits	Electrical circuits that help synchronize the movement of the electron beam across the screen.
Low-voltage power supply	Supplies the operating voltages required by the various circuits inside the monitor.

Monochrome and Color Monitors

One of the differences between a monochrome (single-color) monitor and a color monitor is in the construction of the CRT. The differences are illustrated in Figure 11.

As shown in the figure, the color CRT contains a triad of color phosphor dots. Even though this consists of only three color phosphors, all the colors you see on a color monitor are produced by means of these three colors (including white, which is produced by controlling the intensity of the three colors: red is 30%, green is 59%, and blue is 11%). This process, called **additive color mixing,** is illustrated in Figure 12.

The other differences between monochrome and color monitors are the circuits inside these systems as well as their adapter cards. Some of these differences are the high voltages in a color monitor that are several times higher than those found in a monochrome monitor. Usually, these voltages are on the order of 30,000 V or more. You should note that this high voltage can be stored by the color CRT and still be present even when the set is unplugged from the AC outlet. A special probe is used to discharge the CRT.

Energy Efficiency

Energy-efficient PCs are designed with energy efficiency in mind. The system BIOS, monitor video card, and other hardware must support either the Advanced Power Management

FIGURE 11   Monochrome and color CRTs

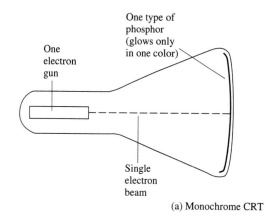

One electron gun

One type of phosphor (glows only in one color)

Single electron beam

(a) Monochrome CRT

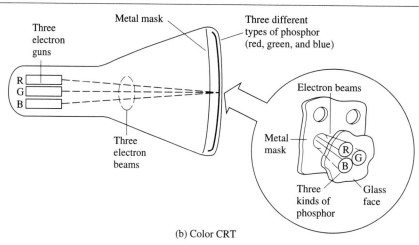

Three electron guns

Metal mask

Three different types of phosphor (red, green, and blue)

Three electron beams

Electron beams

Metal mask

Three kinds of phosphor

Glass face

(b) Color CRT

FIGURE 12   Additive color mixing

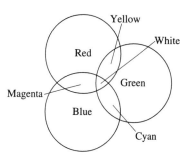

Yellow

White

Red

Green

Magenta

Blue

Cyan

(APM) or VESA BIOS extensions for power management (VBE/PM) standards. Some computers may support limited power management or energy saving features.

It is estimated by the U.S. Environmental Protection Agency (EPA) that the average office desktop computer or workstation uses around $105 of electrical power annually. When all desktops are considered, the total consumption adds up to around 5 percent of all electrical energy consumed in the United States. The EPA estimates that by using energy-efficient equipment, as much as $90 a year per computer can be saved.

The EPA has proposed a set of guidelines for energy-efficient use of computers, workstations, monitors, and printers. The EPA *Energy Star* program requires the computer and monitor to use less than 30 watts each when they are not being used (for a total of 60 watts including both the system unit and the monitor). Personal computers adhering to the Energy Star recommendations are also called *green* PCs.

Each computer can be set up to automatically reduce energy usage using the standby and sleep modes. The standby mode is activated after a user-specified period of inactivity. The sleep mode is automatically activated after the standby time has expired. If the computer is used during standby energy saving mode, it takes just a short period of time before the monitor is usable. Sleep mode is similar to a power-down of the monitor and requires some additional time before the monitor is usable.

The EPA Web site located at http://www.epa.gov/energy_star maintains a list of all energy-efficient computer products. Look for the Energy Star trademark on product packaging and the marketing materials supplied by most manufacturers.

Video Controls

Table 2 lists some of the major video controls and their purposes.

Pixels and Aspect Ratio

Figure 13 illustrates two important characteristics of computer monitors. As shown in the figure, a pixel (or pel) is the smallest area on the screen whose intensity can be controlled. The more pixels available on the screen, the greater the detail that can be displayed. The number of pixels varies among different types of monitors; the more pixels, the more expensive the monitor. The aspect ratio indicates that the face of the CRT is not a perfect square. It is, instead, a rectangle. This is important to remember, especially if you are developing software for drawing squares and circles; you may wind up with rectangles and ellipses. The size of a pixel is referred to as its *dot pitch* and is a function of the number of pixels on a scan line and the distance across the display screen.

TABLE 2 Major video controls

Control	Purpose
Contrast	A gain control for the circuits that determine the strength of the signal used to place images on the screen. It affects the amount of difference between light and dark.
Brightness	Controls the amount of high voltage applied to the CRT, which controls the strength of the beam. The higher the voltage, the stronger the beam and the brighter the picture.
Vertical size	Controls the output of the vertical circuit, changing the amount of the vertical sweep of the CRT and thus changing the vertical size of the displayed image.
Horizontal size	Controls the output of the horizontal circuit, changing the amount of horizontal sweep of the CRT and thus changing the horizontal size of the displayed image.
Vertical hold	Helps adjust the synchronous circuits so the image is stable in the vertical direction.
Horizontal hold	Helps adjust the synchronous circuits so the image is stable in the horizontal direction.

FIGURE 13 Pixels and aspect ratio

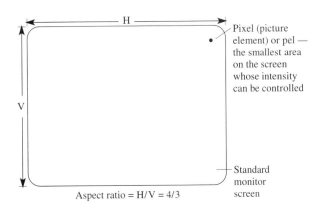

Monitor Modes

There are two fundamental modes in which the monitor operates: the **text mode** and the **graphics mode.** Figure 14 illustrates the difference.

In the text mode, the CRT display gets its information from a built-in ROM chip referred to as the **character ROM.** This may not be a separate ROM chip but part of another, larger one. This ROM contains all the characters on your keyboard, plus many more. This group of characters is known as the **extended character set** and may, among other things, be used in combination to form squares, boxes, and other shapes while your computer is still in the text mode. To get any of these extended characters on the screen (or to get *any* character on the screen), simply hold down the SHIFT and ALT keys at the same time and then type in the character number. For example, to get the character ö, hold down the SHIFT and ALT keys and type in 148 on the keypad; when you lift up on the SHIFT and ALT keys, the character appears (and can also appear on the printer, depending on the type of printer).

The advantage of the text mode is that it doesn't take much memory and the visual results are predictable and easy to achieve (you need only to press a key on the keyboard). The size of the text screen is 80×25 or 40×25. The text screen is sometimes referred to as the **alphanumeric mode.**

When the monitor and its circuits are in the graphics mode, an entirely different use of memory is required. RAM is used because a program has complete control over the intensity and (in the case of color) the color of each pixel. The more pixels available on the monitor, the more memory required; the more memory required, the longer it takes to display a complete picture on the face of the CRT, which in turn means that your whole computer must be able to operate at a very high speed. In order to display detailed graphics, you must have a big and powerful machine, which means a more expensive system as well as a more expensive monitor.

Just to give you an idea of the memory requirements for graphics, if your monitor has 640 horizontal pixels and 480 vertical pixels, the total number of pixels that must be addressed by RAM is $640 \times 480 = 307,200$, which is more than a third of a megabyte for just one screen. If color is not used in the graphics mode, less memory is required (because the computer needs to store less information about each pixel).

Types of Monitors

In order to understand the differences among the most common types of computer monitors, you must first understand the definitions of the terms used to describe them. Table 3 lists the major terms used to distinguish one monitor from another.

Now that you know the definitions of some of the major terms used to distinguish one monitor from another, you can be introduced to the most common types of monitors in use today. Table 4 lists the various types of monitors and their distinguishing characteristics.

VGA Monitor   The **video graphics array (VGA) monitor** is one of the most popular color monitors; it provides high color resolution at a reasonable price. More and more software with graphics is making use of this type of monitor. The associated cards have a high scanning rate, resulting in less eye fatigue both in text and in graphics modes.

FIGURE 14   Text and graphics modes

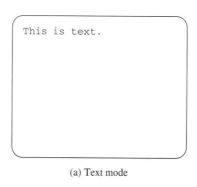

(a) Text mode

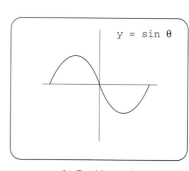

(b) Graphics mode

TABLE 3   Computer monitor terminology

Term	Definition
Resolution	The number of pixels available on the monitor. A resolution of 640 × 480 means that there are 640 pixels horizontally and 480 pixels vertically.
Colors	The number of different colors that may be displayed at one time in the graphics mode. For some color monitors, more colors can be displayed in the text mode than in the graphics mode. This is possible because of the reduced memory requirements of the text mode.
Palette	A measure of the full number of colors available on the monitor. However, not all the available palette colors can be displayed at the same time (again, because of memory requirements). You can usually get a large number of colors with low resolution (fewer pixels) or a smaller number of colors (sometimes only one) with much higher resolution—again, because of memory limitations.
Display (digital or analog)	There are basically two different types of monitor displays, **digital** and **analog.** Some of the first computer monitors used poor-quality analog monitors. Then digital monitors, with their better overall display quality, became more popular. Now, however, the trend is back to analog monitors because of the increasing demand for high-quality graphics, where colors and shades can be varied continuously to give a more realistic appearance.

TABLE 4   Common types of computer monitors

Type	Resolution*	Colors	Palette	Display
Monochrome composite	640 × 200	1	1	Analog
Color composite	640 × 200	4	4	Analog
Monochrome display	720 × 350	1	1	Digital
RGB (CGA)	640 × 200	4	16	Digital
EGA	640 × 350	16	64	Digital
PGA	640 × 480	Unlimited	Unlimited	Analog
VGA	640 × 480	256	262,144	Analog
SVGA	1280 × 1024	Varies	Varies	Digital/ analog
Multiscan	Varies	Unlimited	Unlimited	Digital/ analog

*In general, the higher the resolution, the higher the scan frequency. For example, the typical scan frequencies of EGA and VGA monitors are 21.5 KHz and 31.5 KHz, respectively.

SVGA Monitor   Higher screen resolution and new graphics modes make the **Super VGA (SVGA) monitor** even more popular than the VGA monitor.

Multiscan Monitor   The **multiscan monitor** was one of the first monitors that could be used with a wide variety of monitor adapter cards. Since this type of monitor can accommodate a variety of adapter cards, it is sometimes referred to as the *multidisplay* or *multisync* monitor.

Display Adapters

As previously stated, a computer monitor must be compatible with its adapter card. If it is not, damage to the monitor or adapter card, or both, could result.

VGA Adapter   The **VGA (video graphics array card)** was the fastest-growing graphics card in terms of popularity until the SVGA card became available. The VGA adapter card uses a 15-pin high-density pin-out, as shown in Figure 15. The VGA 15-pin adapter can be wired to fit the standard 9-pin graphics adapter, as shown in Figure 16.

SVGA Adapter   The Super VGA graphics interface uses the same connector that VGA monitors use. However, more display modes are possible with SVGA than with VGA.

VESA   The **VESA** (Video Electronics Standards Association) specification has been developed to guide the operation of new video cards and displays beyond VGA. New BIOS software that supports the VESA conventions is contained in an EPROM mounted on the display card. The software also supports the defined VESA video modes. Some of these new modes are 1024×768, 1280×1024, and 1600×1200, with up to 16 million possible colors.

Graphics Accelerator Adapters

A graphics accelerator is a video adapter containing a microprocessor designed specifically to handle the graphics processing workload. This eliminates the need for the system processor to handle the graphics information, allowing it to process other instructions (nongraphics related) instead.

Aside from the graphics processor, there are other features offered by graphics accelerators. These features include additional video memory, which is reserved for storing graphical representations, and a wide bus capable of moving 64 or 128 bits of data at a time. Video

FIGURE 15   Pin diagram for VGA adapter

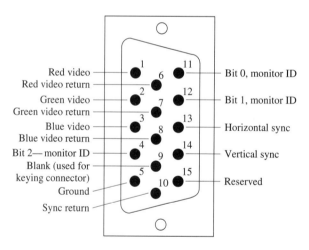

FIGURE 16   Nine-pin adapter cable for VGA

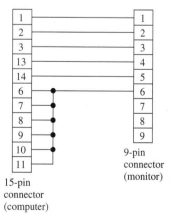

TABLE 5 Common monitor support

Resolution	Colors	Memory	Refresh Rates
640 × 480	256 65K 16M	2MB 2MB 2MB	60, 72, 75, 85
800 × 600	256 65K 16M	2MB 2MB 2MB	56, 60, 72, 75, 85
1024 × 768	256 65K 16M	2MB 2MB 4MB	43 (interlaced), 60, 72, 75, 85
1280 × 1024	256 65K 16M	2MB 4MB 4MB	43 (interlaced), 60, 75, 85

memory is also called VRAM and can be accessed much faster than conventional memory.

Many new multimedia applications require a *graphics accelerator* to provide the necessary graphics throughput in order to gain realism in multimedia applications. Table 5 illustrates the settings available for supporting many different monitor types and refresh rates.

Most graphics accelerators are compatible with the new standards such as Microsoft DirectX, which provides an application programming interface, or API, to the graphics subsystem. Usually, the graphics accelerators are also compatible with OpenGL for the Windows NT environment.

AGP Adapter

The **Accelerated Graphics Port (AGP)** is a new interface specification developed by Intel. The AGP adapter is based on the PCI design but uses a special point-to-point channel so that the graphics controller can directly access the main system memory. The AGP channel is 32 bits wide and runs at 66 MHz. This provides a bandwidth of 266 MBps as opposed to the PCI bandwidth of 133 MBps.

AGP optionally supports two faster modes, with throughput of 533MB and 1.07GB. Sending either one (AGP 1X), two (AGP 2X), or four (AGP 4X) data transfers per clock cycle accomplishes these data rates. Table 6 shows the different AGP modes. Other optional features include AGP texturing, sideband addressing, and pipelining. Each of these options provides additional performance enhancements.

AGP graphics support is provided by the new NLX motherboards, which also support the Pentium II microprocessor (and above). It allows for the graphic subsystem to work much closer with the processor than previously available by providing new paths for data to flow between the processor, memory, and video memory. Figure 17 shows this relationship.

AGP offers many advantages over traditional video adapters. You are encouraged to become familiar with the details of the AGP adapter.

THE COMPUTER PRINTER

Two fundamental types of printers are used with personal computers: the **impact printer** and the **nonimpact printer.** The impact printer uses some kind of mechanical device to

TABLE 6 AGP graphics mode

Mode	Throughput (MB/s)	Data Transfers per Cycle
1x	266	1
2x	533	2
4x	1066	4

FIGURE 17 AGP configuration

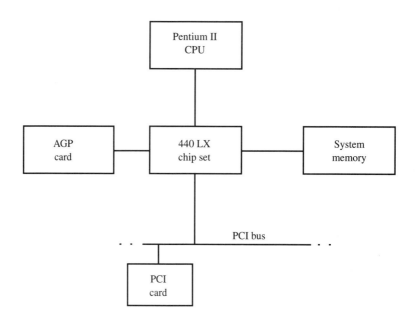

impart an impression to the paper through an inked ribbon. The nonimpact printer uses heat, a jet of ink, electrostatic discharge, or laser light. Nonimpact printers form printed images without making physical contact with the paper. These two types of printers are illustrated in Figure 18.

Impact Printers

The most common type of impact printer is the **dot-matrix printer.** The dot-matrix printer makes up its characters by means of a series of tiny mechanical pins that move in and out to form the various characters printed on the paper.

The Dot-Matrix Printer   The dot-matrix printer, one of the most popular types of printers, uses a mechanical printing head that physically moves across the paper to be printed. This mechanical head consists of tiny movable wires that strike an inked ribbon to form characters on the paper. There are two popular kinds of dot-matrix print heads: one consists of 9 pins (the movable wires) and the other consists of 24 pins. A 9-pin print head is shown in Figure 19.

The 24-pin dot-matrix printer is more expensive than the 9-pin model. However, because both types have modes of operation that allow for an *overstrike* of the image (with the head moving slightly and the image being struck again), the 9-pin model can produce close to what is known as letter-quality printing. The 24-pin model can produce an even sharper character when operated in the same overstrike mode. Because of the manner in which characters are formed in this type of printer, the printing of graphic images is possible.

Nonimpact Printers

The most popular nonimpact printers are the *ink-jet printer, bubble-jet printer,* and the *laser printer.* The ink-jet printer uses tiny jets of ink that are electrically controlled. The laser printer uses a laser to form characters. The laser printer resembles an office photocopying machine.

Ink-Jet and Bubble-Jet Printers   An **ink-jet printer** uses electrostatically charged plates to direct jets of ink onto paper. The ink is under pressure and is formed by a mechanical nozzle into tiny droplets that can be deflected to make up the required images on the paper. A **bubble-jet printer** uses heat to form bubbles of ink. As the bubbles cool, they form the droplets applied to the paper. Ink-jet and bubble-jet printers cost more than impact printers but are quieter and can produce high-quality graphic images.

FIGURE 18   Two fundamental
types of printers

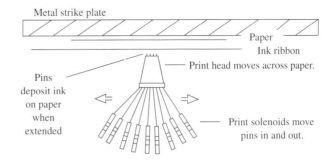

(a) Impact (Dot-matrix Illustration)

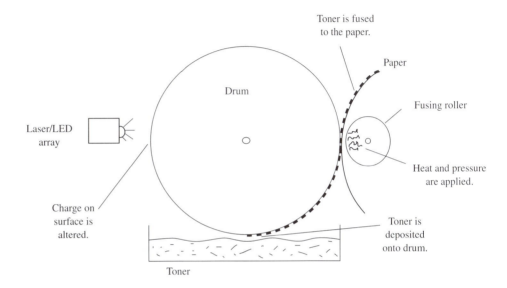

(b) Nonimpact (Laser Illustration)

FIGURE 19   Nine-pin dot-
matrix print head

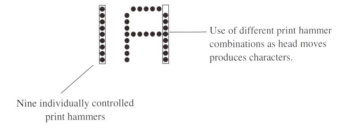

Laser Printers   Through the operation of a laser and mirror (controlled by software), an electrical image is impressed on a photoreceptor drum. This drum picks up powdered toner, which is then transferred to paper by electrostatic discharge. A second drum uses high temperature to bond this image to the paper. The result is a high-quality image capable of excellent characters and graphics.

Because of their high-quality output, laser printers find wide application in desktop publishing, computer-aided design, and other image-intensive computer applications. Laser printers are usually at the high end of the price range for computer printers.

Technical Considerations

Most printer problems are caused by software. What this means is that the software does not match the hardware of the printer. This is especially true when the printing of graphics

301

is involved. Software troubleshooting for printers is presented last in this section. When there is a hardware problem with printers, it is usually the interface cable that goes from the printer adapter card to the printer itself. This is illustrated in Figure 20.

Printer Cables The interface cable is used to connect the printer to the computer. Previously, there was limited communication between the printer and computer. The computer received a few signals from the printer such as the online or offline indicator, the out-of-paper sensor, and the print buffer status. As long as the printer was sending the correct signals to the computer, the computer would continue to send data.

Advances in printer technology now require a two-way communication between the computer and printer. As a result of these changes, a new bidirectional printer cable is required to connect most new printers to the computer. The bidirectional cables may or may not adhere to the new IEEE standard for Bidirectional Parallel Peripheral Interface. The IEEE 1284 Bitronic printer cable standard requires 28 AWG construction, a Hi-flex jacket, and dual shields for low EMI emissions. The conductors are twisted into pairs to reduce possible cross talk.

FIGURE 20   Problem areas in computer printers

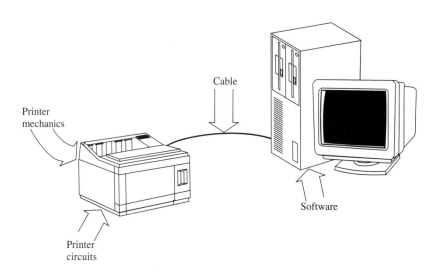

FIGURE 21   Typical printer cables

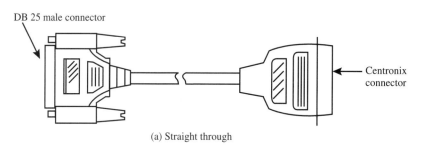

(a) Straight through

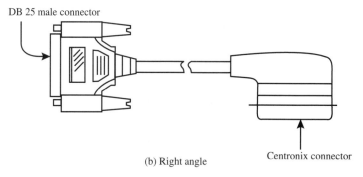

(b) Right angle

FIGURE 22 Print Test Page
option

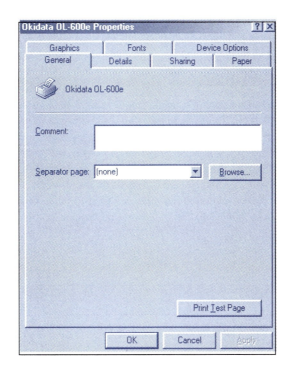

FIGURE 23 Printer test page
confirmation

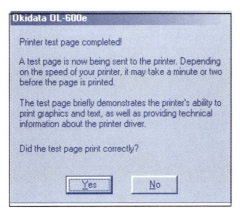

FIGURE 24 Printer test page
output

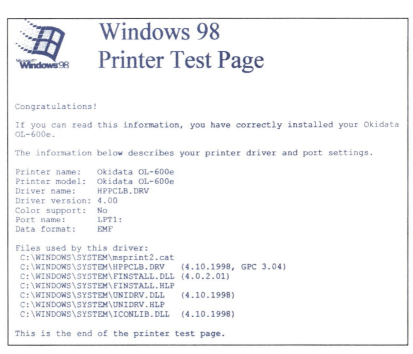

Check the requirements for each printer to determine the proper cable type. Figure 21 shows two popular parallel printer cable styles. Many different lengths of printer cables are available. It is usually best to use the shortest cable possible in order to reduce the possibility of communication errors.

Printer Hardware   A printer requires periodic maintenance. This includes vacuuming out the paper chaff left inside the printer. A soft dry cloth should be used to keep the paper and ribbon paths clean. It is a good idea to use plastic gloves when cleaning a printer, because the ink or toner is usually difficult to remove from the skin. With dot-matrix printers, be careful of the print heads. These heads can get quite hot after extended use. Make sure you do not turn the print platen rollers when the power is on because a stepper motor is engaged when power is applied. This little motor is trying to hold the print platen roller in place. If you force it to move, you could damage the stepper motor.

TABLE 7   Printer troubleshooting methods

Checks	Comments
Check if printer is plugged in and turned on.	The printer must have external AC power to operate.
Check if printer is online and has paper.	Printers must be *online,* meaning that their control switches have been set so that they will print (check the instruction manual). Some printers will not operate if they do not have paper inserted.
Print a test page.	Select the Print Test Page option as shown in Figure 22. Confirm that the page printed properly, see Figure 23. Figure 24 shows the printer test page output.
Do a printer self-test.	Most printers have a self-test mode. In this mode, the printer will repeat its character set over and over again. You must refer to the documentation that comes with the printer to see how this is done.
Do a Print Screen.	If the printer self-test works, then with some characters on the computer monitor, hold down the SHIFT key and press the PRINTSCRN key at the same time. What is on the monitor should now be printed. Do not use a program (such as a word-processing program) because the software in the program may not be compatible with the printer.
Exchange printer cable.	If none of the above tests work, the problem may be in the printer cable. At this point, you should swap the cable with a known good one.
Replace the printer adapter card.	Try replacing the printer adapter card with a known good one. Be sure to refer to the printer manual to make sure you are using the correct adapter card.
Check parameters for a serial interface.	If you are using a serial interface printer from a serial port, make sure you have the transmission rate set correctly, along with the parity, number of data bits, and number of stop bits. Refer to the instruction manual that comes with the printer and use the correct form of the DOS MODE command.
Check the configuration settings.	Check all the configuration settings available on the printer.
Check the software installation.	When software is installed (such as word-processing and spreadsheet programs), the user may have had the wrong printer driver installed (the software that actually operates the printer from the program).

Laser Hardware Essentially, laser printers require very little maintenance. If you follow the instructions that come with the printer, the process of changing the cartridge (after about 3500 copies) also performs the required periodic maintenance on the printer.

When using a laser printer, remember that such a machine uses a large amount of electrical energy and thus produces heat. So make sure that the printer has adequate ventilation and a good source of reliable electrical power. This means that you should not use an electrical expansion plug from your wall outlet to run your computer, monitor, and laser printer. Doing so may overload your power outlet.

When shipping a laser printer, be sure to remove the toner cartridge. If you don't remove it, it could open up and spill toner (a black powder) over the inside of the printer, causing a mess that is difficult to clean up.

Testing Printers

When faced with a printer problem, first determine if the printer ever worked at all or if this is a new installation that never worked. If it is a new installation and has never worked, a careful reading of the manual that comes with the printer is usually required to make sure that the device is compatible with the printer adapter card. Table 7 lists some of the most direct methods for troubleshooting a computer printer.

System Software

Many types of commands are sent to the printer while it is printing. Some of these commands tell the printer what character to print; others tell the printer what to do, such as performing a carriage return, making a new line, or doing a form feed. This is all accomplished by groups of 1s and 0s formed into a standard code that represents all of the printable characters and the other commands that tell the printer what to print and how to print it.

The ASCII Code Table 8 lists all the printable characters for a standard printer. The code used to transmit this information is called the ASCII code. ASCII stands for American Standard Code for Information Interchange.

As you can see in Table 8, each keyboard character is given a unique number value. For example, a space is number 32 (which is actually represented by the binary value 0010 0000 when transmitted from the computer to the printer). The number values that are less than 32 are used for controlling the operations of the printer. These are called **printer-control codes,** or simply **control codes.** These codes are shown in Table 9.

The definitions of the control code abbreviations are as follows:

ACK	Acknowledge	GS	Group separator
BEL	Bell	HT	Horizontal tab
BS	Backspace	LF	Line feed
CAN	Cancel	NAK	Negative acknowledge
CR	Carriage return	NUL	Null
DC_1–DC_4	Device control	RS	Record separator
DEL	Delete	SI	Shift in
DLE	Data link escape	SO	Shift out
EM	End of medium	SOH	Start of heading
ENQ	Enquiry	SP	Space
EOT	End of transmission	STX	Start text
ESC	Escape	SUB	Substitute
ETB	End of transmission block	SYN	Synchronous idle
ETX	End text	US	Unit separator
FF	Form feed	VT	Vertical tab
FS	Form separator		

The way you can enter an ASCII control code is by holding down the ALT key and the SHIFT key at the same time and using the numeric keypad to enter the ASCII control code. You will see the corresponding character on the screen when you release the ALT and SHIFT keys. (The top row of numbers on your keyboard will not work—only the ones on the numeric keypad portion of the keyboard.)

For example, to create a text file that will cause the printer to eject a sheet of paper (a form feed), using an editor, you would press

ALT-SHIFT-12

What will appear on the screen when you release the ALT-SHIFT keys is

^L

This may not be exactly what you expected, but it is your monitor's way of interpreting what you have just entered. If you put this into a text file (say it is called FORMFED.TXT), then when it is sent to the printer, it will be interpreted as a form feed, and the printer will feed a new sheet of paper. You can then make a batch file from this form-feed text file (called FORMFED.BAT) as follows:

COPY FORMFED.TXT FORMFED.BAT

TABLE 8 Standard printable ASCII codes

Dec Hex Char	Dec Hex Char	Dec Hex Char
32 20	64 40 @	96 60 '
33 21 !	65 41 A	97 61 a
34 22 "	66 42 B	98 62 b
35 23 #	67 43 C	99 63 c
36 24 $	68 44 D	100 64 d
37 25 %	69 45 E	101 65 e
38 26 &	70 46 F	102 66 f
39 27 '	71 47 G	103 67 g
40 28 (	72 48 H	104 68 h
41 29)	73 49 I	105 69 i
42 2A *	74 4A J	106 6A j
43 2B +	75 4B K	107 6B k
44 2C ,	76 4C L	108 6C l
45 2D –	77 4D M	109 6D m
46 2E .	78 4E N	110 6E n
47 2F /	79 4F O	111 6F o
48 30 0	80 50 P	112 70 p
49 31 1	81 51 Q	113 71 q
50 32 2	82 52 R	114 72 r
51 33 3	83 53 S	115 73 s
52 34 4	84 54 T	116 74 t
53 35 5	85 55 U	117 75 u
54 36 6	86 56 V	118 76 v
55 37 7	87 57 W	119 77 w
56 38 8	88 58 X	120 78 x
57 39 9	89 59 Y	121 79 y
58 3A :	90 5A Z	122 7A z
59 3B ;	91 5B [	123 7B {
60 3C <	92 5C \	124 7C	
61 3D =	93 5D]	125 7D }
62 3E >	94 5E ^	126 7E ~
63 3F ?	95 5F –	

To test the printer's form feed, you would simply enter

```
A> FORMFED
```

and the printer (if on and ready) should then feed a sheet of paper through. You could also make up your own custom batch files to perform other types of printer tests or put your name in them (so the name appears printed on the sheet).

Extended ASCII Codes

If you set up your printer to act as a graphics printer (by setting the appropriate configuration; refer to the user manual that comes with the printer), you can extend the character set to include many other forms of printable characters. These characters are shown in Table 10.

You can write these extended character codes to your printer by creating text files. To do this, again hold down the ALT and SHIFT keys and type the number code into the numeric keypad on your keyboard. For example, to get the Greek letter Σ, simply press ALT-SHIFT-228; when you lift up on the ALT-SHIFT keys, a Σ will appear on the monitor. If you include this in a text file (or do a Print Screen), you can transfer it to the printer. The extended characters 176 through 223 are used for creating boxes, rectangles, and other shapes on the monitor or printer while it is still in the text mode. If you can't get these extended characters on the printer, it is either because you haven't set the printer to the graphics mode or your printer simply can't perform the functions required by this mode.

TABLE 9   ASCII control codes

Dec	Hex	Char	
0	0	^@	NUL
1	1	☺	SOH
2	2	●	STX
3	3	♥	ETX
4	4	♦	EOT
5	5	♣	ENQ
6	6	♠	ACK
7	7	•	BEL
8	8	◘	BS
9	9	○	HT
10	A	■	LF
11	B	♂	VT
12	C	♀	FF
13	D	♪	CR
14	E	♫	SO
15	F	☼	SI
16	10	►	DLE
17	11	◄	DC_1
18	12	↕	DC_2
19	13	‼	DC_3
20	14	¶	DC_4
21	15	§	NAK
22	16	▬	SYN
23	17	↨	ETB
24	18	↑	CAN
25	19	↓	EM
26	1A	→	SUB
27	1B	←	ESC
28	1C	∟	FS
29	1D	↔	GS
30	1E	▲	RS
31	1F	▼	US
32	20		SP

TABLE 10   Extended ASCII character set

Dec	Hex	Char	Dec	Hex	Char	Dec	Hex	Char	Dec	Hex	Char
128	80	Ç	160	A0	á	192	C0	∟	224	E0	α
129	81	ü	161	A1	í	193	C1	⊥	225	E1	β
130	82	é	162	A2	ó	194	C2	⊤	226	E2	Γ
131	83	â	163	A3	ú	195	C3	├	227	E3	π
132	84	ä	164	A4	ń	196	C4	─	228	E4	Σ
133	85	à	165	A5	Ń	197	C5	┼	229	E5	σ
134	86	å	166	A6	ª	198	C6	╞	230	E6	μ
135	87	ç	167	A7	º	199	C7	╟	231	E7	τ
136	88	ê	168	A8	¿	200	C8	╚	232	E8	φ
137	89	ë	169	A9	⌐	201	C9	╔	233	E9	θ
138	8A	è	170	AA	¬	202	CA	╩	234	EA	Ω
139	8B	ï	171	AB	½	203	CB	╦	235	EB	δ
140	8C	î	172	AC	¼	204	CC	╠	236	EC	∞
141	8D	ì	173	AD	¡	205	CD	═	237	ED	Ø
142	8E	Ä	174	AE	"	206	CE	╬	238	EE	−
143	8F	Å	175	AF	"	207	CF	╧	239	EF	∩
144	90	É	176	B0	░	208	D0	╨	240	F0	≡
145	91	æ	177	B1	▒	209	D1	╤	241	F1	±
146	92	Æ	178	B2	▓	210	D2	╥	242	F2	≥
147	93	ô	179	B3	│	211	D3	╙	243	F3	≤
148	94	ö	180	B4	┤	212	D4	╘	244	F4	⌠
149	95	ò	181	B5	╡	213	D5	╒	245	F5	⌡
150	96	û	182	B6	╢	214	D6	╓	246	F6	÷
151	97	ù	183	B7	╖	215	D7	╫	247	F7	≈
152	98	ÿ	184	B8	╕	216	D8	╪	248	F8	°
153	99	Ö	185	B9	╣	217	D9	┘	249	F9	•
154	9A	Ü	186	BA	║	218	DA	┌	250	FA	·
155	9B	¢	187	BB	╗	219	DB	█	251	FB	√
156	9C	£	188	BC	╝	220	DC	▄	252	FC	η
157	9D	¥	189	BD	╜	221	DD	▌	253	FD	²
158	9E	₧	190	BE	╛	222	DE	▐	254	FE	■
159	9F	ƒ	191	BF	┐	223	DF	▀	255	FF	

Other Printer Features

Recall that most printers allow you to get different kinds of text (such as 80 or 132 characters of text across the page) or change the page orientation from portrait to landscape. You can also create batch files to test the capabilities of the printer to print the following:

- Bold text
- Underscores
- Overscores
- Superscripts
- Subscripts
- Compressed or expanded text
- Italics

For example, suppose a new printer does not do bold text. You could move the printer to a different computer to see if the problem is in the printer, or you could have a batch file you created that will quickly test if the printer really is capable of producing bold text. If the printer can do it, the problem is probably in the software, because a new printer driver needs to be installed (which comes from the software vendor). To do this, you need to understand what printer manufacturers do in order to get their printers to create features such as **bold text,** subscripts, and superscripts.

Printer Escape Codes

The ESC (escape) character is used by printer manufacturers as a preface. It is an easy way for them to get a whole new set of printer commands. The character ESC generally doesn't do anything by itself; what it does is to tell the printer that the character or set of characters that follows is to be treated in a special way. As an example, an <ESC> E means to begin bold text and <ESC>F means to end the bold text. The exact escape sequence is different for different printer manufacturers, and you need to find the sequence for your printer in the user's manual.

You could have a batch file calling a text file that tests for bold printing, such as

Mickey Brown's Printer test:
This is normal text.
<ESC>E
This is now bold text.
<ESC>F
This is back to normal text.

The problem in creating this kind of text file is to actually enter the ESC key into it (just pressing the ESC key doesn't do it). The secret to doing this is to enter a CTRL-V (hold down the CTRL key while pressing the V key) and then follow it with the [(left bracket). So when you see the text

```
<ESC>E
```

it really means CTRL-V[E, which will start boldface printing. Remember, for the printer you are using, the escape code may be different. All you need to do is to use the operator's manual that comes with the printer to determine the proper escape code for each printer's unique features.

Multifunction Print Devices

It is becoming more and more common to see printers bundled with other common products, like a fax machine. For example, a fax machine usually prints any faxes received. With some modifications, it can print data received from a computer. These types of printers generally use either ink-jet or bubble-jet printer technology.

Similarly, when sending a fax, the image or text that is sent must be scanned. Again, by making some additional modifications, the scanner can provide the scanned data to a computer instead of a fax. These three features—printing, faxing, and scanning—are available on most multifunction printers. Other features such as an answering machine may also be included. Multifunction devices can save a lot of money while offering the convenience of many products in one package.

Energy Efficiency

Like computers and monitors, printers can waste a tremendous amount of energy. This is because printers are usually left on 24 hours a day but are active only a small portion of the time. The EPA Energy Star program recommends that a printer automatically enter a sleep mode when not in use. In sleep mode, a printer may consume between 15 and 45 watts of power. This feature may cut a printer's electricity use by more than 65 percent.

Other efficiency options recommended by the EPA include printer sharing, duplex printing, and advanced power management features. Printer sharing reduces the need for an additional printer. Power management features can reduce the amount of heat produced by a printer, contributing to a more comfortable workspace and reduced air-conditioning costs. Consider turning off a printer at night, on weekends, or during extended periods of inactivity.

DEFINITION OF THE MOTHERBOARD

The main system board of the computer is commonly referred to as the **motherboard.** A typical motherboard is shown in Figure 25. Sometimes the motherboard is referred to as the **system board,** or the **planar.**

CONTENTS OF THE MOTHERBOARD

The motherboard holds and electrically interconnects all the major components of a PC. The motherboard contains the following:

• The microprocessor
• The math coprocessor (only on older 386 motherboards)
• BIOS ROM
• RAM (Dynamic RAM, or DRAM, as well as level-2 cache)
• The expansion slots
• Connectors for IDE drives, floppies, and COM ports

    Table 11 lists these major parts and gives a brief overview of the purpose of each part.
    Figure 26 shows a typical motherboard layout and the locations of the major motherboard parts.
    In this section, you will have the opportunity to learn more details about the microprocessor and the coprocessor. You will also learn about the other areas of the motherboard.

THE MICROPROCESSOR

You can think of the **microprocessor** in a computer as the central processing unit (CPU), or the "brain," so to speak, of the computer. The microprocessor sets the stage for everything else in the computer system. Several major features distinguish one microprocessor from another. These features are listed in Table 12.

FIGURE 25   Typical PC motherboard (_photo by John T. Butchko_)

TABLE 11 Purposes of major motherboard parts

Part	Purpose
Microprocessor	Interprets the instructions for the computer and performs the required process for each of these instructions.
Math coprocessor	Used to take over arithmetic functions from the microprocessor.
BIOS ROM	Read-only memory. Memory programmed at the factory that cannot be changed or altered by the user.
RAM	Read/write memory. Memory used to store computer programs and interact with them.
Expansion slots	Connectors used for the purpose of interconnecting adapter cards to the motherboard.
Connectors	Integrated controller on motherboard provides signals for IDE and floppy drives, the printer, and the COM ports.

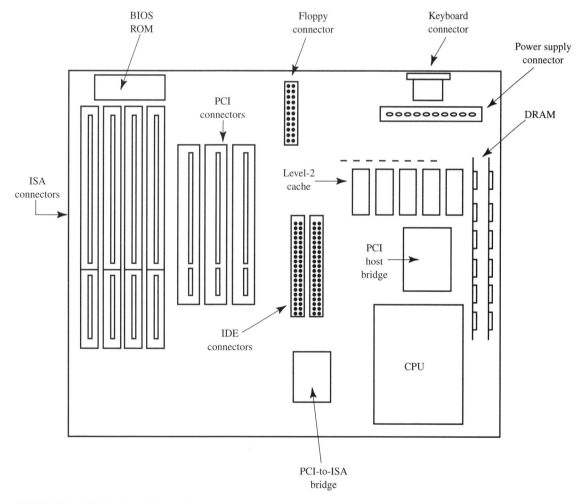

FIGURE 26 Motherboard layout

311

Feature	Description
Bus structure	The number of connectors used for specific tasks.
Word size	The largest number that can be used by the microprocessor in one operation.
Data path size	The largest number that can be copied to or from the microprocessor in one operation.
Maximum memory	The largest amount of memory that can be used by the microprocessor.
Speed	The number of operations that can be performed per unit time.
Code efficiency	The number of steps required for the microprocessor to perform its processes.

TABLE 12 Microprocessor features

You can think of a **bus** as nothing more than a group of wires all dedicated to a specific task. For example, all microprocessors have the following buses:

Data bus	Group of wires for handling data. This determines the data path size.
Address bus	Group of wires for getting and placing data in different locations. This helps determine the maximum memory that can be used by the microprocessor.
Control bus	Group of wires for exercising different controls over the microprocessor.
Power bus	Group of wires for supplying electrical power to the microprocessor.

Figure 27 shows the bus structure of a typical microprocessor.

Since all the data that goes in and out of a microprocessor is in the form of 1s and 0s, the more wires used in the data bus, the more information the microprocessor can handle at one time. For example, some microprocessors have eight lines (wires or pins) in their data buses, others have 16, and some have 32 or 64.

The number of lines used for the address bus determines how many different places the microprocessor can use for getting and placing data. The *places* that the microprocessor uses for getting and placing data are referred to as **memory locations.** The relationship between the data and the address is shown in Figure 28.

The greater the number of lines used in the address bus of a microprocessor, the greater the number of memory locations the microprocessor can use. Table 13 lists the common microprocessors used in the PC. All of these microprocessors are manufactured by Intel Inc.

Note from Table 13 that the greater the number of address lines, the more memory the microprocessor is capable of addressing. In the table, 1MB = 1,048,576 memory locations, and 4GB = 4,294,967,296 memory locations.

FIGURE 27 Typical microprocessor bus structure

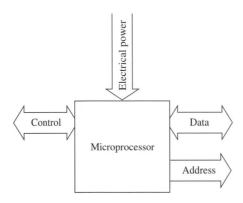

FIGURE 28 Relationship between data and address

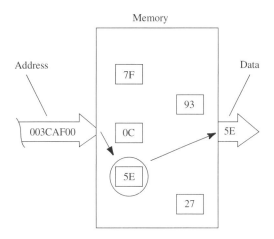

TABLE 13 Types of microprocessors used in the PC

| Microprocessor | Lines | | Maximum Clock Speed | Addressable Memory |
	Data	Address		
8088	8	20	8 MHz	1MB
8086	16	20	8 MHz	1MB
80286	16	24	20 MHz	16MB
80386SX	16	24	20 MHz	16MB
80386	32	32	31 MHz	4GB
80486	32	32	66 MHz	4GB
Pentium	64	32	233+ MHz	4GB
Pentium Pro	64	36	200+ MHz	64GB
Pentium II/III	64	36	400+ MHz	64GB

Compatible CPUs

A number of companies manufacture processors that compete with Intel for use in PC motherboards and other applications. Two of these companies are AMD and Cyrix. Table 14 shows recent sets of compatible CPUs.

Having more than one processor to choose from allows you to examine pricing, chip features, and other factors of importance when making a decision.

About the 80x86 Architecture

The advanced nature of the Pentium microprocessor requires us to think differently about the nature of computing. The Pentium architecture contains exotic techniques such as branch prediction, pipelining, and superscalar processing to pave the way for improved performance. Let us take a quick look at some other improvements from Intel:

- Intel has added MMX technology to its line of Pentium processors (Pentium, Pentium Pro, and Pentium II/III). A total of 57 new instructions enhance the processors' ability to manipulate audio, graphic, and video data. Intel accomplished this major architectural addition by *reusing* the 80-bit floating-point registers in the FPU. Using a method called **SIMD** (single instruction multiple data), one MMX instruction is capable of operating on 64 bits of data stored in an FPU register.
- The Pentium Pro processor (and also the Pentium II/III) use a technique called *speculative execution.* In this technique, multiple instructions are fetched and executed, possibly out of order, in order to keep the pipeline busy. The results of each instruction

313

TABLE 14   Comparing CPUs

Intel	AMD	Cyrix
Pentium	K5	5x86*
Pentium II	K6	6x86MX

*Pentium performance, pin compatible with the 80486.

are speculative until the processor determines that they are needed (based on the result of branch instructions and other program variables). Overall, a high level of parallelism is maintained.

- First used in the Pentium Pro, a bus technology called Dual Independent Bus architecture uses two data buses to transfer data between the processor and main memory (including the level-2 cache). One bus is for main memory, the second is for the level-2 cache. The buses may be used independently or in parallel, significantly improving the bus performance over that of a single-bus machine.
- The five-stage Pentium pipeline was redesigned for the Pentium Pro into a *superpipelined* 14-stage pipeline. By adding more stages, less logic can be used in each stage, which allows the pipeline to be clocked at a higher speed. Although there are drawbacks to superpipelining, such as bigger branch penalties during an incorrect prediction, its benefits are well worth the price.

Spend some time on the Web reading material about these changes, and others. It will be time well invested.

THE COPROCESSOR

Each Intel microprocessor released before the 80486 has a companion to help it do arithmetic calculations. This companion is called a **coprocessor.** For most software, the coprocessor is optional. However, some programs (such as CAD, computer-aided design, programs) have so many math calculations to perform that they need the assistance of the math coprocessor; the main microprocessor simply cannot keep up with the math demand.

These **math chips,** as they are sometimes called, are capable of performing mathematical calculations 10 to 100 times faster than their companion microprocessors and with a higher degree of accuracy. This doesn't mean that if your system is without a coprocessor it can't do math; it simply means that your microprocessor will be handling all the math along with everything else, such as displaying graphics and reading the keyboard.

Table 15 lists the math coprocessors that go with various microprocessors. Note that the 80486 and higher processors have built-in coprocessors.

TABLE 15   Matching math coprocessors

Microprocessor	Math Coprocessor
8086	8087
8088	8087
80286	80287
80386	80387
80386SX	80387SX
80486DX	Built-in coprocessor enabled
80486SX	Coprocessor disabled
Pentium, Pentium Pro, and Pentium II/III	Built-in coprocessor enabled always

For a math coprocessor chip to be used by software, the software must be specifically designed to look for the chip and use it if it is there. Some spreadsheet programs look for the presence of this chip and use the microprocessor for math if the coprocessor is not present. If the coprocessor is present, the software uses it instead. Some programs, such as word-processing programs, have no use for the math functions of the coprocessor and do not use the coprocessor at all. Therefore, the fact that a system has a coprocessor doesn't necessarily mean that the coprocessor will improve the overall system performance. Improvement will take place only if the software is specifically designed to use the coprocessor and there are many complex math functions involved in the program.

COMPUTER MEMORY

Computer memory consists of any device capable of copying a pattern of 1s and 0s that represent some meaningful information to the computer. Computer memory can be contained in *hardware,* such as in chips, or in **magneticware,** such as floppy and hard disks (or other magnetic material such as magnetic tape). Computer memory is not limited to just these two major areas. For example, a laser disk uses light to read large amounts of information into the computer; this too is a form of computer memory. For the purpose of discussion here, computer memory will be divided into two major areas: hardware memory and magneticware memory.

The hardware memory of a computer is referred to as **primary storage.** The magneticware of a computer is referred to as **secondary storage,** or **mass storage.** Here are some facts about each.

Primary Storage

• Immediately accessible to the computer.
• Any part of the memory may be immediately accessed.
• Short-term storage.
• Limited capacity.

Secondary Storage

• Holds very large amounts of information.
• Not immediately accessible.
• May be sequentially accessed.
• To be used, must be transferred to primary storage.
• Long-term storage.

In this section, you will see how primary and secondary computer memories are used (see Figure 29) and how they can work with each other to produce an almost unlimited amount of computer memory. First, let us learn about primary storage.

FIGURE 29 Two major areas of computer memory

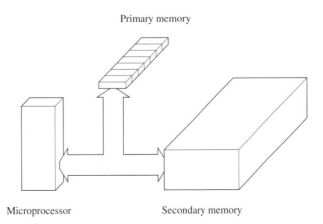

Primary memory

Microprocessor   Secondary memory

PRIMARY STORAGE—RAM AND ROM

There are two basic kinds of primary storage: one kind that the computer can quickly store information in and retrieve information from and another kind that the computer can only receive information from. Figure 30 shows the two basic kinds of primary storage memory.

The kind of memory that the computer can get information (read) from but cannot store information (write) to is called **read-only memory** (ROM). The advantage of having ROM is that it can contain programs that the computer needs when it is first turned on; these programs (called the **Basic Input/Output System,** or BIOS) are needed by the computer so it knows what to do each time it turns on (such as reading the disk and starting the booting process). Obviously, these programs should not be able to be changed by the computer user, because doing so could jeopardize the operation of the system. Therefore, ROM consists of chips that are programmed at the factory. The programs in these chips are permanent and stay that way even when the computer is turned off; they are there when the computer is turned on again.

The kind of memory that the computer can write to as well as read from is called **read/write memory.** The acronym for read/write memory is RWM, which is hard to say. Because of this, read/write memory is called RAM, which stands for **random access memory.** Both ROM and read/write memory are randomly accessible, meaning that the computer can get information from any location without first going through other memory locations. However, read/write memory is traditionally referred to as RAM.

Unlike ROM, RAM loses anything that is stored in it when the power is turned off. Because the information in RAM is not permanent, it is referred to as **volatile memory.** Figure 31 shows this difference.

FIGURE 30   Two basic kinds of primary storage memory

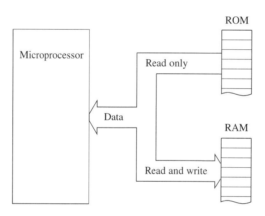

FIGURE 31   ROM and RAM

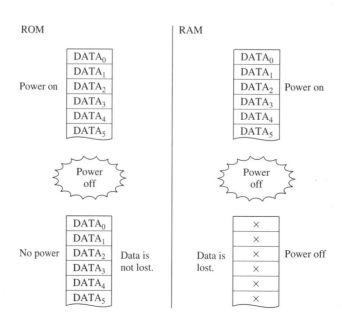

316

The system ROM chip for the PC contains two main programs, the **Power-On Self-Test** (POST) and the **Basic Input/Output System** (BIOS). The programs in the ROM chip set the personality of the computer. As a matter of fact, how compatible a computer is can be determined primarily by the programs in these ROM chips. The ROM chips have changed over time as systems have been improved and upgraded. There have been, for example, more than 20 changes in the ROM BIOS programs by IBM for its different PCs.

You may need to update your old BIOS to use new hardware in your system (large IDE hard drives, for example). To upgrade your BIOS ROM, you may (1) replace the ROM with a new one or (2) run a special upgrade program (typically available for download off the Web) that makes changes to a *flash EPROM,* an EPROM that can be electrically reprogrammed.

BITS, BYTES, AND WORDS

Recall that a bit is a single binary digit. It has only two possible conditions: ON and OFF. Everything in your computer is stored and computed with ONs and OFFs. The bits inside your computer are arranged in such a way as to work in units. The most basic unit, or group, of bits is called a **byte.** A byte consists of 8 bits. Mathematically, 8 bits have 256 unique ON and OFF combinations. You can figure this out with your pocket calculator—just calculate 2^8, which is 2 multiplied by itself eight times. A **word** is 16 bits, or 2 bytes. When 4 bytes are taken together, such as in 32-bit microprocessors, they are called a **double word.** These different arrangements are shown in Figure 32.

In PCs a method called **parity checking** is used to help detect errors. There are times when, in the process of working with computer bits, a bit within a byte may accidentally change from ON to OFF or from OFF to ON. To check for such an error, parity checking uses an extra bit called the **parity bit.** IBM and most compatibles use what is called **even parity** to check their bits. Even parity means that there will always be an even number of ONs for each byte, including the parity bit. Even parity checking is illustrated in Figure 33.

FIGURE 32   Arrangement of computer data

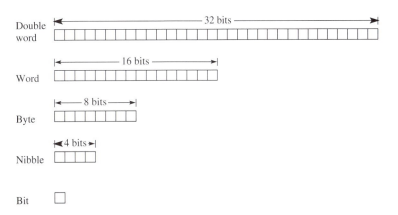

FIGURE 33   Even parity checking

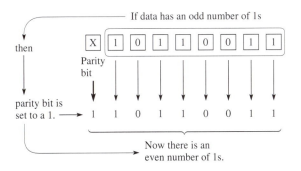
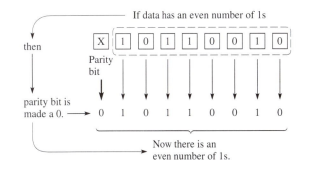

317

FIGURE 34   Single-In-Line
Memory Module (SIMM)

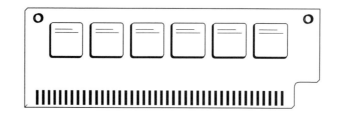

SIMM

The **Single-In-Line Memory Module,** or SIMM, is another way of physically organizing memory. It is a small "boardlet" with several memory chips soldered to it. This boardlet is inserted into a system slot. Figure 34 shows a SIMM.

SIMMs came about in an attempt to solve two problems. The first problem was "chip creep." Chip creep occurs when a chip works its way out of a socket as a result of thermal expansion and contraction. The old solution to this problem—soldering memory chips into the board—wasn't a good solution, because it made them harder to replace. So the SIMM was created. The only problem with the SIMM is that if only 1 bit in any of its chips goes bad, the whole SIMM must be replaced. This is more expensive than replacing only one chip. SIMMs come in 256KB, 1MB, 4MB, and 16MB sizes. A similar type of memory module, called a SIPP, contains metal pins that allow the SIPP to be soldered directly onto the motherboard.

Regarding parity bits in a SIMM, a 32-bit SIMM is nonparity, and a 36-pin SIMM stores one parity bit for each byte of data. Pentium processors incorporate parity in their address and data buses.

DIMM

The Dual In-Line Memory Module (DIMM) was created to fill the need of Pentium-class processors containing 64-bit data buses. A DIMM is like having two SIMMs side by side, and come in 168-pin packages (more than twice that of a 72-pin SIMM). Ordinarily, SIMMs must be added in pairs on a Pentium motherboard to get the 64-bit bus width required by the Pentium.

SDRAM

Synchronous DRAM (SDRAM) is very fast (up to 100-MHz operation) and is designed to synchronize with the system clock to provide high-speed data transfers.

EDO DRAM

Extended Data Out DRAM (EDO DRAM) is used with bus speeds at or below 66 MHz and is capable of starting a new access while the previous one is being completed. This ties in nicely with the bus architecture of the Pentium, which is capable of back-to-back pipelined bus cycles. Burst EDO (BEDO RAM) contains pipelining hardware to support pipelined burst transfers.

VRAM

Video RAM (VRAM) is a special *dual-ported* RAM that allows two accesses at the same time. In a display adapter, the video electronics needs access to the VRAM (to display the Windows desktop, for example) and so does the processor (to open a new window on the desktop). This type of memory is typically local to the display adapter card.

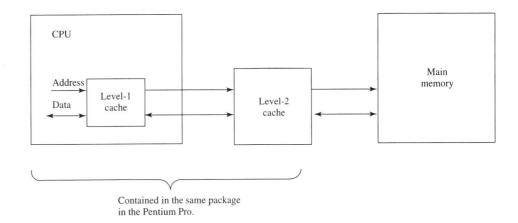

FIGURE 35 Using cache in a memory system

CPU

Address

Data

Level-1 cache

Level-2 cache

Main memory

Contained in the same package in the Pentium Pro.

LEVEL-2 CACHE

Cache is a special high-speed memory capable of providing data within one clock cycle and is typically ten times faster than regular DRAM. Although the processor itself contains a small amount of internal cache (8KB for instructions and 8KB for data in the original Pentium), you can add additional level-2 cache on the motherboard, between the CPU and main memory, as indicated in Figure 35. Level-2 cache adds 64KB to 2MB of external cache to complement the small internal cache of the processor. The basic operation of the cache is to speed up the average access time by storing copies of frequently accessed data.

CHIP SPEED

When replacing a bad memory module, you must pay attention to the *speed* requirements of that module. If you do not, the replacement will not work. You can use memory chips that have a higher speed or the same speed as the replacement, but not a lower speed. The faster the memory, the more it costs. Adding faster memory to your system may not improve overall performance at all because the speed of your computer is determined by the system clock, among other things.

Memory chip speed is measured in **nanoseconds.** A nanosecond is 0.000000001 second. To check the speed rating of a RAM chip, look at the coding on the top of the chip. Typical DRAM speeds are 60 ns and 70 ns.

HOW MEMORY IS ORGANIZED

It is important that you have an understanding of the organization of memory in the computer. The 8088 and 8086 microprocessors are able to address up to 1MB of memory. Since some of the first PCs used the 8088 and 8086 microprocessors, when the 80286, 80386, 80486, and Pentium microprocessors were introduced, they were made **downward compatible** with their predecessors. This meant that software that worked on an older PC would still work on the newer systems.

In order to keep downward compatibility, the newer microprocessors (80286, 80386, 80486, and Pentium) come with two modes of operation (there is one other mode that will be presented later). One mode is called the **real mode,** in which the microprocessors behave like their earlier models (and are limited to 1MB of addressable memory). The other mode, called the **protected mode,** allows the microprocessors to use the newer power designed into them (such as addressing up to 16MB for the 80286 and 4GB for the 80386, 80486, and Pentium).

All PCs and compatibles have what is called a **base memory,** which is the 1MB of memory that is addressable by the 8088, 8086, and newer microprocessors running in real mode. A common way of viewing the organization of memory is through the use of a **memory map.** A memory map is simply a way of graphically showing what is located at different addresses in memory. Figure 36 shows the memory map of the PC in real mode.

319

FIGURE 36 Memory map of PC
in real mode

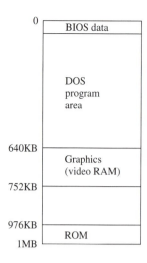

TABLE 16 Purpose of
allocated memory

Assignment	Definition
Base memory	This refers to the amount of memory actually in-stalled in the conventional memory area.
Conventional or user memory	This is the 640KB of memory that is usable by DOS-based programs.
Video or graphics memory	This area of memory (128KB) is reserved for storing text and graphics material for display on the monitor.
Motherboard ROM	This is space reserved for the use of the ROM chips on the motherboard.

As you can see from the memory map, several areas of memory are designated for particular functions; not all the 1MB of memory is available for your programs. As a matter of fact, only 640KB can be used by DOS-operated systems. Table 16 lists the definitions of the various memory sections.

The memory above the conventional 640KB of memory is referred to as **upper memory.**

HOW MEMORY IS USED

There are three ways memory can be allocated: as **conventional memory, extended memory,** or **expanded memory.** Figure 37 shows the relationships among the three types of memories.

Table 17 explains the uses of the three different memory allocation methods. It is important to note that DOS and systems that use DOS are limited to 1MB of addressable space. The reason for this is that DOS is made for microprocessors running in the real mode. This memory limit exists because computer and program designers thought that 1MB of memory would be all that anyone would ever need on a PC for years to come. Since 340KB of the 1MB of addressable DOS memory is reserved by the system, DOS really has only 640KB left for user programs. Thus, DOS is said to have a 640KB limit. However, as you will see, there are ways of allowing DOS to store data in an addressable memory location that is beyond this DOS limit.

VIRTUAL MEMORY

Another method of extending memory is through the use of **virtual memory.** Virtual memory is memory that is not made up of real, physical memory chips. Virtual memory is memory made up of mass storage devices such as disks. In the use of virtual memory, the

FIGURE 37   Relationships among conventional, extended, and expanded memory

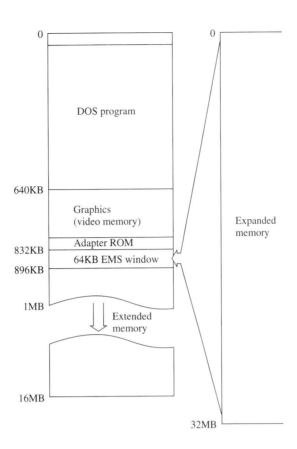

TABLE 17   Memory allocation methods

Memory Type	Comments
Conventional	1. Memory between 0KB and 1MB, with 640KB usable and 34KB reserved. 2. Completely usable by DOS-based systems. 3. Uses real mode of 8086, 8088, 80286, 80386, 80486, and Pentium.
Extended	1. Uses protected mode of 80286 (up to 16MB), 80386, 80486, and Pentium (up to 4GB). 2. Cannot be used by DOS-based systems (which are limited to 1MB of memory). 3. Can all be accessed by the IBM OS/2 operating system. 4. Can be used by a virtual disk in DOS systems. 5. Is the type of memory to use with the 80386, 80486, and Pentium microprocessors.
Expanded	1. Uses "bank-switching" techniques. 2. Requires special hardware and software. 3. Is not a continuous memory but consists of chunks of memory that can be switched in and out of conventional memory. 4. Sometimes referred to as EMS (expanded memory specification) memory.

computer senses when its usable real memory is used up, stores what it deems necessary onto a disk (usually the hard disk), and then uses what it needs of the freed-up real memory. If it again needs the data it stored on the disk, it simply frees up some more real memory (by placing its contents on the disk) and then reads what it needs from the disk back into the freed-up memory. The concept of virtual memory is illustrated in Figure 38. Windows uses demand-paging virtual memory to manage memory, a technique supported by features of protected mode.

FIGURE 38   Virtual memory

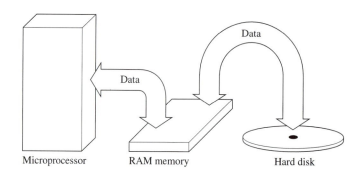

Data

Data

Microprocessor   RAM memory   Hard disk

MEMORY USAGE IN WINDOWS

It is not difficult to determine why an operating system performs better if it has 32MB of RAM available, rather than only 8MB. With 32MB of RAM, the operating system will be able to support more simultaneous processes without having to use the hard drive for virtual memory backup. Additional memory will also be available for the graphical user interface (multiple overlapped windows open at the same time).

It is no secret that the Windows 3.x architecture did not use memory efficiently, typically requiring at least 8MB or 16MB to get a reasonable amount of performance on a 386 or 486 CPU. Windows 95/98 also performs much better when given a large amount of RAM to work with. A minimum of 16MB or 32MB is recommended.

What does Windows 95/98 use memory for? Conventional memory, the first 640KB of RAM, still plays an important role supporting real-mode device drivers and DOS applications. For instance, if you want DOSKEY installed as part of your DOS environment under Windows 95, place its command line in AUTOEXEC.BAT as you normally would.

Upper memory, the next 360KB of the first 1MB of RAM, can be used to place DOS and other memory-resident applications above the 640KB limit, freeing more RAM for DOS applications.

Extended memory, everything above the first 1MB of RAM (4096MB total), is where Windows 95/98 runs most applications, using an addressing scheme called *flat addressing*. The flat addressing model uses 32-bit addresses to access any location in physical memory, without the need to worry about the segmented memory scheme normally used.

Windows 95/98 provides the Resource Meter (in the System Tools folder under Accessories) to monitor system resources. As indicated in Figure 39, the amount of resources available is shown graphically. The display is updated as resources are used and freed up.

Another useful tool is the System Monitor in Windows 95/98 and the Performance Monitor in Windows NT (shown in Figure 40), which display a running tally of resource usage over a period of time. The display format is selectable (bar, line, or numeric charts), as are the colors and type of information displayed.

SECONDARY STORAGE, PART 1: THE FLOPPY DISK DRIVE

A floppy disk drive is a device that enables a computer to read and write information on a floppy disk.

FIGURE 39   Resource Meter display

Resource Meter

System resources: 76% free

User resources: 76% free

GDI resources: 80% free

OK

FIGURE 40 (a) Windows 95/98 System Monitor display, and (b) Windows NT Performance Monitor display

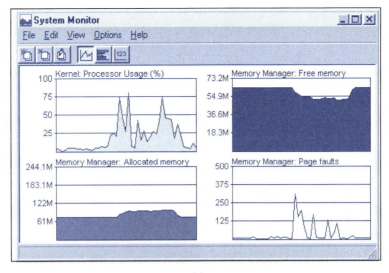

(a)

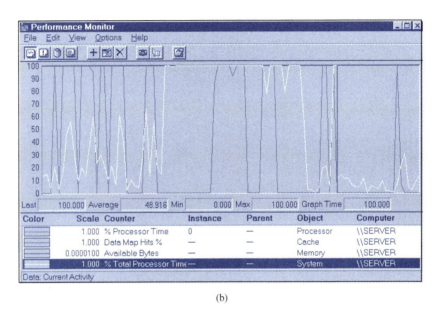

(b)

Floppy disk drives (FDDs) are located at the front panel of the computer. The most common is the $3^1/2$" drive (the old $5^1/4$" drive is almost obsolete). These two disk drives are illustrated in Figure 41.

HOW A FLOPPY DISK DRIVE WORKS

Figure 42 is a simplified drawing of an FDD with its major components. Table 18 summarizes the purpose of each major component of the FDD.

OPERATING SEQUENCE

The operating sequence of a typical $3^1/2$" drive is as follows. Pushing the floppy disk into the drive causes the disk to be properly seated. The initial start-up for the drive consists of determining where track 0 is located. This is usually accomplished by a mechanical device, which is activated once the drive head is over track 0. When information is read, the stepping motor moves the read/write heads to their proper location. When information is written, the disk's write-protect status is checked and then new information is added to the disk.

323

FIGURE 41   Typical floppy disk drives

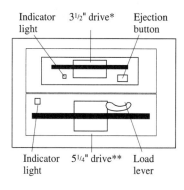

Indicator light   3½" drive*   Ejection button

Indicator light   5¼" drive**   Load lever

FRONT OF COMPUTER

*Most frequently used now.

**Rarely used now.

FIGURE 42   Major components of a floppy disk drive

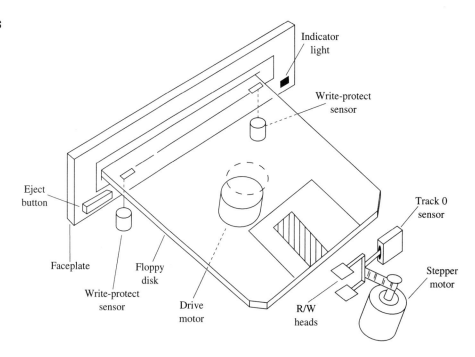

TABLE 18   Main parts of an FDD

Part	Purpose
Eject button	Used to eject a disk from the drive.
Write-protect sensor	Checks the condition of the floppy disk's write-protect system.
Read/write heads	Read and write information magnetically on the floppy disk. The heads move together, each working from its own side of the disk.
Track 0 sensor	Indicates when the read/write head is over track 0 of the floppy disk.
Drive motor	Spins the floppy disk inside the FDD.
Stepper motor	Moves the read/write head to different positions on the floppy disk.
Indicator light	Indicates if the disk drive is active.

DISK DRIVE SUPPORT SYSTEM

For proper operation, each part of the FDD support system must function properly.

1. *OS.* The operating system must be compatible with the media on the floppy disks.
2. *Floppy disk.* The disk itself must contain accurately recorded information in the proper format.

3. *Disk drive controller.* The drive controller conditions the signals between the motherboard and the FDD. Originally, a controller card that plugged into the motherboard was used to control the floppy drive. Most motherboards now have the floppy controllers built in.
4. *Disk drive electronic assembly.* This assembly consists of circuit boards that control the logical operations of the FDD. They act as an electrical interface between the disk drive controller and the electromechanical parts of the FDD.
5. *Disk drive mechanical assembly.* This assembly ensures proper alignment of the disk and read/write heads for reading and writing information.
6. *System power supply.* The power supply provides electrical power for all parts of the FDD, including the motors.
7. *Interconnecting cable.* Ribbon cable is used to transfer signals between the disk drive electrical assembly and the disk drive controller card.
8. *Power cable.* The DC power cable supplies electrical power to all parts of the FDD.

Let us take a closer look at many of these important components.

The OS

The operating system plays an important role in the operation of the floppy drive. Beginning with the system BIOS, all drive parameters must be known by the operating system so that data can be properly exchanged. Windows contains many applications designed specifically for disk drive operations. For example, right-clicking the drive A: icon in the My Computer window produces a Properties window similar to that shown in Figure 43. A pie chart is used to graphically illustrate used/free space on the drive. The volume label can be changed by entering a new one in the text box.

The Floppy Disk

There are several built-in Windows tools available for working with floppy disks. These tools are contained in the Tools submenu, as indicated in Figure 44.

The disk can be scanned for errors (using ScanDisk), backed up, or *defragmented.* A disk that has had many files created and deleted on it eventually becomes fragmented, with the files broken up into groups of sectors and scattered all over the disk (but still logically connected through the use of the FAT). This fragmentation increases the amount of time required to read or write entire files to the disk. By defragmenting the disk, all the files are reorganized, stored in consecutive groups of sectors at the beginning of the storage space on the disk.

FIGURE 43   Drive A: Properties window

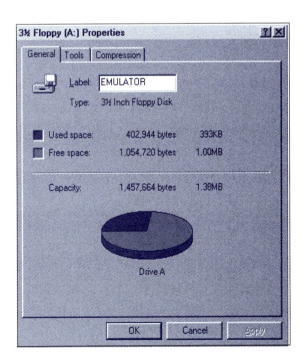

FIGURE 44   Floppy disk tools

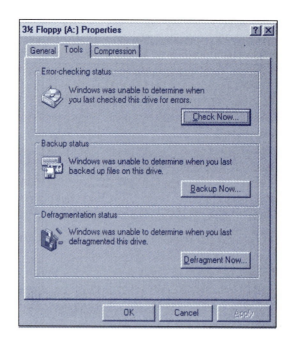

FIGURE 45   Compression submenu

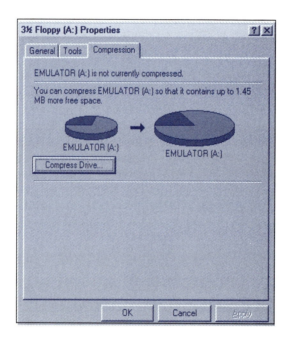

A fourth tool is included that allows you to *compress* the data on your floppy, increasing the amount of free space available. The Compression submenu shown in Figure 45 indicates what will be gained by compressing the current disk.

The Disk Drive Controller

The disk drive controller used to be an individual chip on a controller card. Now, the controller is just one part of a multifunction peripheral IC designed to operate the floppy and hard drives, the printer, and the serial ports. As always, I/O ports and interrupts are used to control the floppy disk drive. The settings used by Windows 95/98 can be examined/changed by using the Device Manager submenu of System Properties. The settings used by Windows NT can be examined from the Windows NT Diagnostics menu. Figure 46 shows the hardware configuration of a typical A: drive. These settings can be changed if necessary.

FIGURE 46 (a) Hardware settings for floppy drive, and (b) Windows NT Floppy Properties

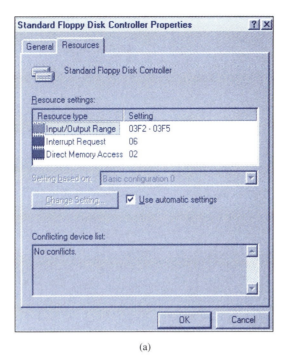

(a)

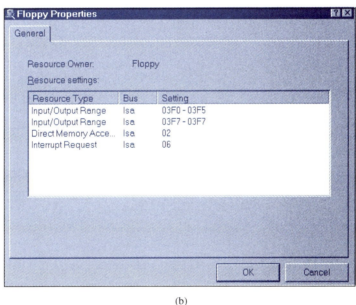

(b)

The Drive Cable

A 34-conductor ribbon cable is used to connect one or two disk drives to the controller. A twist in the cable between the two drive connectors reverses the signals on pins 9 through 16 at each connector. This twist differentiates the two drive connectors, forcing them to be used specifically for drive A: or drive B:. Figure 47 shows the cable details.

The meanings of the signals on the drive cable are shown in Table 19. Note that the signals affected by the twist in the cable are the select and enable signals for each drive.

The Power Cable

Like any other peripheral, the disk drive needs power to operate the drive and stepper motors, the read/write amplifiers and logic, and the other drive electronics. Figure 48 shows the pinout of the standard power connector used on the floppy drive. The connector is keyed so that it only plugs in one way.

FIGURE 47 **Floppy drive cable**

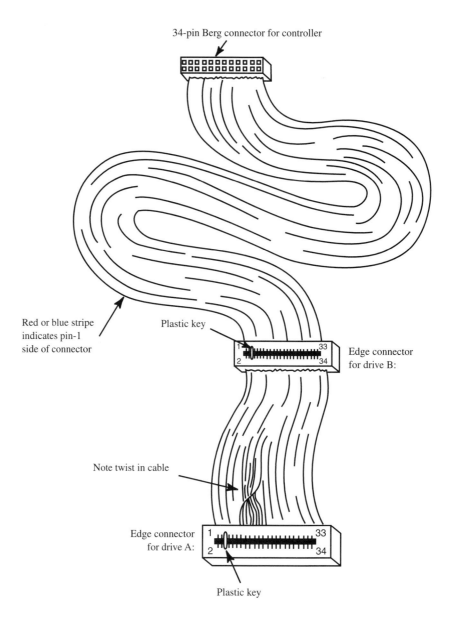

34-pin Berg connector for controller

Red or blue stripe
indicates pin-1
side of connector

Plastic key

Edge connector
for drive B:

Note twist in cable

Edge connector
for drive A:

Plastic key

Zip Drives

A device similar to the floppy drive is the Zip drive, manufactured by Iomega. Zip drives connect to the printer port, have removable 100MB cartridges, and boast a data transfer rate of 60MB/minute (using an SCSI connection). The 100MB disks spin at 2941 RPM, have an average access time of 29 milliseconds, and are relatively inexpensive. Newer 250MB cartridges are also available.

The software driver for the Zip drive uses the signal assignments shown in Figure 49 to control the Zip drive through the printer port. Using the printer port to control the Zip drive allows you to easily exchange data between two computers.

Jaz Drives

Similar to the Zip drive, the Jaz drive uses a 1GB removable cartridge that spins at 5400 RPM, has an average seek time around 10 milliseconds, and has a sustained data transfer rate of more than 6MB/second. A Jaz drive operates similarly to a hard drive, except the drive media is removable.

The 120MB SuperDisk

The SuperDisk is a new type of floppy drive with a 120MB capacity. SuperDisk drives can read/write both 120MB SuperDisk floppies and 1.44/2.88 MB 3^1/$_2$" disks.

TABLE 19 Floppy drive cable signals

Conductor (Pin)	Signal
1–33 odd	Ground
2	Unused
4	Unused
6	Unused
8	Index
10	Motor Enable A
12	Drive Select B
14	Drive Select A
16	Motor Enable B
18	Stepper Motor Direction
20	Step Pulse
22	Write Data
24	Write Enable
26	Track 0
28	Write Protect
30	Read Data
32	Select Head 1
34	Ground

FIGURE 48 Floppy drive power connector

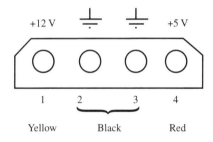

FIGURE 49 Zip drive parallel interface

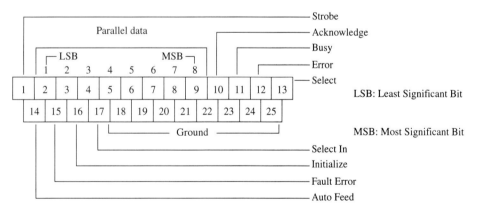

Working with Floppies

A few tips to keep in mind when working with floppy drives:

- Sometimes disks that are formatted on one system cannot be read by another. This may be due to differences in the read/write head alignments between both drives.
- If a floppy gives unexpected read errors, try ejecting the floppy and reinserting it.
- Run ScanDisk or some other suitable disk tool (such as Norton Utilities) to check a troublesome floppy.

- Always beware of disks given to you by someone else. Before you begin using them, scan them for viruses. This is especially important in an educational setting, where students and instructors often exchange disks as a normal part of class or lab.

SECONDARY STORAGE, PART 2: THE HARD DRIVE

The construction of a hard disk system is much different from that of a floppy disk system. With the floppy disk system, data is stored on each side of the disk, but in a hard disk system, there is usually more than one disk, or **platter.** Figure 50 shows the structure of a two-platter hard disk system, in which there are four sides for storing data.

As you can see from Figure 50, the four sides are labeled 0 through 3. Figure 51 shows how a floppy disk organizes its data in single concentric tracks, as compared with a hard disk system, which organizes its data in a combination of tracks called **cylinders.**

FIGURE 50   Typical hard disk structure

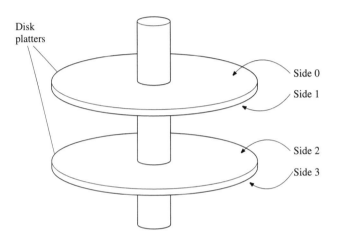

FIGURE 51   Floppy and hard disk organization

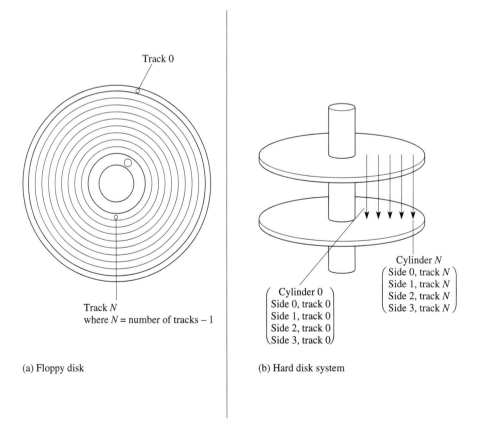

(a) Floppy disk

(b) Hard disk system

PARTITIONS

A hard disk can be formatted so that it acts as two or more independent systems. As an example, it is possible for a hard disk to operate under two entirely different operating systems, such as Windows and UNIX. Doing this is called **partitioning** the disk, as shown in Figure 52. This is sometimes necessary when a computer is part of a collection of computers connected together over a network.

In the early versions of DOS, no disk partition could be larger than 32MB, and no one operating system could access more than one partition. This means that no matter how much data the hard disk could hold, early versions could make use of only 32MB of the disk surface. DOS 4.0 broke the 32MB limitation on hard disks.

When a disk is partitioned, its **primary partition** (the one from which it boots) is called the C: drive, and the remaining partitions are referred to as D, E, F, and so on.

Every disk may be partitioned differently, but there are a few rules that must be followed. For example, Windows 95 Version A can recognize disks as large as 2.1GB. Windows 95 Version B and above can address disks as large as 4TB (terabytes). The operating system determines the maximum size that can be handled. Updates to an operating system add capabilities for new technology.

The FDISK tool supplied by DOS and Windows is used to create partitions on the hard disk. Figure 53 shows the Windows 98 FDISK menu. It looks very similar to the old DOS

FIGURE 52   Single and multiple partitions of a hard disk

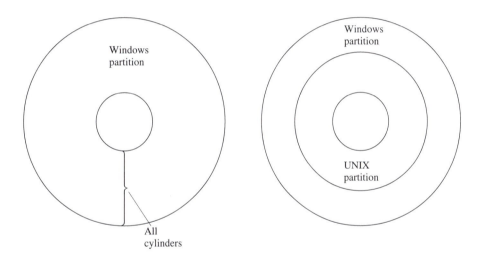

FIGURE 53   Windows 98 FDISK utility

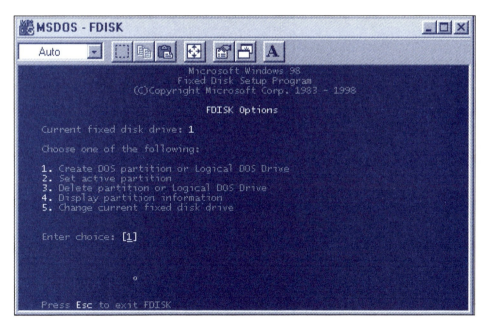

FDISK program. FDISK is used to create, modify, or delete partitions on a hard drive. Extreme caution must be observed when working with the FDISK program.

There are also many specialty programs designed to make the disk partitioning process easier and more flexible than the FDISK program. For example, PartitionMagic by Power-Quest allows for a hard drive to be partitioned dynamically, saving time and disk space. Figure 54 shows the PartitionMagic main window. Information about the default drive is displayed automatically, showing the size of each partition and associated disk format. By selecting the Info Options button, the Partition Information window is displayed showing the default disk usage statistics, as shown in Figure 55.

The Cluster Waste tab shows the current amount of disk space that is wasted. This waste is attributed to the smallest amount of disk space that can be allocated by the oper-ating system. For example, if we want to store one character in a file, it will be stored in a

FIGURE 54   PartitionMagic main window

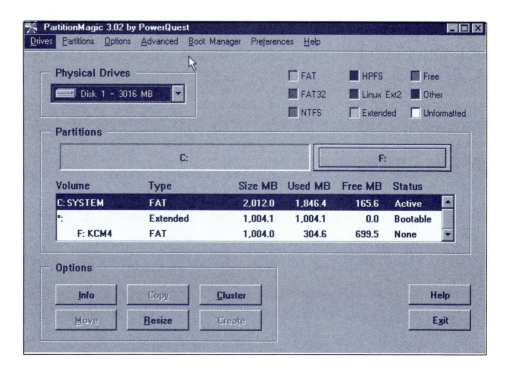

FIGURE 55   Disk Usage Partition Information

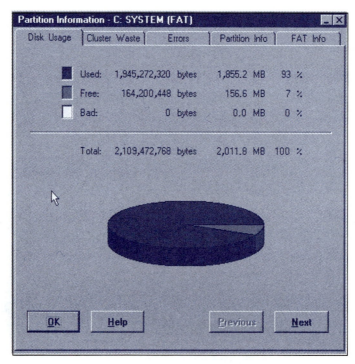

32K chunk of space on a computer with 32K clusters (shown in Figure 56). The cluster size is determined by the type of file structure used on the disk, such as FAT16 or FAT32, which will be discussed shortly.

PartitionMagic can also display information about the physical layout of a partition, as shown in Figure 57. The first, last, and total physical sectors are displayed along with the corresponding cylinder and head information. The disk physical geometry is also indicated on the Partition tab.

Details about the FAT are available on the FAT Info tab shown in Figure 58. This window contains the details of the FAT structure, such as the number of FATs, root directory capacity, First FAT sector, First Data sector, and other interesting information.

FIGURE 56   Cluster Waste Partition Information

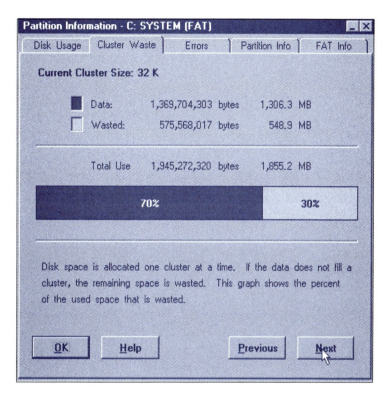

FIGURE 57   Physical Partition Information

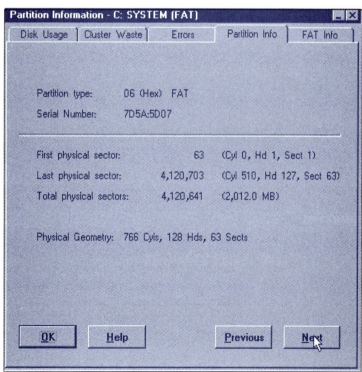

FIGURE 58   FAT Info tab

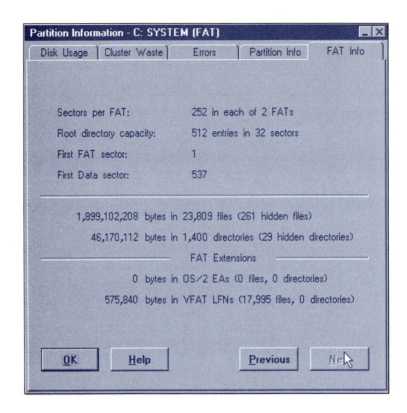

FIGURE 59   Cluster Analysis window

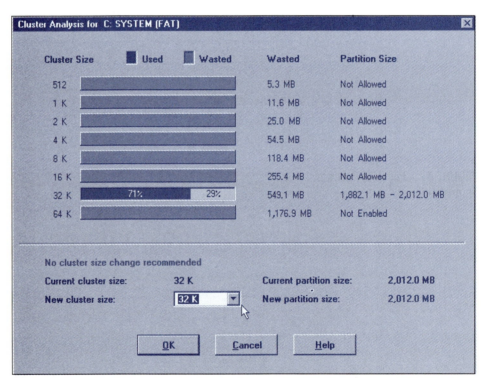

PartitionMagic can also change partition information dynamically, such as changing the cluster size. Figure 59 shows the common cluster sizes and associated wasted space. In this case, since the disk is close to capacity, PartitionMagic cannot recommend any type of changes; otherwise, the user may select a new cluster size and change the partition size. If it is necessary to change disk partition size frequently, it may be a good idea to invest in a software package.

FIGURE 60 Windows NT Disk Administrator window

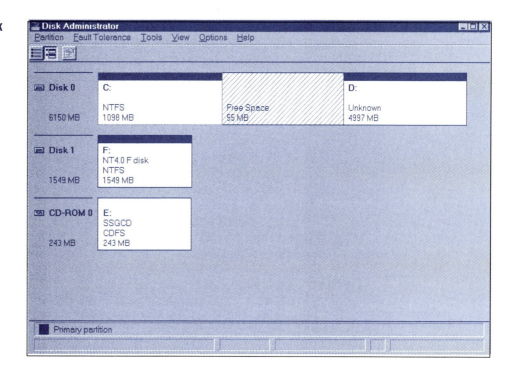

WINDOWS NT DISK ADMINISTRATOR

Windows NT provides a different method to deal with the chore of managing disks and disk partitions. The Disk Administrator utility, located in the Windows NT Administrative Tools (Common) menu, offers many advantages over running the traditional FDISK program in a DOS window. Figure 60 shows a graphical display of the physical partitions on a computer system. Notice how Disk 0 contains an NTFS partition at the beginning of the disk labeled "C:" and an unknown partition at the end of Disk 0 labeled "D:" with 55MB of free space remaining. Also note from the figure that Disk 1 contains an NTFS partition and a label of "F:" with no free space and the CD-ROM 0 reports information about the CD currently in the drive. On this computer, the D: disk is actually the Windows 98 operating system, but because the format is FAT32, Windows NT cannot read it. Likewise, Windows 98 cannot see the NTFS partitions.

You are encouraged to explore the capabilities of the Disk Administrator utility.

LBA

Logical block addressing (LBA), is a method to access IDE (Integrated Drive Electronics) hard disk drives. Using LBA, disks larger than 504MB (1024 cylinders) can be partitioned using FDISK or PartitionMagic. Actually, LBA has been around for quite some time now, and has been incorporated into system BIOS on most PCs. Before this, BIOS limitations prevented FDISK from using the entire drive and it was necessary to use custom software, called a Dynamic Drive Overlay. The last option was to simply stay under the limit.

LBA may be implemented in four ways. For example:

1. ROM BIOS support for INT 13h.
2. Hard disk controller support for INT 13h.
3. Use only 1024 cylinders per partition.
4. Real-mode device driver support for geometry translation.

Windows 95 and Windows 98 support the first three methods directly. The last method requires a special version of the Dynamic Drive Overlay software.

IDE disks using the ATA interface also use BIOS INT 13h services. The disk drive identifies itself to the system BIOS specifying the number of cylinders, heads, and sectors per track. The number of bytes in each sector is always 512.

FAT32

As hard drives grew in storage capacity, they quickly reached the maximum size supported by DOS and Windows (initially only 32MB, then 504MB, then 2.1GB). This limitation was based on the number of bits used to store a cluster number. The original FAT12 used a 12-bit FAT entry. FAT16 added four more bits, allowing for up to 65,536 clusters. With each cluster representing 16 sectors on the disk, and each sector storing 512 bytes, a cluster would contain 8KB. The total disk space available with 65,536 clusters of 8KB is 512MB, a small hard drive by today's standards.

One way to support larger partitions is to increase the size of a cluster. Storing 32KB in a cluster allows a 2048MB (2.048GB) hard drive, but also increases the amount of wasted space on the hard drive when files smaller than 32KB are stored. For example, a file of only 100 bytes is still allocated 32KB of disk space when it is created because that is the smallest allocation unit (one cluster). You would agree that most of the cluster is wasted space. Some disk compression utilities reclaim this wasted file space for use by other files. In general, however, large cluster sizes are not the solution to the limitation of the FAT16 file system.

FAT32 uses 32-bit FAT entries, allowing 2200GB hard drives without having to result to using large cluster sizes. In fact, FAT32 typically uses 4KB clusters, which helps keep the size of the FAT small and lowers the amount of wasted space. FAT32 is only used by Windows 95 Version B (OEM Service Pack 2) and Windows 98. Several utilities, such as PartitionMagic, are able to convert a FAT16 disk into a FAT32 disk. Windows NT has its own incompatible file system called NTFS.

NTFS

The **NT file system,** or NTFS, is used on Windows NT computers. Using NTFS, it is possible to protect individual items on a disk and therefore prevent them from being examined or copied. This is a feature commonly found on multiuser computers such as Windows NT.

HARD DRIVE INTERFACES

Many companies manufacture hard drives for personal computers. Even though each company may design and build its hard drives differently, the interface connectors on each drive must conform to one of the accepted standards for hard drive interfaces. These interface standards are illustrated in Figure 61.

The first popular hard drive interface scheme was invented by Shugart Technologies. Called **ST506,** it requires two cables (control and data) between the controller card and the hard drive. This is shown in Figure 61(a). Serial data passes back and forth between the controller and hard drive over the data cable. A second hard drive is allowed; the second drive shares the control cable with the first drive, but has its own data cable. Jumpers must be set on each drive for proper operation, and the last drive needs to contain termination resistors.

An improvement on the ST506 standard was developed by Maxtor, another hard drive manufacturer. Called **ESDI** (Enhanced Small Device Interface), this interface uses the same two cables as the ST506, but allows data to be exchanged between the controller and hard drive at a faster rate. Although similar in operation to ST506, ESDI is not electrically compatible with it. Thus, ST506 hard drives require ST506 controllers, and ESDI hard drives require ESDI controllers.

The **IDE** (Integrated Drive Electronics) interface, shown in Figure 61(b), has virtually replaced the older ST506 standard. A single cable is used to exchange parallel data between the adapter card and the hard drive. A second hard drive uses the same cable as the first, with a single jumper on each drive indicating if it is the primary drive. A significant differ-

FIGURE 61 Hard drive connections

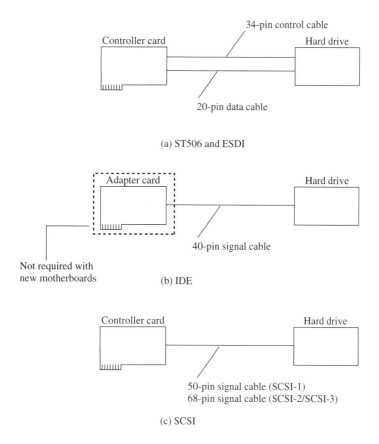

(a) ST506 and ESDI

(b) IDE

Not required with new motherboards

(c) SCSI

ence in the IDE standard is that the controller electronics are located *on the hard drive it-self.* In a two-drive system, the primary drive controls itself and the other hard drive as well. The adapter card plugged into the motherboard merely connects the hard drive to the system buses, using parallel data transfers. This allows typical data transfers with an IDE hard drive of up to 8MB per second.

New motherboards have built-in **EIDE** (enhanced IDE) controllers, which provide signals for four EIDE connectors. This eliminates the need for an adapter card. One pair of connectors are the *primary* connectors, the other pair are the *secondary* connectors. Each pair can support two IDE hard drives (CD-ROM and tape backup drives as well) in a master/slave configuration. Figure 62 shows the pin and signal assignments for the EIDE interface, and a typical IDE hard drive.

The earlier IDE interface lacked the upper eight data lines D_8 through D_{15}. The IDE specification also limited hard drive capacity to just 504MB. The EIDE specification increases drive capacity to more than 8GB, expands the maximum number of drives from two to four, and increases the data transfer rate to over 16MB/second.

Right-clicking on My Computer and then choosing Properties allows you to check the system properties. In Device Manager, double-click on the Hard disk controller entry and then the primary IDE controller entry. You should get a settings window similar to that shown in Figure 63. In Windows NT, this information is available in the Windows NT Diagnostics window.

As indicated in Figure 63, the primary IDE controller uses interrupt 14. A handful of I/O ports are required as well, to issue commands and read data from the controller.

The **SCSI** (Small Computer System Interface) standard, shown in Figure 61(c), offers more expansion capability than any of the three previously mentioned standards. Like the IDE standard, the SCSI standard uses a single cable to control the hard drive. But where all three previous standards allowed at most two hard drives, the SCSI standard allows up to *seven* devices to be daisy-chained on the single 50-pin cable. Each device can be a hard drive, if necessary. This difference is important to computer users who have large storage requirements and might require **gigabytes** (1GB = 1024MB) of hard drive capacity. SCSI is used to connect a wide variety of devices together on a shared bus. For example,

FIGURE 62 (a) Primary EIDE connector pin/signal assignments, and (b) typical IDE hard drive (*photograph by John T. Butchko*)

$\overline{\text{RESET}}$	1	2	GND
D_7	3	4	D_8
D_6	5	6	D_9
D_5	7	8	D_{10}
D_4	9	10	D_{11}
D_3	11	12	D_{12}
D_2	13	14	D_{13}
D_1	15	16	D_{14}
D_0	17	18	D_{15}
Ground	19	20	Key (missing pin)
DRQ3	21	22	GND
$\overline{\text{IOW}}$	23	24	GND
$\overline{\text{IOR}}$	25	26	GND
IOCHRDY	27	28	ALE
DACK3	29	30	GND
IRQ14*	31	32	IO16
A_1	33	34	GND
A0	35	36	A_2
CS0	37	38	CS1
SLV/ACT	39	40	GND

* IRQ15 for the secondary IDE connector

(a)

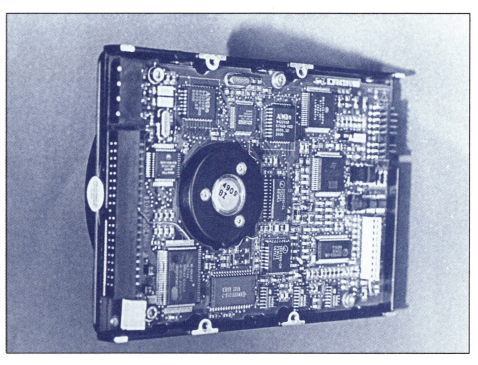

(b)

several disk drives, a tape drive, and a scanner can be connected to one SCSI bus. Each device on the bus and the controller card itself requires an address and the end of the SCSI cable, or the last device on the bus, must be terminated. Figure 64 shows a daisy chain of SCSI devices. The last device contains a terminator.

There are several different types of SCSI buses, each allowing for a specific cable type, transfer rate, bus width, and so on. Table 20 shows the different SCSI standards.

FIGURE 63   Hardware settings for primary IDE controller

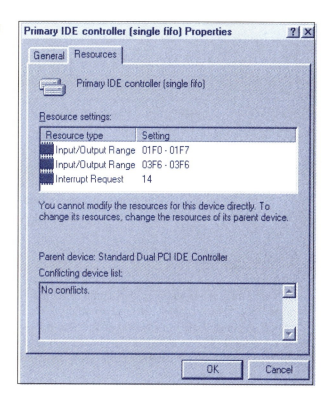

FIGURE 64   A daisy chain of SCSI devices

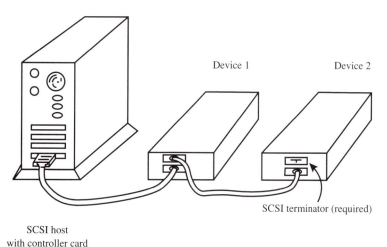

TABLE 20   SCSI bus standards

Standard	Bus Width	Max Transfer Rate (Mbps)	Cable Type
SCSI-1	8	4	Not specified
SCSI-2	8 16	5 10	A B
SCSI-3	16 32	10 20	P P, Q

SCSI buses also have specific length requirements falling into two categories: single-ended and differential. A single-ended SCSI bus is cheap and fast over short distances. Differential SCSI can be used over longer distances. Table 21 shows the SCSI bus length requirements.

339

TABLE 21   SCSI bus
lengths

Bus Type	Single-Ended	Differential
SCSI-1	6 meters	25 meters
SCSI-2	6 meters	25 meters
SCSI-3	3 meters	25 meters

There are also different types of connectors used to connect SCSI devices together. Check the individual requirements for each SCSI device to determine the appropriate type. Note that SCSI devices are generally more expensive than non-SCSI devices, but they provide for combinations of devices not possible with standard PC technology.

DATA STORAGE

Although there are differences between IDE hard drives and SCSI hard drives (and all the other types), there is also something in common: each hard drive uses the flux changes of a magnetic field to store information on the hard drive platter surface. A number of different techniques are used to read and write 0s and 1s using flux transitions. Some of the more common techniques are listed in Table 22.

DISK CACHING

Because of mechanical limitations (rotational speed of the platters; movement and settling time of the read/write head), the rate at which data can be exchanged with the hard drive is limited. It is possible to increase the data transfer rate significantly through a technique called **caching.** A hardware cache is a special high-speed memory whose access time is much shorter than that of ordinary system RAM. A software cache is a program that manages a portion of system RAM, making it operate as a hardware cache. A computer system might use one or both of these types of caches, or none at all.

The main idea behind the use of a cache is to increase the *average* rate at which data is transferred. Let us see how this is done. First, we begin with an empty cache. Now, suppose that a request to the hard drive controller requires 26 sectors to be read. The controller positions the read/write head and waits for the platters to rotate into the correct positions. As the information from each sector is read from the platter surface, a copy is written into the cache. This entire process may take a few *milliseconds* to complete, depending on the drive's mechanical properties. If a future request requires information from the same 26 sectors, the controller reads the copy from the cache instead of waiting for the platter and read/write head to position themselves. This means that data is accessed at the faster rate of the cache (whose access time might be as short as 10 *ns*). This is called a cache *hit.* If the

TABLE 22   Data recording
techniques

Technique	Meaning/Operation
MFM	Modified Frequency Modulation. Magnetic flux transitions are used to store 0s and 1s.
RLL	Run Length Limited. Special flux patterns are used to store *groups* of 0s and 1s.
Advanced RLL	Advanced Run Length Limited. Permits data to be recorded at higher density than RLL.

requested data is not in the cache (a *miss*), it is read from the platter surface and copied into the cache as it is outputted, to avoid a miss in the future. The cache uses an algorithm to help maintain a high hit ratio.

The same method is used for writing. Data intended for the hard drive is written into the cache very quickly (8MB/second), and then from the cache to the platter surface at a slower rate (2.5MB/second) under the guidance of the controller.

Many hard drives now come with 256KB of onboard hardware cache. In addition, a program called SMARTDRV can be used to manage system RAM as a cache for the hard drive. To use SMARTDRV, a line such as

```
C:\DOS\SMARTDRV.EXE 2048
```

must be added to your AUTOEXEC.BAT file. This command instructs SMARTDRV to use 2MB of expanded or extended memory as a cache. Small programs that are run frequently (DOS utilities stored on the hard drive) load and execute much more quickly with the help of SMARTDRV.

DISK STRUCTURE

The information presented here applies equally well to floppy disks as to hard disks. Figure 65 shows how a disk is divided into **sectors.** From the figure, you can see that a sector is a specified pie-slice area on the disk.

Disk sectors and tracks are not physically on the disk, just as data is not physically on the disk. They are simply magnetic patterns placed on the disk by electrical impulses. The number of tracks available on the disk varies. For example, a standard $3^1/_2$-in. floppy disk has 80 tracks, whereas a hard disk may contain 650. The number of sectors a disk has also varies. This is illustrated in Table 23.

DISK STORAGE CAPACITY

You can calculate the storage capacity of a disk as follows:

$$DSC = sides \times tracks \times sectors \times size$$

FIGURE 65   Disk sectors

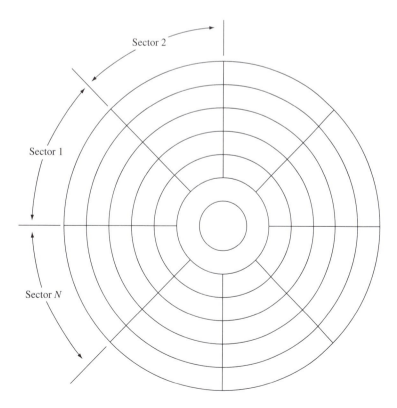

TABLE 23   Floppy disk
configurations

Disk Size	Disk Type	Tracks/Side	Total Sectors*
5.25	Single-sided—8 sectors per track	40	320
5.25	Single-sided—9 sectors per track	40	360
5.25	Double-sided—8 sectors per track	40	640
3.5	Double-sided—9 sectors per track	40	720
3.5	Quad-density—9 sectors per track	80	1440
5.25	Quad-density—15 sectors per track	80	2400

*Common sector sizes for disks are 128, 256, 512, and 1024 bytes.

where

$\quad$ DSC = disk storage capacity
$\quad$ sides = number of disk sides used
$\quad$ tracks = number of disk tracks per side
$\quad$ sectors = number of disk sectors per side
$\quad$ size = size of each sector in bytes (usually 512)

As an example, consider a double-sided, double-density (nine-sector-per-track) disk. From Table 23, you can see that such a disk has 40 tracks per side and nine sectors per track, where each sector stores 512 bytes. Thus, for this type of disk, the total disk storage capacity is

$$DSC = 2 \times 40 \times 9 \times 512 = 368,640 \text{ bytes (or 360KB)}$$

Recall that each disk contains a boot sector. This boot sector is contained in sector 1, side 0, track 0, as illustrated in Figure 66. Table 24 lists the information contained in a boot sector.

Following the boot sector, there is a file allocation table (FAT). This table is used by DOS to record the number of disk sectors on the disk that can be used for storage as well as bad sectors that must not be used for storage. Several sectors are reserved for the FAT. All hard disks come from the factory with a certain number of bad sectors. During final product testing, these bad sectors are usually found and are usually then labeled on the hard drive unit itself.

FIGURE 66   Location of the
boot sector

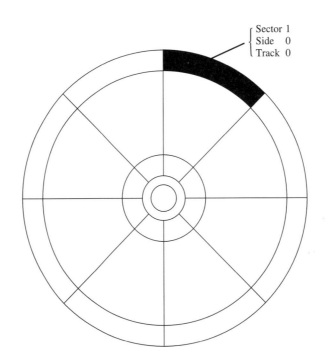

Sector 1
Side   0
Track 0

TABLE 24   Information in a boot sector

- Microprocessor jump instruction
- Manufacturer's name (IBM or Microsoft) and version number
- Number of bytes per disk sector
- Number of sectors per cluster
- Number of reserved sectors
- Number of maximum root directory entries
- Total number of sectors
- Description of the disk media
- Number of sectors per track
- Number of disk heads
- Number of hidden sectors
- The BOOTSTRAP program

512 bytes

In order to ensure reliability, each disk contains a duplicate copy of the FAT. This means that if one copy of the FAT goes bad, the backup copy is available for use.

DISK FRAGMENTATION

Disk fragmentation is the result of one or more disk files being contained in scattered sectors around the disk, as shown in Figure 67. Observe that the read/write heads may take more than one revolution of the disk to read all the file information scattered across the various sectors as a result of disk fragmentation.

Next, consider the file data distributed in contiguous sectors around the disk, as shown in Figure 68. The way the data is distributed here, it is conceivable that it could all be read in one revolution of the disk. The difference between fragmented data and contiguous data is that it takes longer to read fragmented data.

You must keep in mind that disk drives are slow when compared with the rest of the computer system. If information is fragmented over the hard disk, it takes longer to read the disk and significantly slows the entire computer system (because it takes longer to read and write the information). Disk fragmentation occurs when files are repeatedly added and deleted on a disk. Once disk fragmentation occurs (especially on the hard drive), your system will begin to run more slowly when it interacts with the hard drive.

FIGURE 67   Disk fragmentation

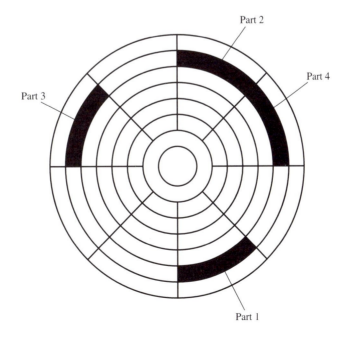

Part 2

Part 4

Part 3

Part 1

FIGURE 68 Contiguous file
data on the same disk track

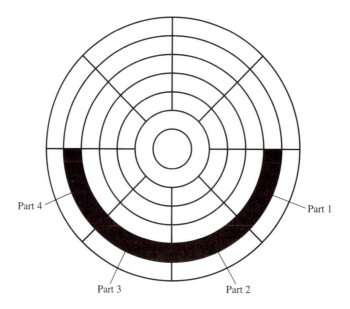

To eliminate this, you must use a hard disk utility program to defragment the disk. It is important now to realize that you must do this periodically in order to maximize system performance.

The defragmentation process can be automated in Windows 95/98 using the system agent. During periods of inactivity, the system agent will periodically run the disk utility tools, keeping the disk in reasonably good condition. The defragmentation process can also be run on demand by selecting Disk Defragmenter from the System Tools submenu (located under Accessories on the Start menu). Figure 69 shows the initial defrag screen presented by Windows.

From this menu, it is very easy to start the defragmentation process, select another drive, check or modify the advanced parameters, or exit. Figure 70 shows the Disk Defragmenter Settings window.

FIGURE 69 Selecting a disk to defragment

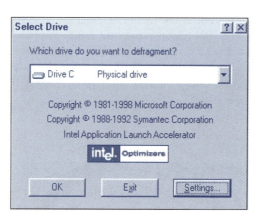

FIGURE 70 Disk Defragmenter Settings window

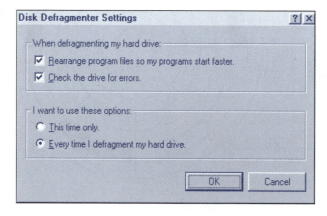

Figure 71 shows the brief status of the defragmentation process. By clicking the Show Details button, we can also view the details of the defragmentation process as shown in Figure 72. Notice how each of the disk clusters is presented on the screen. It is interesting to watch the defragmentation process. Depending on the amount of fragmentation, the process may last just a few minutes or as long as a few hours. Select Legend to view the legend shown in Figure 73 to help you identify the different types of disk clusters. As you can see, there are many different possible states for a disk cluster.

FIGURE 71   Defragmentation status window

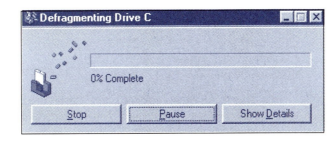

FIGURE 72   Details of the defragmentation process

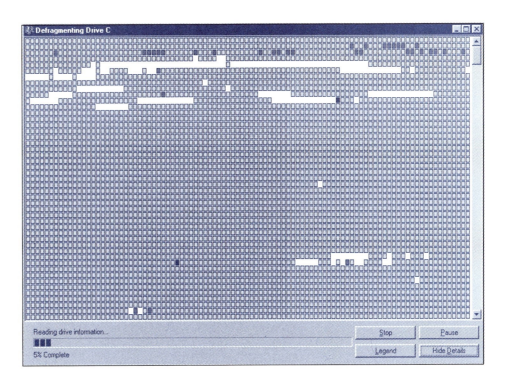

FIGURE 73   Defragmentation details legend

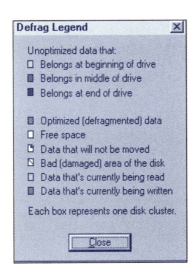

FILE ALLOCATION

Every time DOS has to get space on the disk for a file, it looks at the FAT for unused disk **clusters.** A cluster is a set of contiguous disk sectors. DOS will always try to minimize disk fragmentation, if possible. Every file written to the disk has a directory entry that is a 32-byte record, which contains the following information:

- Name of file
- File extension
- Attribute byte
- File time
- File date
- Number of the starting cluster
- Size of the file

DOS allocates disk space for directory entries in a special directory area that follows the FAT. This means that a disk can hold only a certain number of files, depending on the size and density of the disk. To DOS, a subdirectory is treated simply as a standard DOS file. This means that DOS stores subdirectory information in the same manner as it stores file information. Because of this, the number of subdirectories is also limited by the amount of free file space on the disk.

SECONDARY STORAGE, PART 3: THE CD-ROM DRIVE

A CD-ROM stores binary information in the form of microscopic *pits* on the disk surface. The pits are so small that a CD-ROM typically stores more than 650MB of data. This is equivalent to more than 430 1.44MB floppies. A laser beam is shined on the disk surface and either reflects (no pit) or does not reflect (pit), as you can see in Figure 74.

These two light states (reflection and no reflection) are easily translated into a binary 0 and a binary 1. Since the pits are mechanically pressed into a hard surface and only touched by light, they do not wear out or change as a result of being accessed.

PHYSICAL LAYOUT OF A COMPACT DISK

Figure 75 shows the dimensions and structure of a compact disk. The pits previously described are put into the reflective aluminum layer when the disk is manufactured. Newer recordable CDs use a layer of gold instead of aluminum so that they can be written to using a low-power laser diode.

FIGURE 74   Reading data from a CD

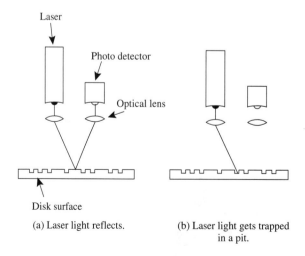

(a) Laser light reflects.

(b) Laser light gets trapped in a pit.

FIGURE 75 Compact disk

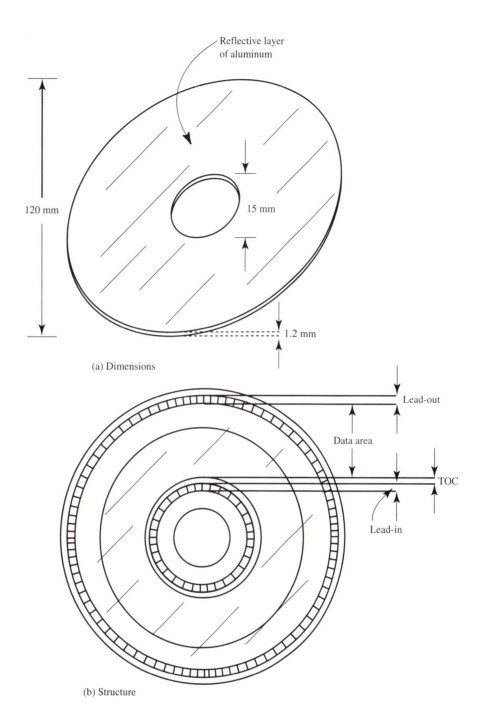

(a) Dimensions

(b) Structure

THE HIGH SIERRA FORMAT

The High Sierra format specifies the way the CD is logically formatted (tracks, sectors, directory structure, file name conventions). This specification is officially called ISO-9660 (International Standards Organization).

CD-ROM STANDARDS

The evolution of the CD-ROM is documented in several *books*. These are described in Table 25. The red book describes the method used to store digital audio on the CD-ROM. Pulse code modulation (PCM) is used to sample the audio 44,100 times/second with 16-bit sampling.

**TABLE 25   CD-ROM
standards**

Book	Feature
Red	CD-DA. Digital audio. PCM encoding.
Green	CD-I. Interactive (text, sound, and video). ADPCM and MPEG-1 encoding.
Orange	CD-R. Recordable.
Yellow	CD-ROM. Original PC CD-ROM format. 150Kbps transfer rate.

The green book specifies the CD interactive format typically used by home video games. Text, audio, and video are interleaved on the CD-ROM, and the MPEG-1 (Motion Picture Entertainment Group) video encoding method requires special hardware inside the player to support real-time video. The CD-ROM-XA (extended architecture) format is similar to CD-I.

The orange book provides the details on recordable CD-ROM drives. Gold-based disks are used to enable data to be written to the CD-ROM, up to 650MB. A *multisession* CD-ROM allows you to write to the CD more than once, and requires a multisession CD-ROM drive. Compact disks that only allow one recording session are known as WORM drives, for write once, read mostly.

The yellow book describes the original PC-based CD-ROM format (single spin), which specifies a data transfer rate of 150KB/s. 2x CD-ROMs transfer at 300KB/s. 4x CD-ROMs transfer data at 600KB/s. Currently there are 48x CD-ROM drives on the market, with faster ones coming.

PHOTO CD

Developed by Kodak, the photo CD provides a way to store high-quality photographic images on a CD (using recordable technology) in the CD-I format. Each image is stored in several different resolutions, from 192×128 to 3072×2048, using 24-bit color. This allows for around 100 images on one photo CD.

ATAPI

ATAPI stands for AT Attachment Packet Interface. ATAPI is an improved version of the IDE hard drive interface, and uses *packets* of data during transfers. The ATAPI specification supports CD-ROM drives, hard drives, tape backup units, and plug-and-play adapters.

I/O, PART 1: EXPANSION BUSES

Expansion buses (slots) serve a very important function in personal computers. They allow you to plug in electronic cards to expand and enhance the operation of your computer. The concept of expansion slots is simple; however, in practical terms, there are many things to consider. Figure 76 is a simple illustration of the function of expansion slots.

An expansion slot must be able to communicate with the computer. This communication usually includes access to the microprocessor. In achieving this access, the expansion bus must not interfere with the normal operation of the microprocessor. This means that the expansion bus must have access not only to the address and data lines used by the microprocessor but also to special control signals.

You need to be familiar with the functions of the various types of expansion buses that provide the means to add new features to your computer. Most of these added features, such as extra memory, additional types of monitor displays, extra or different disk drives, telephone communications, and other enhancements, are added in part by cards that fit into

FIGURE 76   Expansion slots

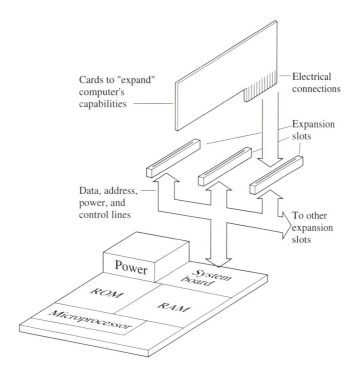

expansion slots on the computer (peripheral cards). However, as you will soon see, not all expansion slots are the same. It is important that you know their differences.

MAKEUP OF AN EXPANSION SLOT

Expansion slots have more similarities than differences. Table 26 lists the purposes of the different lines that are connected to the expansion slots. It should be noted that not all expansion slots use every one of these lines. The terminology used in this table does, however, apply to all expansion slots that use any of these lines.

TABLE 26   Expansion slot terminology

Connections	Purpose
Power lines	Power lines supply the voltages that may be needed by the various expansion cards. The power lines are +5 V DC, −5 V DC, +12 V DC, −12 V DC, and ground.
Data lines	Data lines are used to transfer programming information between the expansion card and the computer. One of the major differences among expansion slots in different computers is the number of data lines available.
Address lines	Address lines are used to select different memory locations. Another major difference among expansion slots is the number of address lines available.
Interrupt request lines	Interrupt request lines are used for hardware signals. These signals come from various devices, including the expansion card itself. Interrupt request signals are used to get the attention of the microprocessor. This is done so that the expansion card can temporarily use the services of the microprocessor.

(continued on the next page)

TABLE 26 *(continued)*

Connections	Purpose
DMA lines	DMA stands for *direct memory access.* DMA lines are control lines that provide direct access to memory (without having to go through the microprocessor, which tends to slow things down). DMA lines are also used to indicate when memory access is temporarily unavailable because it is being used by some other part of the system. DMA lines are used to indicate that direct memory access is being requested (called a DMA request line) and to acknowledge that request (called a DMA acknowledge).
NMI line	NMI stands for *nonmaskable interrupt.* This line is so called because it cannot be "masked," or switched off, by software. It is primarily used when a parity check error occurs in the system.
Memory-read, memory-write lines	The memory-read and memory-write lines are used to indicate that memory is either being written to or read from.
I/O read, I/O write lines	The I/O read and write lines are used to indicate that an input or output device (such as a disk drive) is to be written to or read from.
Special lines	Another one of the major differences among expansion slots in different types of computers is the number (and the types) of specialized lines used. For example, some of the OS/2 systems offer an *audio channel line* for the purpose of carrying a sound signal.

ISA Expansion Slots

Figure 77 shows the ISA (Industry Standard Architecture) expansion slot. True PC compatibles also use the same kind of expansion slot. Pin assignments are shown in Figure 78.

The major features of the ISA expansion slots are listed in Table 27. The features are described in terms of a *bus.* Recall that a bus is nothing more than a group of conductors treated as a unit; as a *data bus,* it is a group of conductors used to carry data. In terms of an expansion slot, a bus can be thought of as a group of connectors that is connected to the bus on the motherboard.

Figure 79 shows the design of an expansion card used in a PC. Note that there are two major types of PC expansion cards: one type goes straight back from the connector and the other has a skirt that dips back down to the board level. This distinction becomes important in the design of expansion slots used in other types of computers to accommodate PC expansion cards.

THE LOCAL BUS

The EISA connector supports 80386, 80486, and Pentium microprocessors by providing a full 32-bit data bus. Three special bus-controlling chips are used to manage data transfers

FIGURE 77 ISA expansion slot and pin numbering

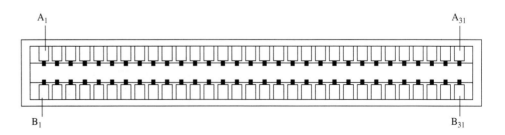

FIGURE 78 ISA expansion slot pin assignments

	B	A	
GND	B1	A1	I/O CH CK
RESETDRV			D7
+5 V			D6
IRQ2			D5
–5 V	B5	A5	D4
DRQ2			D3
–12 V			D2
RESERVED			D1
+12 V			D0
GND	B10	A10	I/O CH RDY
MEMW			AEN
MEMR			A19
IOW			A18
IOR			A17
DACK3	B15	A15	A16
DRQ3			A15
DACK1			A14
DRQ1			A13
DACK0			A12
CLK	B20	A20	A11
IRQ7			A10
IRQ6			A9
IRQ5			A8
IRQ4			A7
IRQ3	B25	A25	A6
DACK2			A5
TC			A4
ALE			A3
+5 V			A2
OSC	B30	A30	A1
GND	B31	A31	A0

TABLE 27 Major features of ISA expansion slots

Type of Bus	Comments
Total pins	62 separate connectors
Data bus	8 data lines
Address bus	20 address lines (1MB addressable memory)

FIGURE 79 Typical ISA expansion card

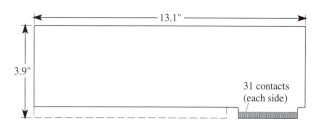

13.1"

3.9"

31 contacts (each side)

through the EISA connectors. Thus, data that gets transferred between an expansion card and the CPU must go through the bus controller chip set. This effectively reduces the rate at which data can be transferred.

To get around this problem, a new bus architecture was introduced, called the **local bus.** A local bus connector provides the fastest communication possible between a plug-in card and the machine by bypassing the EISA chip set and connecting directly to the CPU. Local bus video cards and hard drive controllers are popular because of their high-speed data transfer capability.

One initial attempt to define the new local bus was the VESA local bus. VESA stands for Video Electronics Standards Association, an organization dedicated to improving video display and bus technology. VESA local bus cards typically run at 33-MHz speeds, and were originally designed to interface with 80486 signals. VESA connectors are simply add-ons to existing connectors; no special VESA local bus connector exists.

THE PCI BUS

PCI stands for Peripheral Component Interconnect, and it is Intel's offering in the world of standardized buses. The PCI bus uses a *bridge* to control data transfers between the processor and the system bus, as indicated in Figure 80.

In essence, the PCI bus is not strictly a local bus, since connections to the PCI bus are not connections to the processor, but rather a special PCI-to-host controller chip. Other chips, such as PCI-to-ISA bridges, interface the older ISA bus with the PCI bus, allowing both types of buses on one motherboard, with a single chip controlling them all. The PCI

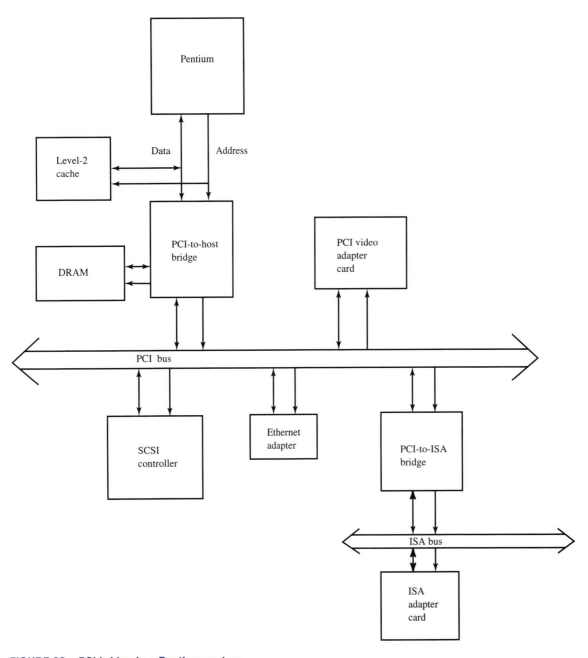

FIGURE 80   PCI bridge in a Pentium system

FIGURE 81   32-bit PCI
connector

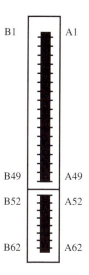

B1   A1

B49   A49

B52   A52

B62   A62

bus is designed to be processor independent, plug-and-play compatible, and capable of 64-bit transfers at 33 MHz and above.

PCI connectors are physically different from all other connectors. Refer to Figure 25, which shows four ISA connectors and three PCI connectors. Figure 81 shows the pinout for a 32-bit PCI connector.

THE PCMCIA BUS

The PCMCIA (Personal Computer Memory Card International Association) bus, now referred to as *PC card bus,* evolved from the need to expand the memory available on early laptop computers. The standard has since expanded to include almost any kind of peripheral you can imagine, from hard drives, to LAN adapters and modem/fax cards. Figure 82 shows a typical PCMCIA Ethernet card.

The PCMCIA bus supports four styles of cards, as shown in Table 28.

All PCMCIA cards allow *hot swapping,* removing and inserting the card with power on.

A type I connector is shown in Figure 83. The signal assignments are illustrated in Tables 29 and 30. The popularity of laptop and notebook computers suggests the continued use of this bus.

FIGURE 82   PCMCIA Ethernet
card (*photograph by John T. Butchko*)

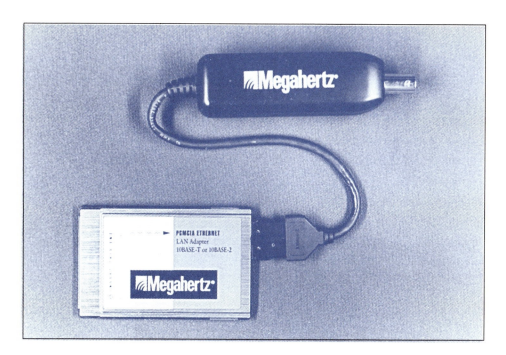

TABLE 28   PCMCIA slot styles

Slot Type	Meaning
I	Original standard. Supports 3.3-mm cards. Memory cards only.
II	Supports 3.3-mm and 5-mm cards.
III	Supports 10.5-mm cards, as well as types I and II.
IV	Greater than 10.5 mm supported.

FIGURE 83   PCMCIA connector

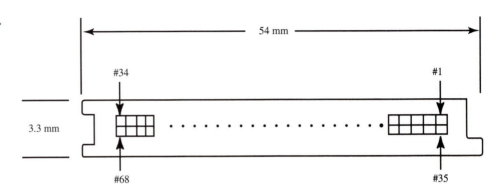

TABLE 29   PCMCIA pin assignments (available at card insertion)

Pin	Signal	Pin	Signal	Pin	Signal
1	GND	24	A_5	47	A_{18}
2	D_3	25	A_4	48	A_{19}
3	D_4	26	A_3	49	A_{20}
4	D_5	27	A_2	50	A_{21}
5	D_6	28	A_1	51	Vcc
6	D_7	29	A_0	52	Vpp2
7	CE1	30	D_0	53	A_{22}
8	A_{10}	31	D_1	54	A_{23}
9	OE	32	D_2	55	A_{24}
10	A_{11}	33	WP	56	A_{25}
11	A_9	34	GND	57	RFU
12	A_8	35	GND	58	RESET
13	A_{13}	36	CD1	59	WAIT
14	A_{14}	37	D_{11}	60	RFU
15	WE/PGM	38	D_{12}	61	REG
16	RDY/BSY	39	D_{13}	62	BVD2
17	Vcc	40	D_{14}	63	BVD1
18	Vpp1	41	D_{15}	64	D_8
19	A_{16}	42	CE2	65	D_9
20	A_{15}	43	RFSH	66	D_{10}
21	A_{12}	44	RFU	67	CD2
22	A_7	45	RFU	68	GND
23	A_6	46	A_{17}		

TABLE 30   PCMCIA signal differences

Pin	Memory Card	I/O Card
16	RDY/BSY	IREQ
33	WP	IOIS16
44	RFU	IORD
45	RFU	IOWR
60	RFU	INPACK
62	BVD2	SPKR
63	BVD1	STSCHG

FIGURE 84   AGP interface

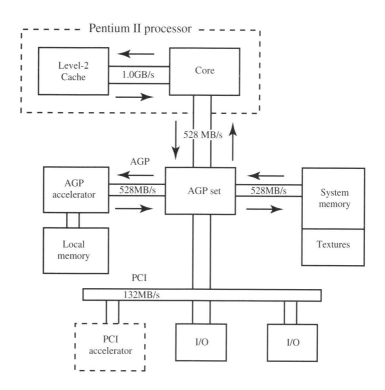

AGP

The Accelerated Graphics Port (AGP) is a new technology that improves multimedia performance on Pentium II/III computers. Figure 84 shows where the AGP technology fits into the other bus hardware.

The heart of the AGP is the 440LX AGPset hardware, a *quad-ported* data switch that controls transfers between the processor, main memory, graphics memory, and the PCI bus. AGP technology uses a connector similar to a PCI connector.

With growing emphasis on multimedia applications, AGP technology sets the stage for improved performance.

I/O, PART 2: THE MODEM

This section has to do with communications between computers. The user of one computer may interact with another computer—which may be located thousands of miles away—as if it were sitting right in the same room.

In order for computers to communicate in this manner, four items must be available, as shown in Figure 85. There must be some kind of link between the computers. The most convenient link to use is the already-established telephone system lines. Using these lines and a properly equipped computer allows communications between any two computers that have access to a telephone. This becomes a very convenient and inexpensive method of communicating between computers.

There is, however, one problem. Telephone lines were designed for the transmission of the human voice, not for the transmission of digital data. Therefore, in order to make use of these telephone lines for transmitting computer data, the ONs and OFFs of the computer must first be converted to sound, sent over the telephone line, and then reconverted from sound to the ONs and OFFs that the computer understands. This concept is shown in Figure 86.

MODEM DEFINITION

The word *modulate* means to change. Thus an electronic circuit that changes digital data into sound data can be called a *modulator.* The word *demodulate* can be thought of as

FIGURE 85 Four items
necessary for computer
communications

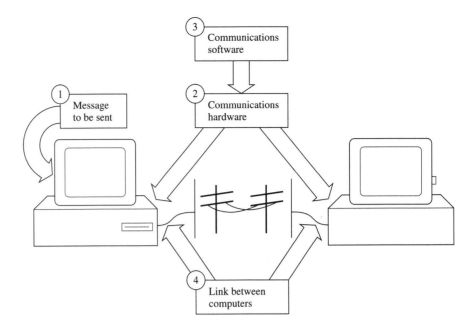

FIGURE 86 Basic needs for the
use of telephone lines in
computer communications

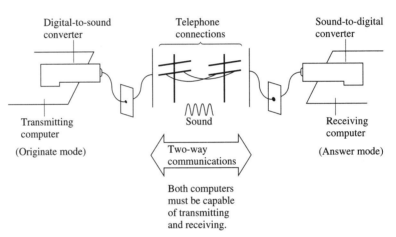

meaning "unchange," or restore to an original condition. Any electronic circuit that converts the sound used to represent the digital signals back to the ONs and OFFs understood by a computer can, therefore, be called a demodulator. Since each computer must be capable of both transmission and reception, each computer must contain an electrical circuit that can modulate as well as demodulate. Such a circuit is commonly called a m̲o̲dulator/ de̲m̲odulator, or **modem.**

For personal computers, a modem may be an internal or an external device—both perform identical functions.

THE RS-232 STANDARD

The EIA (Electronics Industries Association) has published the EIA *Standard Interface Between Data Terminal Equipment Employing Serial Binary Data Interchange*—specifically, EIA-232-C. This is a standard defining 25 conductors that may be used in interfacing **data terminal equipment** (DTE, such as your computer) and **data communications equipment** (DCE, such as a modem) hardware. The standard specifies the function of each conductor, but it does not state the physical connector that is to be used. This standard exists so that different manufacturers of communications equipment can communicate with each other. In other words, the RS-232 standard is an example of an interface, essentially an agreement among equipment manufacturers on how to allow their equipment to communicate.

FIGURE 87   The RS-232
standard

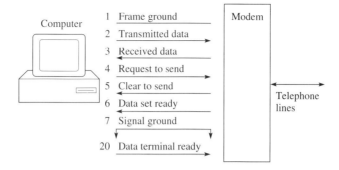

TABLE 31   Standard baud rates

Low Speed	High Speed
300	
600	
1200	14,400
2400	28,800
4800	33,600
9600	56K

The RS-232 standard is designed to allow DTEs to communicate with DCEs. The RS-232 uses a DB-25 connector; the male DB-25 goes on the DTEs and the female goes on the DCEs. The RS-232 standard is shown in Figure 87.

ʾ The RS-232 is a digital interface designed to operate at no more than 50 feet with a 20,000-bit/second bit rate. The **baud,** named after J. M. E. Baudot, actually indicates the number of *discrete* signal changes per second. In the transmission of binary values, one such change represents a single bit. What this means is that the popular usage of the term *baud* has become the same as bits per second (bps). Table 31 shows the standard set of baud rates.

TELEPHONE MODEM SETUP

The most common problem with telephone modems is the correct setting of the software. There are essentially six distinct areas to which you must pay attention when using a telephone modem:

1. Port to be used
2. Baud rate
3. Parity
4. Number of data bits
5. Number of stop bits
6. Local echo ON or OFF

Most telephone modems have a default setting for each of these areas. However, as a user you should understand what each of these areas means. You will have to consult the specific documentation that comes with the modem in order to see how to change any of the above settings. For now it is important that you understand the idea behind each of these areas.

Port to Be Used

The most common ports to be used are COM1 and COM2. Other possible ports are COM3 and COM4. The port you select from the communications software depends on the port to which you have the modem selected. On most communications software, once you set the correct port number, you do not need to set it again.

Baud Rate

Typical values for the baud rate are between 9600 and 56K. Again, these values can be selected from the communications software menu. It is important that both computers be set at the same baud rate.

Parity

Parity is a way of having the data checked. Normally, parity is not used. Depending on your software, there can be up to five options for the parity bit, as follows:

Space: Parity bit is always a 0.
  Odd: Parity bit is 0 if there is an odd number of 1s in the transmission and is a 1 if there is an even number of 1s in the transmission.
  Even: Parity bit is a 1 if there is an odd number of 1s in the transmission and is a 0 if there is an even number of 1s in the transmission.
Mark: Parity bit is always a 1.
None: No parity bit is transmitted.

Again, what is important is that both the sending and receiving units are set up to agree on the status of the parity bit.

Number of Data Bits

The number of data bits to be used is usually set at 8. There are options that allow the number of data bits to be set at 7. It is important that both computers expect the same number of data bits.

Number of Stop Bits

The number of stop bits used is normally 1. However, depending on the system, the number of stop bits may be 2. Stop bits are used to mark the end of each character transmitted. Both computers must have their communications software set to agree on the number of stop bits used.

WINDOWS MODEM SOFTWARE

Windows has built-in modem software, accessed through the Control Panel. Clicking on the Mouse icon displays the window shown in Figure 88. Notice that Windows indicates the presence of an external Sportster modem. To test the modem, click the Diagnostics tab. Figure 89 shows the Diagnostics window.

Selecting COM2 (the Sportster modem) and then clicking More Info will cause Windows to talk to the modem for a few moments, interrogate it, and then display the results in a new window, shown in Figure 90.

Specific information about the modem port is displayed, along with the responses to several AT commands. The *AT command set* is a standard set of commands that can be sent to the modem to configure, test, and control it. Table 32 lists the typical **Hayes compatible** commands (first used by Hayes in its modem products). An example of an AT command is:

ATDT 778 8108

which stands for AT (attention) DT (dial using tones). This AT command causes the modem to touch-tone dial the indicated phone number. Many modems require an initial AT command string to be properly initialized. This string is automatically output to the modem when a modem application is executed.

TELEPHONE MODEM TERMINOLOGY

In using technical documentation concerning a telephone modem, you will encounter some specialized terminology. Figure 91 illustrates some of the ideas behind some basic

FIGURE 88   Modems Properties
window

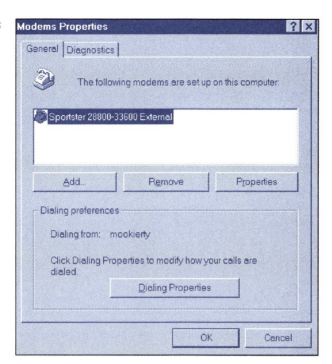

FIGURE 89   Modems
Diagnostics window

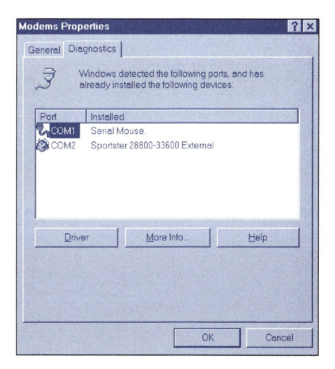

communication methods. As you can see from the figure, **simplex** is a term that refers to a communications channel in which information flows in one direction only. An example of this is a radio or a television station.

Duplex

The **duplex** mode refers to two-way communication between two systems. This term is further refined as follows. **Full duplex** describes a communication link that can pass data in two directions at the same time. This mode is analogous to an everyday conversation between two people either face-to-face or over the telephone. The other mode, which is not commonly available with telephone modems, is the **multiplex** mode. Multiplex refers to a communications link in which multiple transmissions are possible.

FIGURE 90   Modem diagnostic information

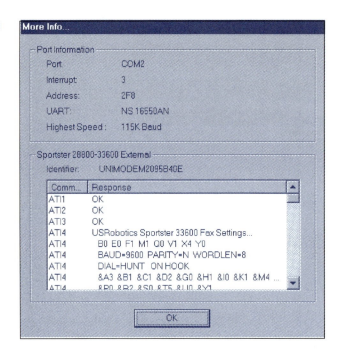

TABLE 32   Selected AT commands

Command	Function	Command	Function
A/	Repeat last command	Xn	Result code type
A	Answer	Yn	Long space disconnect
Bn	Select CCITT or Bell	Zn	Recall stored profile
Cn	Carrier control option	&Cn	DCD option
D	Dial command	&Dn	DTR option
En	Command echo	&F	Load factory defaults
Fn	Online echo	&Gn	Guard tone option
Hn	Switch hook control	&Jn	Auxiliary relay control
In	Identification/checksum	&M0	Communication mode option
Kn	SRAM buffer control	&Pn	Dial pulse ratio
Ln	Speaker volume control	&Q0	Communication mode option
Mn	Speaker control	&Sn	DSR option
Nn	Connection data rate control	&Tn	Self-test commands
On	Go online	&Vn	View active and stored configuration
P	Select pulse dialing	&Un	Disable Trellis coding
Qn	Result code display control	&Wn	Stored active profile
Sn	Select an S-register	&Yn	Select stored profile on power-on
Sn=x	Write to an S-register	&Zn=x	Store telephone number
Sn?	Read from an S-register	%En	Auto-retrain control
?	Read last accessed S-register	%G0	Rate renegotiation
T	Select DTMF dialing	%Q	Line signal quality
Vn	Result code form	-Cn	Generate data modem calling tone

Echo

Terminology used here has to do with how the characters you send to the other terminal are displayed on your monitor screen. The term **echo** refers to the method used to display characters on the monitor screen. First, there is a **local echo**. A local echo means that the sending modem immediately returns or echoes each character back to the screen as it is entered into the keyboard. This mode is required before transmission, so that you can see what instructions you are giving the communications software. Next there is **remote echo.** Remote echo means that during the communications between two

FIGURE 91   Some basic communication methods

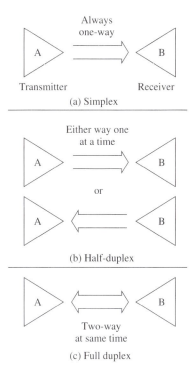

computers, the remote computer (the one being transmitted to) sends back the character it is receiving. The character that then appears on your screen is the result of a transmission from the remote unit. This is a method of verifying what you are sending. To use the remote-echo mode, you must be in the full-duplex mode. This idea is illustrated in Figure 92.

MODULATION METHODS

Many different techniques are used to encode digital data into analog form (for use by the modem). Several of these techniques are:

- AM (amplitude modulation)
- FSK (frequency shift keying)
- Phase modulation
- Group coding

Figure 93 shows how the first three of these techniques encode their digital data.

To get a high data rate (in bits per second) over ordinary telephone lines, group coding techniques are used. In this method, one cycle of the transmitted signal encodes two or more bits of data. For example, using *quadrature modulation,* the binary patterns 00, 01, 10, and 11 encode one of four different phase shifts for the current output signal. Thus, a signal that changes at a rate of 2400 baud actually represents 9600 bps!

Another technique, called *Trellis modulation,* combines two or more other techniques, such as AM and quadrature modulation, to increase the data rate.

MNP STANDARDS

MNP (Microcom Networking Protocol) is a set of protocols used to provide error detection and correction, as well as compression, to the modem data stream. Table 33 lists the MNP classes and their characteristics.

MNP classes 4 and above are used with newer, high-speed modems. When two modems initially connect, they will negotiate the best type of connection possible, based on line properties, and the features and capabilities of each modem. The CCITT

FIGURE 92 Echo modes

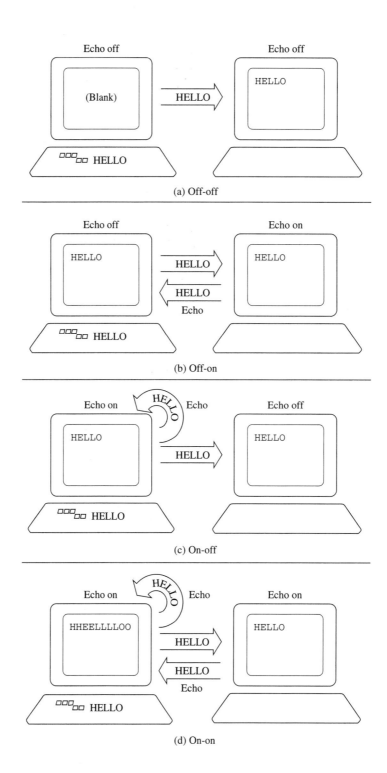

(a) Off-off

(b) Off-on

(c) On-off

(d) On-on

standards supported by the modem are also part of the negotiation. Let us look at these standards as well.

CCITT STANDARDS

CCITT (French abbreviation for International Telegraph and Telephone Consultive Committee) standards define the maximum operating speed (as well as other features) available in a modem (which is a function of the modulation techniques used). Table 34 lists the CCITT standards.

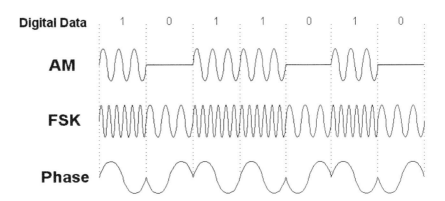

FIGURE 93 Modulation techniques

TABLE 33 MNP standards

Class	Feature
1	Asynchronous, half-duplex, byte-oriented
2	Asynchronous, full-duplex, byte-oriented
3	Synchronous, full-duplex, byte-oriented
4	Error correction, packet-oriented
5	Data compression
6	Negotiation
7	Huffman data compression
9	Improved error correction
10	Line monitoring

Note: There is no MNP-8 standard.

TABLE 34 CCITT standards

Standard	Data Rate (bps)
V.22	1200
V.22 bis	2400
V.32	9600
V.32 bis	14,400
V.32 terbo	19,200
V.34	28,800/33,600
V.90	56K

Earlier, low-speed standards not shown are the Bell 103 (300 bps using FSK) and Bell 212A (1200 bps using quadrature modulation). V.22 is similar in operation to Bell 212A, and is more widely accepted outside the United States.

The V.90 standard, finalized in early 1998, outlines the details of modem communication at 56K bps, currently the fastest speed available for regular modems. Fax modems have their own set of standards.

ISDN MODEMS

ISDN (Integrated Services Digital Network) is a special connection available from the telephone company that provides 64K bps digital service. An ISDN modem will typically connect to a *basic rate ISDN* (BRI) line, which contains two full-duplex 64Kbps B channels (for voice/data) and a 16Kbps D channel (for control). This allows up to 128Kbps communication. ISDN modems are more expensive than ordinary modems, and require you to have an ISDN line installed before you can use it.

CABLE MODEMS

One of the most inexpensive, high-speed connections available today is the cable modem. A cable modem connects between the television cable supplying your home and a network interface card in your computer. Two unused cable channels are used to provide data rates in the hundreds of thousands of bits per second. For example, downloading a 6MB file over a cable modem takes less than 20 seconds (during several tests of a new cable modem installation). That corresponds to 2,400,000 bps! Of course, the actual data rate available depends on many factors, such as the speed the data is transmitted from the other end and any communication delays. But unlike all other modems, the cable modem has the capability to be staggeringly fast, due to the high bandwidth available on the cable. In addition, a cable modem is typically part of the entire package from the cable company, and is returned when you terminate service. The cost is roughly the same as the cost of basic cable service.

FAX/DATA MODEMS

It is difficult to find a modem manufactured today that does not have fax capabilities built into it. Since fax/data modems are relatively inexpensive, it does not make sense to purchase a separate fax machine (unless it is imperative that you be able to scan a document before transmission). Word-processing programs (such as WordPerfect) now support the use of a fax/data modem, helping to make the personal computer almost an entire office by itself.

PROTOCOLS

A **protocol** is a prearranged communication procedure agreed upon by two or more parties. When two modems are communicating over telephone lines (during a file transfer from a computer bulletin board or an America Online session), each modem has to agree on the technique used for transmission and reception of data. Table 35 shows some of the more common protocols. The modem software that is supplied with a new modem usually allows the user to specify a particular protocol.

I/O, PART 3: THE SOUND CARD

Along with CD-ROM drives, sound cards for PCs have also increased in popularity. Currently, 16-bit sound cards are available that provide multiple audio channels and FM-quality sound, and are compatible with the MIDI (Musical Instrument Data Interface) specification.

The basic operation of the sound card is shown in Figure 94. Digital information representing samples of an analog waveform are inputted to a *digital-to-analog* converter, which translates the binary patterns into corresponding analog voltages. These analog voltages are then passed to a *low-pass filter* to smooth out the differences between the individual voltage samples, resulting in a continuous analog waveform. All of the digital/analog signal processing is done in a custom **digital signal processor** chip included on the sound card.

Sound cards also come with a microphone input that allows the user to record any desired audio signal.

TABLE 35   Modem
communication protocols

Protocol	Operation
Xmodem	Blocks of 128 bytes are transmitted. A checksum byte is used to validate received data. Bad data is retransmitted.
Xmodem CRC	Xmodem using Cyclic Redundancy Check to detect errors.
Xmodem-1K	Essentially Xmodem CRC with 1024-byte blocks.
Ymodem	Similar to Xmodem-1K. Multiple files may be transferred with one command.
Zmodem	Uses 512-byte blocks and CRC for error detection. Can resume an interrupted transmission from where it left off.
Kermit	Transmits data in packets whose sizes are adjusted to fit the needs of the other machine's protocol.

FIGURE 94   How binary data is converted into an analog waveform

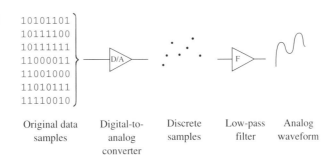

| Original data samples | Digital-to-analog converter | Discrete samples | Low-pass filter | Analog waveform |

MIDI

MIDI stands for Musical Instrument Digital Interface. A MIDI-capable device (electronic keyboard, synthesizer) will use a MIDI-in and MIDI-out serial connection to send messages between a *controller* and a *sequencer.* The PC operates as the sequencer when connected to a MIDI device. MIDI messages specify the type of note to play and how to play it, among other things. Using MIDI, a total of 128 pitched instruments can generate 24 notes in 16 channels. This can be accomplished in a PC sound card by using frequency modulation or *wave table synthesis,* the latter method utilizing prerecorded samples of notes stored in a data table. The output of a note is controlled by several parameters. Figure 95 illustrates the use of attack, delay, and release times to shape the output waveform envelope. Each of the four parameters can be set to a value from 0 to 15.

FIGURE 95   Note envelope

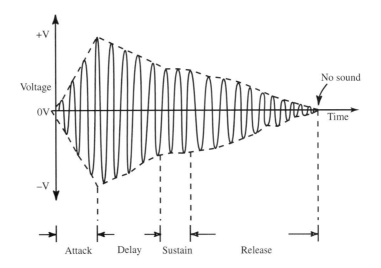

TROUBLESHOOTING THE MOUSE

Usually the biggest problem with a mouse-to-computer connection is improper installation of the mouse software. To correct this problem, read the literature that comes with the mouse. In order for the mouse to interface effectively with the computer, the software that comes with the mouse must be installed in the system as directed by the manufacturer. If you are running DOS, you also want to make sure that the system's CONFIG.SYS and AUTOEXEC.BAT files have been properly set up so that the mouse driver is automatically installed each time the system is booted up. Again, this information is included in the literature that comes with the mouse. The important point here is that you read the literature and follow the directions.

Usually, there is an optional utility program with the software that comes with the mouse that helps you test and adjust the mouse interface (by means of software). These utilities are perhaps one of the best tests of mouse performance.

FIGURE 96   Initial Mouse Properties window

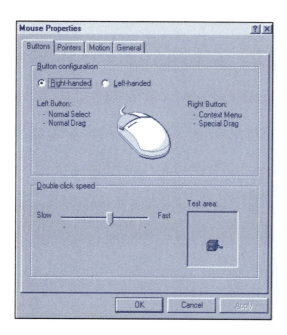

FIGURE 97   Mouse pointer types

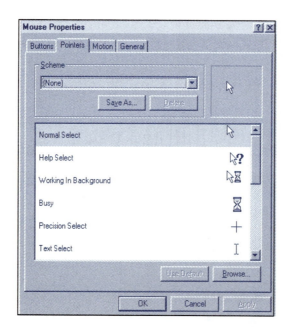

FIGURE 98   Additional pointer
controls

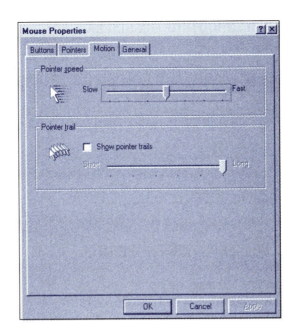

FIGURE 99   Examining the
mouse driver information

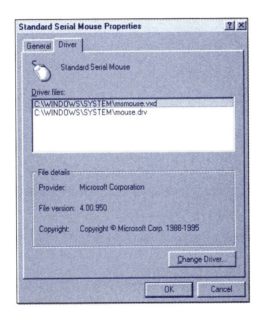

Windows supplies its own mouse drivers, eliminating the need for any setup in CONFIG.SYS and AUTOEXEC.BAT, and provides a great amount of control over how the mouse appears and operates.

Clicking on the Mouse icon in Control Panel brings up the Mouse Properties window shown in Figure 96. Here you can adjust such important parameters as the double-click speed and left/right-handed operation. Figures 97 and 98 show two additional control windows dealing with the appearance of the mouse pointer. If the mouse is not responding correctly, it may be necessary to change its driver or hardware properties. In Control Panel, double-clicking the System icon and then selecting the Device Manager tab will allow you to double-click Mouse and check the driver information. This information is illustrated in Figure 99. Mouse information is available under the Devices icon in the Windows NT Control Panel. Figure 100 shows the associated Windows NT mouse information.

TROUBLESHOOTING THE DISPLAY

Most monitors produced today can be controlled by software. In order to take advantage of this feature, the monitor must be recognized by the operating system. Windows 95/98 will display the monitor's specific information when the Change Display Type button is selected from the Display Properties screen as shown in Figure 101.

The Change Display Type window shows the current settings of the display adapter and the monitor type, as illustrated in Figure 102. Note the additional check box setting used to inform Windows the monitor is Energy Star compliant. If enabled, the monitor may be shut down during a period of inactivity. Be sure to verify the system, video card, and monitor can support Energy Star features before enabling them. Windows NT provides similar screens to accomplish the same tasks.

TROUBLESHOOTING THE PRINTER

The most common problems with printers usually involve the quality of the output. Many problems are associated with the supply of ink or toner. Printers also contain many mechanical components that are a common point of failure. In the case of a dot-matrix printer, the ribbon may need to be replaced, the print head may need to be replaced, or the pin feeds may occasionally require some adjustment. For an ink-jet printer, the ink cartridge may become clogged with dried ink and may need to be cleaned to restore the print quality. There is no set schedule for these events to occur.

The best course of action is to be prepared for common problems that can be encountered. For example, it is a good idea to keep printer supplies on hand, so when a problem occurs, it can be remedied quickly. Table 36 contains a list of items that should be kept on hand. Remember, many of these items have a certain shelf life. Rotate the stock regularly.

TROUBLESHOOTING MEMORY

One of the simplest ways to determine whether your Windows system has enough RAM to handle its workload is to watch the hard drive light. No or little activity, except when opening or closing an application, is a good sign.

If the hard drive activates sporadically, doing a little work every now and then, the system is borderline. If the activity increases when additional applications are opened, there is a definite lack of RAM.

Frustrated with hard drive activity when only a few applications were open, one user increased the amount of RAM in his system from 32MB to 128MB (taking advantage of a drop in memory prices at that time). Now, even with a taskbar full of applications, the hard drive remains inactive.

TABLE 36 Common printer types and supplies

Printer Type	Supplies
Dot matrix	Ribbons, pin-feed paper
Ink jet and bubble jet	Black ink cartridges, color ink cartridges, single-sheet ink-jet paper
Laser	Toner cartridges, single-sheet laser-quality paper
Color laser	Cyan, yellow, magenta, and black toner cartridges, single-sheet color laser-quality paper

FIGURE 100 Windows NT Mouse Driver status

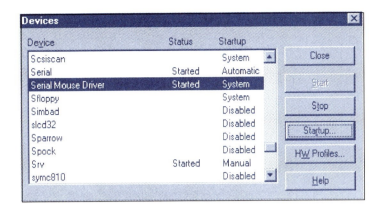

FIGURE 101 Access to change the display type

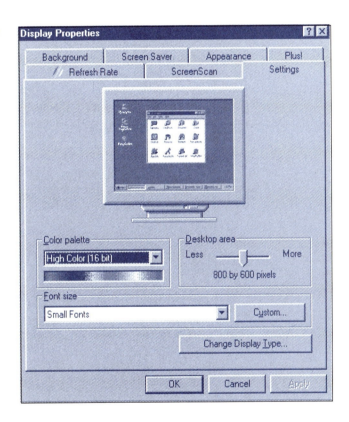

FIGURE 102 Setting the monitor type

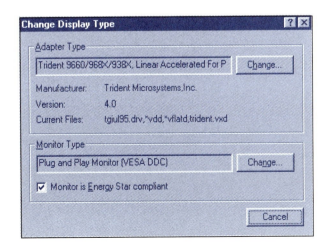

TROUBLESHOOTING FLOPPY DRIVES

Probable causes of what appears to be an FDD failure may be in one of the areas shown in Figure 103. This figure illustrates the relationship of the FDD to the entire computer system. At one end is the software on the disk; at the other extreme is the power cord connection to the electrical outlet. Every part of this system must be functioning properly for the disk drive to do its part. What is important here is to ensure that what appears to be a disk drive problem is not actually a problem caused by one of these other areas.

Troubleshooting Logic

The first step in troubleshooting the disk drive is to classify the problem as occurring in one of the areas shown in Figure 103. Once the area at fault is determined, corrective action may be taken. Figure 104 is a troubleshooting diagram for determining which of these areas may be at fault.

Troubleshooting Steps

1. *Ask questions.* With all computer servicing, inquire about the history of the system. Was it recently modified? Did anyone attempt to make any changes or repairs? Did the user buy the system used? For example, if the disk drive was recently installed, it may be that certain settings need to be set differently. In this case, you will need to refer to the documentation for the system as well as the newly installed FDD. Asking questions may help you quickly spot the problem, saving you time and money.
2. *Try again.* Sometimes reseating the floppy in the drive fixes any read errors encountered. You could also try using the floppy in a different drive.
3. *Visual inspection.* A good visual inspection may reveal a burned component, an improperly seated cable, a dirty read/write head, or other mechanical evidence of the

FIGURE 103   System relationship to floppy disk drive

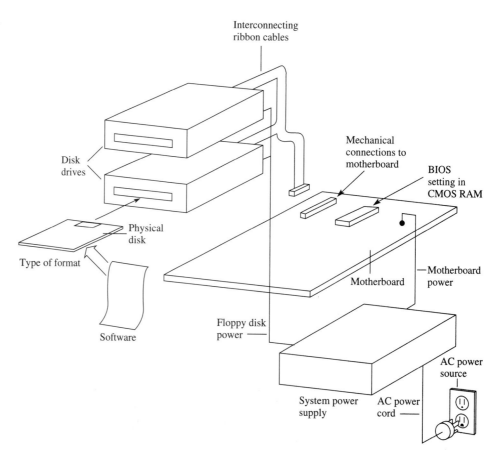

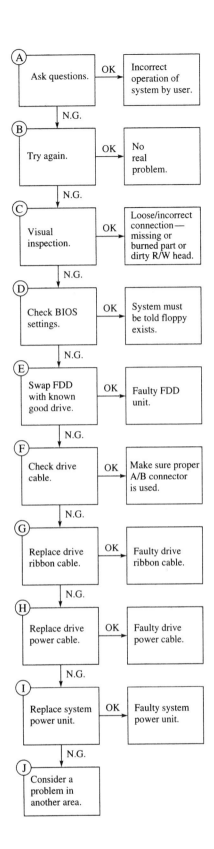

FIGURE 104 FDD troubleshooting chart

problem. Remember, a good visual inspection is an important part of any troubleshooting process.

4. *Check BIOS settings.* The system must be aware that a floppy drive exists. This is accomplished by running the BIOS setup program at boot time.

5. *Swap FDD with known good drive.* You need to take some precautions when doing this. It may be that the A: drive has a terminating resistor. You need to refer to documentation for the disk drive in question.

6. *Check drive cable.* In a one-drive system, make sure the connector with the twist is connected to drive A:. In a two-drive system, make sure both connectors are in the correct drives.

7. *Replace drive ribbon cable.* Here, another good visual inspection may be needed. Make sure you refer to the servicing manual to ensure that the cable was correctly installed in the first place. Replacing the cable with a known good one will help determine if the original cable or its connectors are at fault.

8. *Replace drive power cable.* This may be difficult on some systems because the power cable may be permanently attached to the power supply. If this is the case, skip this step and go on to the next one.

9. *Replace system power unit.* Doing this will help eliminate the problem of a power unit supplying the correct voltages when the power demands on it are small, but failing, as a result of the increased power demand, when a disk drive motor is activated (such as when the FDD is attempting to read a disk). Make a note of the power rating of the power supply; if it is below 100 W, substitute another known good power unit with a higher power rating.

10. *Consider a problem in another area.* If all the preceding steps fail to locate the problem, the problem is in another area of the unit. The most likely area in this case is the motherboard.

TROUBLESHOOTING MODEMS

Table 37 lists some of the most common problems encountered in telephone modems. As you will see, most of the problems are software related.

Other common problems encountered involve very simple hardware considerations. For example, telephone modems usually come with two separate telephone line connectors.

The purpose of the phone input is to connect a telephone, not the output line from the modem, to the telephone wall jack. The phone input is simply a convenience. It allows the telephone to be used without having to disconnect a telephone line from the computer to the wall telephone jack. If you mistakenly connect the line from the wall telephone jack to the phone input, you will be able to dial out from your communications software, but your system will hang up on you. Make sure that the telephone line that goes to the telephone wall jack comes from the *line output* and not the *phone output* jack of your modem.

TABLE 37   Common telephone modem problems

Symptom	Possible Cause(s)
Can't connect	Usually this means that your baud rates or numbers of data bits are not matched. This is especially true if you see garbage on the screen, especially the { character.
Can't see input	You are typing in information but it doesn't appear on the screen. However, if the person on the other side can see what you are typing, it means that you need to turn your local echo on. In this way, what you type will be echoed back to you, and you will see it on your screen.
Get double characters	Here you are typing information and getting double characters. This means that if you type HELLO, you get HHEELLLLOO; at the same time, what the other computer is getting appears normal. This means that you need to turn your local echo off. In this way, you will not be echoing back the extra character. With some systems *half-duplex* refers to local echo on, whereas *full duplex* refers to local echo off.

Another common hardware problem is a problem in your telephone line. This can be quickly checked by simply using your phone to get through to the other party. If you can't do this, then neither can your computer.

A problem that is frequently encountered in an office or school building involves the phone system used within the building. You may have to issue extra commands on your software in order to get your call out of the building. In this case you need to check with your telecommunications manager or the local phone company.

Sometimes your problem is simply a noisy line. This may have to do with your communications provider or it may have to do with how your telephone line is installed. You may have to switch to a long-distance telephone company that can provide service over more reliable communication lines. Or you may have to physically trace where your telephone line goes from the wall telephone jack. If this is an old installation, your telephone line could be running in the wall right next to the AC power lines. If this is the case, you need to reroute the phone line.

TROUBLESHOOTING SOUND CARDS

One of the most common reasons a new CD-ROM drive or sound card does not work has to do with the way its interrupts and/or DMA channels are assigned.

Figure 105 shows the location of the sound card in the hardware list provided by Device Manager. The AWE-32 indicates that the sound card is capable of advanced wave effects using 32 voices.

Figure 106 shows the interrupt and DMA assignments for the sound card. Typically, interrupt 5 is used (some network interface cards also use interrupt 5), as well as DMA channels 1 and 5. If the standard settings do not work, you need to experiment until you find the right combination.

FIGURE 105   Selecting the sound card

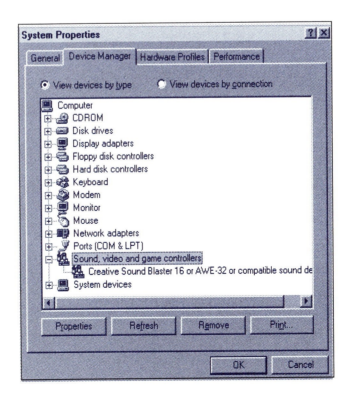

FIGURE 106   Sound card settings

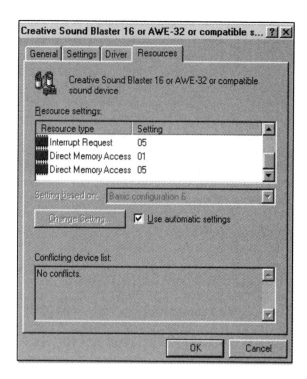

IDENTIFYING THE PROCESSOR

Windows identifies the processor it is running on. Use System Properties in Control Panel to check the processor type. This kind of information is not easily found in existing documentation. It is good to talk directly to the manufacturer of the motherboard to determine how your performance might be affected.

Answers to Odd-Numbered Self-Test Questions

EXERCISE 1

1. True 3. False 5. True 7. b 9. b 11. h, i 13. f 15. a
17. output 19. mouse

EXERCISE 2

1. False 3. False 5. False 7. False 9. True 11. c 13. a 15. b
17. TCP/IP 19. Desktop 21. Registry 23. Shutdown

EXERCISE 3

1. True 3. False 5. False 7. True 9. False 11. b 13. b 15. c 17. c
19. Shutdown 21. Application 23. Diagnostics 25. workgroup 27. one

EXERCISE 4

1. True 3. False 5. False 7. True 9. False 11. b 13. c 15. b 17. b
19. b 21. Taskbar 23. tiled 25. Briefcase 27. screen saver 29. Settings

EXERCISE 5

1. False 3. True 5. True 7. c 9. d 11. MPEG 13. Properties 15. WAV

EXERCISE 6

1. False 3. False 5. False 7. d 9. d
11. My Computer, Network Neighborhood, Recycle Bin
13. pull-down 15. Find Now

EXERCISE 7

1. False   3. False   5. False   7. b   9. c   11. b   13. dithering
15. local   17. default

EXERCISE 8

1. False   3. False   5. True   7. False   9. d   11. b   13. d
15. a   17. fragmented   19. WordPad

EXERCISE 9

1. False   3. True   5. False   7. a   9. a   11. a   13. Installation
15. Application setting

EXERCISE 10

1. True   3. False   5. b   7. b   9. a   11. c
13. fully connected   15. data-link

EXERCISE 11

1. False   3. True   5. False   7. a   9. a   11. Extended User Interface
13. Accessories   15. Internet service provider

EXERCISE 12

1. False   3. False   5. d   7. b   9. checksums; signatures

EXERCISE 13

1. False   3. True   5. False   7. c   9. a   11. HTML   13. active   15. HTTP

EXERCISE 14

1. True   3. False   5. False   7. b   9. a   11. View
13. view   15. grammatical

EXERCISE 15

1. True   3. True   5. True   7. c   9. a   11. selected
13. Wizard   15. A1

EXERCISE 16

1. False   3. True   5. True   7. a   9. b
11. slides   13. spell   15. templates

EXERCISE 17

1. True   3. False   5. False   7. a   9. c   11. Planning   13. Number   15. Map

EXERCISE 18

1. False   3. True   5. True   7. c   9. b   11. Multimedia Internet Mail Extensions
13. Cc   15. Server

EXERCISE 19

1. True   3. False   5. False   7. c   9. b   11. Switchboard   13. relational
15. form

EXERCISE 20

1. False   3. True   5. True   7. b   9. a   11. Capacitor   13. Semiconductors
15. Ground

EXERCISE 21

1. False   3. False   5. False   7. b   9. c   11. Content   13. Select   15. Overlap
   or obscure

Index